CW00709313

Mammalian Host Defense Peptides

Cationic antimicrobial peptides are multifunctional peptides of the innate immune system, which not only act directly against microorganisms, but also signal between early and late immune responses and modulate inflammatory responses. The significance of these host defense peptides in combating infection and in host–microbe homeostasis has become increasingly clear, through advances made by microbiologists, biochemists, biophysicists, immunologists, molecular biologists, and a range of medical and pharmaceutical researchers. This book, drawing together contributions from leading scientists, reviews significant recent advances in our knowledge of mammalian antimicrobial peptides. In addition to providing up-to-date overviews of their structure, expression, and biology, their multiple activities and interactions with microbial populations as well as their potential application as novel therapeutic agents are summarized. Chapters that describe developments by use of animal models and investigations of the roles of these host defense peptides in microbial infections are complemented by chapters that address their mechanisms of action and of microbial resistance.

DEIRDRE A. DEVINE is a senior lecturer in Microbiology in the Division of Oral Biology at the Dental Institute, University of Leeds, UK. Her research focuses on the role of antimicrobial peptides in control of the resident microflora and in preventing disease. She also works on oral anaerobes, environmental regulation of bacterial virulence factors, biofilms, and new antimicrobial approaches. She is the coeditor of *Medical Implications of Biofilms* (2003).

ROBERT E. W. HANCOCK is professor of Microbiology and Immunology at the University of British Columbia (UBC) and is a Canada Research Chair holder. He is head of the UBC Centre for Microbial Diseases and Immunity Research. His research interests include antibiotic uptake and resistance, functional genomics, and the development of small cationic peptides as novel antimicrobials and immune modulators.

Published Titles

1. *Bacterial Adhesion to Host Tissues.* Edited by Michael Wilson 0521801079
2. *Bacterial Evasion of Host Immune Responses.* Edited by Brian Henderson and Petra Oyston 0521801737
3. *Dormancy and Low-Growth States in Microbial Disease.* Edited by Anthony R. M. Coates 0521809401
4. *Susceptibility to Infectious Diseases.* Edited by Richard Bellamy 0521815258
5. *Bacterial Invasion of Host Cells.* Edited by Richard Lamont 0521809541

Forthcoming Titles in the Series

The Dynamic Bacterial Genome. Edited by Peter Mullany 0521821576
Bacterial Protein Toxins. Edited by Alistair Lax 052182091X
The Influence of Bacterial Communities on Host Biology. Edited by Margaret McFall Ngai, Brian Henderson, and Edward Ruby 0521834651
The Yeast Cell Cycle. Edited by Jeremy Hyams 0521835569
Salmonella Infection. Edited by Pietro Mastroeni and Duncan Maskell 0521835046

Over the past decade, the rapid development of an array of techniques in the fields of cellular and molecular biology have transformed whole areas of research across the biological sciences. Microbiology has perhaps been influenced most of all. Our understanding of microbial diversity and evolutionary biology and of how pathogenic bacteria and viruses interact with their animal and plant hosts at the molecular level, for example, have been revolutionized. Perhaps the most exciting recent advance in microbiology has been the development of the interface discipline of cellular microbiology, a fusion of classic microbiology, microbial molecular biology, and eukaryotic cellular and molecular biology. Cellular microbiology is revealing how pathogenic bacteria interact with host cells in what is turning out to be a complex evolutionary battle of competing gene products. Molecular biology and cellular biology are no longer discrete subject areas but vital tools and an integrated part of current microbiological research. As part of this revolution in molecular biology, the genomes of a growing number of pathogenic and model bacteria have been fully sequenced, with immense implications for our future understanding of microorganisms at the molecular level.

Advances in Molecular and Cellular Microbiology is a series edited by researchers active in these exciting and rapidly expanding fields. Each volume focuses on a particular aspect of cellular or molecular microbiology and provides an overview of the area, as well as examining current research. This series will enable graduate students and researchers to keep up with the rapidly diversifying literature in current microbiological research.

ADVANCES IN MOLECULAR AND CELLULAR MICROBIOLOGY

Advances in Molecular and Cellular Microbiology 6

Mammalian Host Defense Peptides

EDITED BY
Deirdre A. Devine
University of Leeds

Robert E. W. Hancock
University of British Columbia

PUBLISHED BY THE PRESS SYNDICATE OF THE UNIVERSITY OF CAMBRIDGE
The Pitt Building, Trumpington Street, Cambridge, United Kingdom

CAMBRIDGE UNIVERSITY PRESS
The Edinburgh Building, Cambridge CB2 2RU, UK
40 West 20th Street, New York, NY 10011-4211, USA
477 Williamstown Road, Port Melbourne, VIC 3207, Australia
Ruiz de Alarcón 13, 28014 Madrid, Spain
Dock House, The Waterfront, Cape Town 8001, South Africa

http://www.cambridge.org

First published 2004

Printed in the United States of America

Typefaces FF Scala 9.5/13 pt., Formata and Quadraat Sans   *System* LaTeX 2ε [TB]

A catalog record for this book is available from the British Library.

Library of Congress Cataloging in Publication Data

Mammalian host defense peptides / edited by Deirdre A Devine,
Robert E.W. Hancock.

p. cm.

Includes bibliographical references and index.

ISBN 0-521-82220-3 (hbk.)

1. Peptides. 2. Mammals–Physiology. I. Devine, Deirdre, 1961–
II. Hancock, R. E. W. (Robert E. W.)

QP552.P4M366 2004
571.9′619–dc22

2003060609

ISBN 0 521 82220 3 hardback

Contents

Contributors

Birgitta Agerberth
Karolinska Institute
Stockholm, Sweden

Mark Ackermann
Agricultural Research Service
United States Department of Agriculture
Ames, Iowa

Robert Bals
Klinikum der Philipps – Universität Marburg
Germany

Charles L. Bevins
University of California Davis School
of Medicine
Davis, California

Kim A. Brogden
Agricultural Research Service
United States Department of Agriculture
Ames, Iowa

Deirdre A. Devine
University of Leeds Dental Institute
Leeds, United Kingdom

Gill Diamond
UMDNJ – New Jersey Medical School
Newark, New Jersey

Timothy J. Falla
Helix Biomedix Inc.
Bothell, Washington

Tomas Ganz
David Geffen School of Medicine at UCLA
Los Angeles, California

Gudmundur H. Gudmundsson
Karolinska Institute
Stockholm, Sweden

John S. Gunn
The Ohio State University
Columbus, Ohio

Robert E. W. Hancock
University of British Columbia
Vancouver, Canada

Eva J. Helmerhorst
Boston University Goldman School of Dental Medicine
Boston, Massachusetts

Marcia Klein-Patel
UMDNJ – New Jersey Medical School
Newark, New Jersey

Danielle Laube
UMDNJ – New Jersey Medical School
Newark, New Jersey

Robert I. Lehrer
UCLA School of Medicine
Los Angeles, California

Frank G. Oppenheim
Boston University
Boston, Massachusetts

Joost J. Oppenheim
Frederick Cancer Research and Development Centre
Frederick, Maryland

Amanda C. Portillo
University of Texas Health Science Center at San Antonio
San Antonio, Texas

Rita Tamayo
University of Texas Health Science Center at San Antonio
San Antonio, Texas

Michael J. Welsh
Agricultural Research Service
United States Department of Agriculture
Ames, Iowa

De Yang
Frederick Cancer Research and Development Centre
Frederick, Maryland

Michael R. Yeaman
Harbor – UCLA Medical Centre
Torrance, California

Joseph Zabner
Agricultural Research Service
United States Department of Agriculture
Ames, Iowa

Lijuan Zhang
Helix Biomedix Inc.
Bothell, Washington

Mammalian Host Defense Peptides

Editorial: "Antimicrobial" or "host defense" peptides

Robert E. W. Hancock and Deirdre A. Devine

1

Short cationic amphipathic peptides were first demonstrated in the 1970s to be present in amphibians, insects, and human phagocytes. When examined by use of in vitro assays of antimicrobial activity, they could be demonstrated to kill bacteria and other microorganisms and were thus accorded the general names "cationic antimicrobial peptides" or "antibiotic peptides" and were lauded as "Nature's antibiotics." As is clear from recent research summarized in this book and in leading journal review articles (e.g., Boman, 1995; Andreu and Rivas, 1998; Gudmundsson and Agerberth, 1999; Hancock and Diamond, 2000), they have many other activities that are relevant to the anti-infective host defense process known as innate immunity. We would like to propose here that, with some prominent exceptions, most of these peptides have no *relevant* antibiotic activities at physiological concentrations and conditions and, because they have multiple impacts on innate immunity, they should be classed as "host (innate) defense peptides" or "peptides of the innate immune system."

The prevailing conditions in vivo do not favor the antimicrobial activity of cationic peptides. Often these activities are assessed ex vivo by either a 10-mM phosphate buffer or, for example, a tenfold diluted bacterial growth medium. Of necessity, such conditions are artificial and certainly do not reflect most mammalian tissue environments. Some papers in the literature have considered the higher levels of salt in vivo; however, sodium and chloride ions have a relatively modest effect on antimicrobial peptide activity. Indeed, divalent cations have a much stronger effect and, at the millimolar concentration found in vivo (e.g., blood has approximately 2-mM Ca^{2+} and 1-mM Mg^{2+}), can completely ablate the activity of many or most natural peptides. This happens only with 200-mM monovalent cations (Friedrich et al., 1999). Other highly antagonistic agents include polyanionic saccharides (e.g., mucins) and cell surfaces, and serum factors, including lipoproteins and proteases. Such considerations have been made for the cationic aminoglycosides, for which it has been shown that in vivo conditions can be partly reflected by supplementation of in vitro minimum inhibitory concentration (MIC) assays (Reller et al., 1974); however, they rarely are for cationic peptides. For example, human LL-37 is classed by many as antimicrobial in that it has MICs of around 1–8 μg/ml in diluted broth containing up to 100-mM NaCl (Turner et al., 1998). However, in normal Mueller Hinton medium, which has moderate divalent cation levels, MICs of ≥32 μg/ml are observed (Turner et al., 1998) as confirmed in the Hancock laboratory, far higher than the 2–5 μg/ml found at mucosal surfaces. It will be important as this field moves forward to rate as "antimicrobial" only those peptides that are functionally antimicrobial at physiologically meaningful concentrations and under physiological conditions.

We do not intend to imply that direct antimicrobial activity never occurs with such peptides, but rather that many of these peptides do not act as antimicrobials in most locations where they are found in the host. For example, the work of Lehrer and colleagues has indicated that cationic α-defensins constitute 5% or more of total neutrophil proteins (Spitznagel, 1990), and this means that the concentration would be around 10–100 mg/ml in the compartments where they are found (azurophilic granules and, during phagocytosis, phagolysosomes). Also, estimates of defensin concentrations in intestinal crypts are around 25 mg/ml (Charles Bevins, personal communication). Similarly, peptides can be found at concentrations of >100 μg/ml at sites of chronic inflammation (Hancock and Diamond, 2000). Other peptides, such as polyphemusins from horseshoe crabs (Zhang et al., 2000) and protegrins from pigs, and so on (Steinberg et al., 1997), are far more active than most of the peptides discussed here. In addition, synergy between individual peptides is possible, although such studies have not been performed under physiologically meaningful conditions. Nevertheless, direct killing of microbes would be a part of the host defenses constituting innate immunity, and we submit that a more accurate description for this class of molecules is "host defense peptides." It is likely that some peptides have antimicrobial functions at one body site (e.g., in a phagosome) and other host defence roles at other sites (e.g., at epithelial surfaces when released by degranulation). Also, these peptides may play different roles at heavily colonized sites compared with those that are normally sterile (e.g., intestinal compared with lung epithelia).

The nonantimicrobial activities of these peptides include stimulation of chemotaxis of phagocytic cells, vasodilation (through encouragement of histamine release from mast cells), neutralization of bacterial-signaling molecules such as lipopolysaccharide (LPS) and lipoteichoic acid (LTA), cell differentiation, and so forth (Boman, 1995; Andreu and Rivas, 1998; Gudmundsson and Agerberth, 1999; Hancock and Diamond, 2000). Given these activities, we must explain why the mammalian host contains measurable activities of many peptides at a range of body sites, which at the same time harbor large numbers of bacteria that constitute the normal resident microbiota. Added to this is the fact that these resident populations produce most of the same surface molecules that signal Toll-like (pattern-recognition) receptors (TLRs). We hypothesize that the background expression of innate immunity peptides in the normal host provides a homeostatic balance to signaling by the natural flora, preventing undesirable induction of innate immunity. When this situation is locally perturbed by the introduction of new microbes onto a mucosal/epidermal surface, by increases in certain populations, or by released microbial components above threshold levels, TLRs are activated, leading to local upregulation of innate immunity. At the same time, signaling through TLRs leads to an increased expression of host (innate) defense peptides. These peptides themselves induce novel gene responses that block the upregulation of gene responses signaled by bacterial surface molecules, permitting reestablishment of homeostasis. If this model is correct, then peptides have a central role in the process of innate immunity and may also assist in the decision to induce both chronic inflammation and adaptive immunity.

The reviews presented in this book discuss a variety of the aspects previously discussed. Many interesting perspectives are presented and, especially, we invite reviewers to read, consider, and make up their own minds about how these peptides might function.

ACKNOWLEDGMENT

The terms host defense peptides and peptides of the innate immune system were suggested to us by Alex Tossi and Tim Falla, respectively.

REFERENCES

Andreu, D. and Rivas, L. (1998). Animal antimicrobial peptides: An overview. *Biopolymers*, 47, 415–33.

Boman, H. G. (1995). Peptide antibiotics and their role in innate immunity. *Annual Reviews of Immunology*, 13, 61–92.

Friedrich, C., Scott, M. G., Karunaratne, N., Yan, H., and Hancock, R. E. W. (1999). Salt-resistant alpha-helical cationic antimicrobial peptides. *Antimicrobial Agents and Chemotherapy*, 43, 1542–8.

Gudmundsson, G. H. and Agerberth, B. (1999). Neutrophil antibacterial peptides, multifunctional effector molecules in the mammalian immune system. *Journal of Immunological Methods*, 232, 45–54.

Hancock, R. E. W. and Diamond, G. (2000). The role of cationic antimicrobial peptides in innate host defences. *Trends in Microbiology*, 8, 402–10.

Reller, L. B., Schoenknecht, F. D., Kenny, M. A., and Sherris, J. C. (1974). Antibiotic susceptibility testing of *Pseudomonas aeruginosa*: Selection of a control strain and criteria for magnesium and calcium content in media. *Journal of Infectious Diseases*, 130, 454–63.

Spitznagel, J. K. (1990). Antibiotic proteins of human neutrophils. *Journal of Clinical Investigation*, 86, 1381–6.

Steinberg, D. A., Hurst, M. A., Fujii, C. A., Kung, A. H., Ho, J. F., Cheng, F. C., Loury, D. J., and Fiddles, J. C. (1997). Protegrin-1: A broad-spectrum, rapidly microbicidal peptide with in vivo activity. *Antimicrobial Agents and Chemotherapy*, 41, 1738–42.

Turner, J., Cho, Y., Dinh, N. N., Waring, A. J., and Lehrer, R. I. (1998). Activities of LL-37, a cathelin-associated antimicrobial peptide of human neutrophils. *Antimicrobial Agents and Chemotherapy*, 42, 2206–14.

Zhang, L., Scott, M. G., Yan, H., Mayer, L. D., and Hancock, R. E. W. (2000). Interaction of polyphemusin I and structural analogs with bacterial membranes, lipopolysaccharide and lipid monolayers. *Biochemistry*, 39, 14504–14.

CHAPTER 1

Overview: Antimicrobial peptides, as seen from a rearview mirror

R. I. Lehrer

Here I sit, having just celebrated my sixty-fifth birthday, wondering why I agreed to write this overview and also why I never learned to type. I will be brief. If these reflections seem uninteresting, remember that nobody is forcing you to read them. The other chapters in this volume will provide an up-to-date and "serious" introduction to the antimicrobial peptides of mammals.

The gene-encoded antimicrobial peptides of mammals are very old, because such peptides also exist in archaea, eubacteria, protists, plants, and invertebrates. Nevertheless their study is relatively new. Consequently it may be helpful to recall the following dialogue. After William Gladstone (1809–98), Chancellor of the Exchequer, witnessed a demonstration of the generation of electricity by Michael Faraday (1791–1867), Gladstone said "It is very interesting, Mr. Faraday, but what practical worth is it?" Faraday replied "One day, sir, you may tax it." To date, mammalian antimicrobial peptides have been tax exempt.

I complete this overview by recounting how the field began and how I got into it and by mentioning some other early investigators. The search for endogenous antimicrobial molecules arose in the middle third of the nineteenth century. Eli Metchnikoff (1845–1916), an insightful Russian émigré who spent his later years at the Pasteur Institute, first recognized the vital role of phagocytes in host defense and also inquired into their microbicidal mechanisms. In those pre-Sigma Catalogue days only trypsin and pepsin preparations were readily available to him. Finding that these did not kill bacteria, Metchnikoff surmised that other leukocyte enzymes might do so. His speculation was proven correct when, over 30 years later, Alexander Fleming described lysozyme. According to the accounts of Lady Fleming, lysozyme's discovery was largely ignored by the medical community of the day because it was effective only against nonpathogens. When Fleming later described penicillin, this discovery also received little attention, and the industrial development of penicillin had to wait for the exigencies of World War II.

Recognizing the implications of the nascent science of bacteriology, a Scottish surgeon named Joseph Lister (1827–1912) revolutionized surgical practice by using aerosolized phenol (carbolic acid) to prevent infection and by using phenol-soaked lint to dress wounds. No less than the introduction of ether anesthesia in 1846, a generation before, disinfection and antisepsis revolutionized surgical practice. Although Lister knew of Metchnikoff's work, neither knew that phagocytes used disinfectants that were less cytotoxic than phenol. They produced these substances "on demand" through the agencies of two tightly regulated enzyme complexes: nicotinamide-adenine dinucleotide phosphate (NADPH) oxidase and inducible nitric oxide synthase.

Mammalian neutrophils contain myeloperoxidase, an enzyme that converts hydrogen peroxide, a product of NADPH oxidase, into more potent microbicidal oxidants that include hypochlorite and chloramines. During World War I, Henry Drysdale Dakin (1880–1952), an English-born biochemist who once worked at the Lister Institute, joined Alexis Carrel in introducing dilute sodium hypochlorite irrigations to treat wound infections. "Carrel–Dakins solution" was highly effective, and, unlike Lister's phenol, it retained activity in blood. Sodium hypochlorite is also the active ingredient in Clorox, a common household bleach and disinfectant that was "invented" in 1916.

Leukocytes also have much to teach about antimicrobial peptides. The antimicrobial properties of crude leukocyte extracts were noted in the 1940s and 1950s. Although memorable names, such as leukins or phagocytin, were created to describe the phenomenon, precise molecular characterization of the active principle was not yet feasible. The modern era of antimicrobial peptide research began in the mid–1960s when Hussein Zeya and John Spitznagel described highly cationic polypeptides ("lysosomal cationic proteins") in leukocytes from rabbits and guinea pigs. Considering that their most powerful preparative tools were cellulose and free boundary electrophoresis, they had remarkable success in characterizing these peptides. Unfortunately, their progress stopped when most workers in the field became enthralled with an inherited condition called chronic granulomatous disease (CGD).

Indeed, there were many reasons to be interested in CGD. Although the condition was rare, it was serious; most of the affected children sustained frequent infections, and many died by their late teens. The blood neutrophils and monocytes of CGD patients could ingest various bacteria and fungi normally, but showed defective killing of many of them because of deficient production of hydrogen peroxide and related oxidants by their NADPH oxidase.

Over the next two decades, many laboratories worked to define NADPH oxidase, to ascertain the details of its regulation and structure, and to identify the molecular defects responsible for CGD. During this time, NADPH oxidase was a Holy Grail, and only heretics or skeptics began other quests.

I was also involved in these mainstream issues, but as I tested the neutrophils and monocytes of individuals with CGD or hereditary myeloperoxidase deficiency, I found that they killed many bacteria and fungi with normal or near-normal efficacy. Hence I began to look for other antimicrobial components in leukocytes. By 1974, I had learned how to obtain large numbers of "activated" rabbit alveolar macrophages in considerable purity by using a technique developed by Eva S. Leake and Quentin N. Myrvik. I extracted these macrophages with acid, and subjected the clarified extracts to nondenaturing polyacrylamide gel electrophoresis (PAGE) in pencil-sized tube gels. After the gels were hemisected longitudinally, one half was stained and the other half was sliced at 1-mm intervals with an array of single-edged razor blades. The 60 or so little gel pieces were transferred to test tubes, pulverized in a small volume of distilled water, and the eluted contents were tested against various bacteria and fungi. This simple and direct preparative procedure identified two highly cationic antibacterial and antifungal components. With this preliminary data in hand, I applied for National Institutes of Health funding and six years and three proposals later secured it. Although it is amusing to read the reviewer's comments now, it was less amusing then. Fortunately, I had grants to study postphagocytic ion fluxes in neutrophils and the activation of NADPH oxidase, so the work could continue "on the side."

In the early 1980s, work on insect antimicrobial peptides from Hans Boman's lab in Sweden began to appear. At the same time, the UCLA group (including myself, Judith Delafield, Michael Selsted, Tomas Ganz, and the late Sylvia Harwig) began to isolate and characterize the peptides now called α-defensins. Gradually others began to join the search. I recall that, when I found Bob Hancock's 1989 publication on rabbit NP-1, I sent him a letter (I did not then know him) welcoming him to the "defensin club." A recent Medline keyword search on defensins retrieved well over 1,000 hits. Had I continued to write welcoming letters, I would surely have become an expert typist by now.

By the end of that decade, the first β-defensins had been described in the tracheal epithelial cells and leukocytes of cattle, and Michael Zasloff had captured the imagination of the public with his description of magainins. The first cathelicidin peptides had been recognized, largely through the efforts of Dominico Romeo, Margarita Zanetti, and Renato Gennaro. The first three human β-defensin (HBD) peptides, HBD1, HBD2, and HBD3, were isolated and described by Harder and Schroeder between 1996 and 2001. More recently, powerful genomics-based search strategies identified 28 "new" β-defensin genes (DEFB) in humans and 43 "new" DEFB genes in mice. Although these numbers are small compared with odorant receptor genes (approximately 900 in humans and 1,500 in the mouse) and some other mammalian multigene families, they are nevertheless impressive. HBD1 is prominently expressed in the human vagina and multiple β-defensin genes are expressed in the human and murine epididymis, suggesting that these peptides play significant roles in reproductive processes.

In any rapidly developing field, surprises can be expected. I end by mentioning two that come from our recent studies. We recently established that several θ-(and α-) defensins are lectins. This property enables them to bind surface glycoproteins and glycolipids involved in cell entry by HIV-1 and herpes simplex viruses. I suspect that the ability to bind sugars could contribute to many other properties, including pathogen recognition and receptor-mediated signaling. At the least, in the words of Linda Loman, "Attention must be paid!"

We have formed somewhat heretical views about the mechanism of action of two exceptionally potent antimicrobial peptides: protegrins and sheep myeloid antimicrobial peptide (SMAP-29). We have evidence that these peptides kill susceptible microbes by inducing a process akin to fresh water drowning – namely, a massive influx of water that overwhelms the microbe's osmoregulatory apparatus. I named this the HOTTER (an acronym for hydro-osmotic transtesseral extrusion and rupture) mechanism. As soon as I get my typing up to speed, I intend to put the supporting data into a manuscript.

The principal risk in "naming names" comes from leaving some out. Although I expect no complaints from Metchnikoff or Lister, if I did not mention you in the view from my rearview mirror, then perhaps you were and are in front of me. Please excuse the lack of references. I will learn how to use my citation manager after mastering typing. By the time a second edition comes around, I should have it perfected.

CHAPTER 2

Cationic antimicrobial peptides in regulation of commensal and pathogenic microbial populations

Deirdre A. Devine

2.1. INTRODUCTION

Microbial cells that comprise the diverse resident communities colonizing mucosal sites outnumber cells of the human body by 10:1 (Savage, 1977). It seems remarkable that these potentially overwhelming populations coexist with a host, with harmful effect only if the host becomes immunocompromised or organisms reach sites to which they do not normally have access, for example, through trauma. Effective maintenance and control of resident populations is very important to a colonized host, as these populations contribute to host protection through blocking of colonization by pathogens (e.g., Mead and Barrow, 1990; Roos, Håkansson, and Holm, 2000), development of cell structure and function (Hooper, Falk, and Gordon, 2000; Freitas et al., 2002), and development of the immune system (Cebra, 1999). In addition, nonpathogenic bacteria can downregulate or attenuate inflammatory responses (Neish et al., 2000). Disruption of the host–microbe balance and loss of regulation of these populations may have seriously detrimental effects in development of infections (e.g., in immunocompromised patients) or chronic inflammatory disorders (Neish et al., 2000; Wehkamp et al., 2002). The mammalian host is able, under normal circumstances, to allow the survival and long-term tolerance of these essential resident microbial communities without eliciting a damaging chronic inflammatory response.

The mechanisms involved in this host–microbe homeostasis are not well understood, but cationic antimicrobial peptides possess many characteristics that indicate roles in regulating resident populations as well as defending against specific pathogens. Diverse antimicrobial peptides are components of the innate defenses of a wide range of higher and lower host species, and there is evidence that they have evolved under positive pressures exerted by colonizing microorganisms. At each site of production, antimicrobial peptides form part of a cocktail of antimicrobial substances that in vivo work synergistically to combat infection (Gudmundsson and Agerberth, 1999; Hancock and Diamond, 2000). Recent research has provided evidence that antimicrobial peptides have multiple activities in host immunity and have a key role in modulating early immune responses (Yang, Kwak, and Oppenheim, 2002, see also chapter 3). This multiplicity of function, combined with the fact that every host species produces such a range of site-specific antimicrobial peptides, has led to the proposal that cationic antimicrobial peptides are of key importance in host responses to highly diverse resident populations (Boman, 1996; Garabedian et al., 1997; Simmaco et al., 1998). The complexity of these host–microbe relationships is illustrated by consideration of the composition and diversity of resident populations, which vary according to site and host species.

2.2. DIVERSITY AND SITE SPECIFICITY OF RESIDENT MICROBIAL POPULATIONS

The complex mechanisms involved in regulating responses to colonizing mammalian hosts must interact with a vast diversity of microorganisms, mostly bacteria, although some protozoa, fungi, and viruses are members of the resident microbiota (Tannock, 1999). In humans and other mammals, some sites are usually sterile (e.g., lung, bladder), whereas others (e.g., oral cavity and colon) are heavily colonized by largely anaerobic bacterial populations (Fig. 2.1). It is estimated that up to 600 species, only 50% of which can be grown in monoculture by conventional methods, are normal inhabitants of the human mouth (Wilson, Weightman, and Wade, 1997; Paster et al., 2001), and the human gut harbors more than 400 bacterial species (Berg, 1996). In spite of gaining access to the gastrointestinal tract through frequent swallowing, few oral organisms colonize the gut. Recent phylogenetic studies have shown that resident populations in the human gut and oral cavity are equally highly diverse, but substantially different in composition (Martin, 2002). Studies of oral and nasopharyngeal resident populations have also shown that sites that are anatomically close or adjacent can nonetheless harbor very distinct microbiota (e.g., Rasmussen et al., 2000; Hohwy, Reinholdt, and Kilian, 2001; Könönen et al., 2002).

In addition to the microbial species diversity evident in resident populations, single species exhibit substantial genetic diversity (e.g., Jolley et al., 2000; Hohwy et al., 2001). The genetic composition of a species at one site can fluctuate significantly, with resident populations comprising persistent

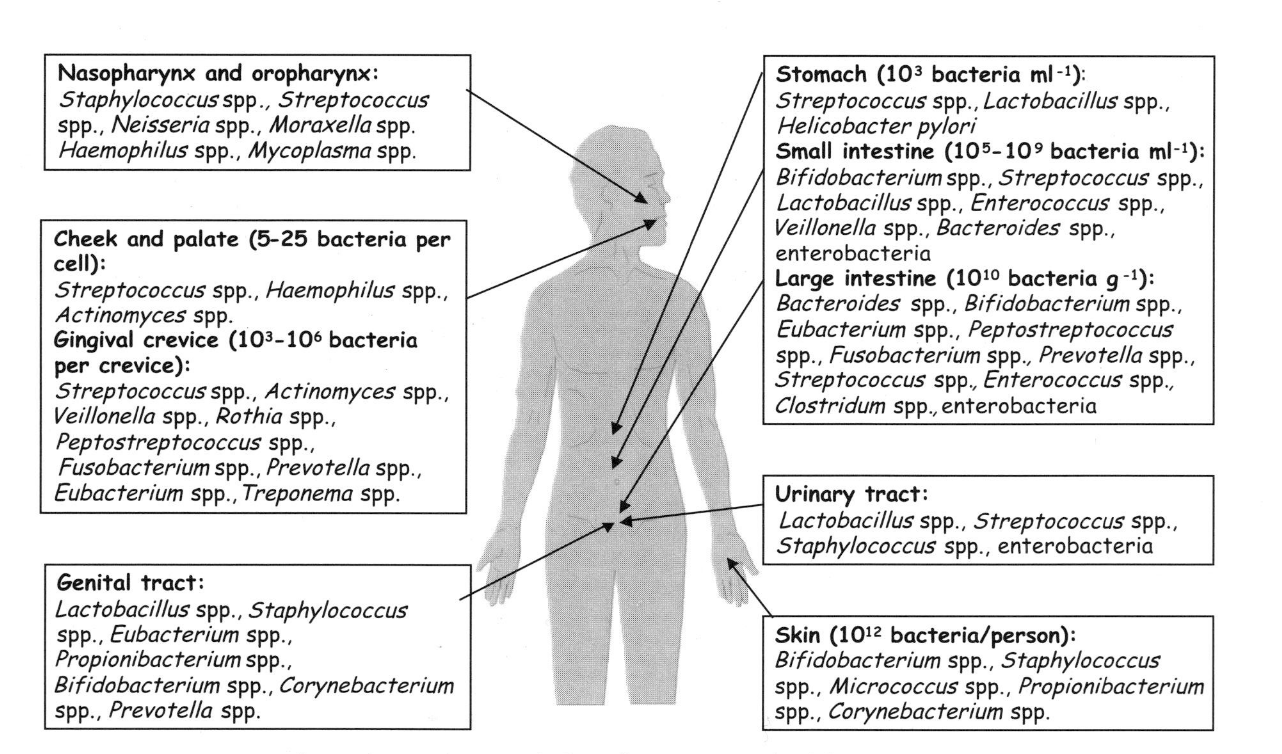

Figure 2.1. Resident microbial populations associated with human tissues.

clones as well as a succession of transient clones (Caugant, Levin, and Selander, 1981; Hohwy et al., 2001), illustrating the dynamic relationship between host and resident populations.

Finally, the complexity of resident populations is further increased through the genetic and phenotypic adaptability of microorganisms. Concentrations of ions such as Fe^{3+}, Mg^{2+}, Ca^{2+}, and H^{+} vary widely among tissues, cells, and serum, and inflammation or bleeding also alter environmental parameters such as temperature, pH, and nutrient availability. Bacteria have exquisite mechanisms for responding to such environmental changes, in many cases comprising two-component signal transduction pathways, for example, PhoP–PhoQ (Groisman, 2001). Bacterial gene expression also alters in response to exposure to host defenses (including antimicrobial peptides) and adhesion to cells or surfaces (Agnani, Tricot-Doleux, and Bonnaure-Mallet, 2000; Donlan and Costerton, 2002; Kolenbrander et al., 2002; Hong et al., 2003). Organisms within resident populations form interacting and codependent communities, often within biofilms, the gene expression of each member responding to signals from others through, for example, quorum sensing mechanisms (Bassler, 1999; de Kievet and Iglewski, 2000; Kolenbrander et al., 2002). Finally, gene alteration and adaptation occur through horizontal gene transfer between closely and distantly related species (Hentschel, Steinert, and Hacker, 2000).

In view of their contributions to colonization resistance and immune and cellular development, it is important that resident populations be established and tolerated. To this end, it is essential that mechanisms regulating host–microbe relationships exhibit a breadth of recognition to accommodate such microbial diversity and genetic flexibility. Also, in view of the potentially catastrophic effects of disruption of host–microbe homeostatic mechanisms, it would make sense for such mechanisms to exhibit some redundancy. Antimicrobial peptides fulfill these requirements.

2.3. DIVERSITY AND SITE SPECIFICITY OF MAMMALIAN CATIONIC ANTIMICROBIAL PEPTIDES

The number of known antimicrobial peptides has increased to over 500 (Boman, 2000). In mammals, the most commonly isolated antimicrobial peptides include the α- and β-defensins, which have been described in humans, cattle, sheep, pigs, rats, mice, rabbits, and guinea pigs (Schutte et al., 2002), and defensinlike molecules have additionally been described (Jia et al., 2001; Morrison et al., 2002b). The discovery of β-defensins and subsequent genome sequence searches have indicated that the defensins constitute a larger family of peptides than previously described (Bals et al., 1999a; Jia et al., 2000, 2001; Garcia et al., 2001; Schutte et al., 2002; Yamaguchi et al., 2002). Molecules from nondefensin structural groups further increase the diversity of mammalian antimicrobial peptides. These include, for example, human LL-37 and histatins, murine cathelin-related antimicrobial peptides (CRAMPs), porcine protegrins, bovine bactenecin, and indolicidin. A novel circular molecule has been described in primates (Tam et al., 1999) and the 8-cysteine hepcidin was isolated from human liver and urine (Park et al., 2001).

Mammalian antimicrobial peptides are expressed by a wide range of tissues and cells (Huttner and Bevins, 1999; Devine and Hancock, 2002; Schutte et al., 2002; Diamond, Laube, and Klein-Patel, Chapter 5 of this book) and, along with factors such as the expression of specific microbial adhesin molecules and host cell surface receptors, probably contribute to the tissue tropism exhibited by commensal and pathogenic microorganisms. Some peptides, such as human β-defensins (HBDs) 1–3 (HBD1, HBD2, and HBD3) and LL-37, provide protection at a number of sites, many of which are normally colonized by large resident microbial populations. Other antimicrobial peptides are specialized in terms of sites of production. For example, peptides expressed specifically in epididymal tissues have been described in human (von Horsten, Derr, and Kirchhoff, 2002; Yamaguchi et al., 2002), rat (Li et al., 2001), and mouse tissues (Yamaguchi et al., 2002); hepcidin has been detected only in human liver and urine; histatins are solely produced within human salivary glands. This specialization may suggest that a particular spectrum of activity against certain pathogenic or resident microorganisms is of primary importance, in addition to the other functions exhibited by antimicrobial peptides. Valore et al. (1998) found a number of β-defensin isoforms in human urine, some of which were more effective against relevant bacteria, especially when tested in urine. Specialized site-specific production may also reflect a need to protect critical cells, for example, secretory cells in salivary ducts, although evidence provided by Garabedian et al. (1997) did not support a contention that production of Paneth cell antimicrobial factors, which include cryptdins, is essential to prevent colonization of intestinal crypts. However, secretion of cryptdins is significantly upregulated by bacteria and bacterial cellular components, and cryptdins may nonetheless contribute to the defense of the environment adjacent to crypts (Ayabe et al., 2000).

Further interindividual antimicrobial peptide diversity is indicated by the fact that multiple polymorphisms in the β-defensin 1 gene have been reported (Dörk and Stuhrmann, 1998; Vatta et al., 2000; Circo et al., 2002). Although some of these would result in neutral amino acid changes or were in noncoding regions, others, if transcribed, would produce molecules that may be predicted to differ in biological activity. The in vivo importance of these polymorphisms is largely unknown, but one has been linked with chronic obstructive pulmonary disease (Matsushita et al., 2002).

Studies of isolated peptides have shown that the range of molecules produced may be increased posttranslationally through the formation of isoforms, resulting from minor truncations of the amino-terminus (Valore et al., 1998; O'Neil et al., 1999; Park et al., 2001). Isoforms of human antimicrobial peptides HBD1, HBD2, hepcidin, and histatins have been described (Valore et al., 1998; Hiratsuka et al., 2000; Zucht et al., 1998; O'Neil et al., 1999; Park et al., 2001; Diamond et al., 2001), and isoforms of defensins in rat, mouse, and rabbit have been isolated (Miyasaki et al., 1990; Kohashi et al., 1992; Ouellette et al., 2000). These minor differences in amino acid sequence can produce significant differences in antimicrobial activity (Raj, Antonyraj, and Karunakaran, 2000), providing an additional mechanism to ensure maximal and optimal antimicrobial cover at a specific site, with little metabolic cost to the host cell. In addition, some peptides derived from larger proteins have cationic antimicrobial peptide properties; for example, fragments of lactoferrin, bactericidal permeability-inducing peptides, histones, ribosomal protein, hemoglobin, and mucin (Odell et al., 1996; Boman, 2000; Levy, 2000; Mak et al., 2001; Bobek and Situ, 2003). Some peptides have antimicrobial functions, although they were originally described because they displayed other important biological functions, for example, chemokines and adrenomedullin (Allaker, Zihni, and Kapas, 1999; Cole et al., 2001).

Thus antimicrobial peptides are diverse and site specific, as are resident commensal microbial populations and pathogens. Evolutionary evidence also indicates a link between antimicrobial peptides and regulation of microbial populations, and this is discussed in the next section.

2.4. EVOLUTION OF MAMMALIAN ANTIMICROBIAL PEPTIDES AND COLONIZING MICROORGANISMS

There is evidence that pathogenic and other bacteria have evolved and diversified in response to host immune mechanisms (Gupta and Maiden, 2001). Additionally, genetic studies of a range of commensal organisms have shown that mucosal populations of these species comprise clones that are in a constant state of change (Caugant et al., 1981; Hohwy et al., 2001). The ecological pressure driving this dynamic replacement of clones is not known.

Many studies have examined the evolution of defensins, and these have indicated an evolutionary relationship between α- and β-defensin families (Hughes, 1999; Morrison et al., 2002b, Yamaguchi et al., 2002). These studies have also indicated that diverse defensins have arisen through rapid evolution, with positive Darwinian pressure resulting in selection of mutations arising after gene-duplication events (Hughes, 1999). Antimicrobial peptide molecules are amenable to considerable diversification; Bauer et al. (2001) suggested that the sequence variability seen in defensins in and between species indicates that the defensin structural fold is able to accommodate a wide range of amino acids at most positions within the molecule. Evidence gained from examining rates of synonymous and nonsynonymous nucleotide substitution in genes for mammalian α-defensins, as well as bovine and ovine β-defensins, has shown that these antimicrobial peptides have evolved under the positive pressure of colonizing microorganisms (Hughes, 1999). Similar analyses of primate (nonhuman) β-defensin 1 genes indicated that these genes diversified without being subjected to such positive pressure (Del Pero et al., 2002), which may imply that the nonantimicrobial (e.g., immunomodulatory) functions of these primate peptides are most important and have therefore evolved in response to distinct selective pressures.

2.5. THE ROLES OF ANTIMICROBIAL PEPTIDES IN HOST–MICROBE RELATIONSHIPS

There is little doubt that antimicrobial peptides are significant in the host response and defense against infection. In humans and mammals, concentrations of neutrophil and epithelial antimicrobial peptides increase to significant levels following infection or injury (Hancock and Diamond, 2000; Dorschner et al., 2001). In some cases, these increased concentrations have been implicated in the pathology of the disease (Devine and Hancock, 2002), and certain diseases or susceptibility to disease may be associated with overexpression or underexpression of antimicrobial peptides (Lawrence, Fiocchi, and Chakravarti, 2001; Dauletbaev et al., 2002; Ong et al., 2002; Wehkamp et al., 2002). Resistance to neutrophil antimicrobial peptides has been cited as a pathogenicity determinant for intracellular or invasive pathogens, such as *Salmonella enterica* serovar Typhimurium (Ernst, Guina, and Miller, 2001) and group A streptococci (Nizet et al., 2001).

It is logical and intuitive to propose that antimicrobial peptides play a central role not only in defending against pathogens and responding to infection, but also in regulating colonization and selection of resident populations. The reasons for such a proposal are as follows: (1) Resident microbial populations and antimicrobial peptides are diverse, species specific, and site specific; (2) there is evidence that antimicrobial peptides have evolved in response to selection pressures exerted by resident and pathogenic microbial populations; and (3) antimicrobial peptides have multiple functions in innate defense and communicate between innate and adaptive immunity, suggesting regulatory or signaling roles in immune responses. However, experimental examination of the role of antimicrobial peptides in host–microbe relationships, with resident populations in particular, is challenging.

One problem associated with elucidating fully the role of antimicrobial peptides in host–microbe relationships is that it is essential to understand these relationships and to use meaningful experimental models. This is by no means straightforward. Some organisms do not cause clinically overt disease in all hosts, and important pathogens may be carried as part of the normal resident microbiota of individuals for considerable periods of time. Usually, disease is the manifestation of the combined activities of microbes and host, which has led to attempts to redefine these interactions and terms such as pathogenicity, with host and microbe contributions taken into account (Casadevall and Pirofski, 1999, 2000; Hentschel et al., 2000). Also, certain infections are not caused by single species and arise at sites that are very heavily colonized, often by fastidious or unculturable organisms. For example, sites in the oral cavity, and the periodontal pocket in particular, are heavily colonized with extremely diverse bacterial populations, so that consortia, rather than individual organisms, are associated with diseases (Haffajee et al., 1999; Marsh and Bradshaw, 1999). In such cases it can be extremely difficult to determine which organisms contribute to the etiology of the disease and which are bystanders in the process. It may also be difficult to identify organisms that directly colonize the mucosa and that have contact with antimicrobial peptides or antimicrobial-peptide-secreting cells. Intestinal populations of *Escherichia coli*, for example, are composed of mixtures of populations derived from mucosal and luminal sites. Bacteria associated with periodontal disease are not primary colonizers and may not interact much, if at all, with healthy mucosa. Resident microorganisms of mucosal tissues often establish commensal relationships with a host early in life, and these primary colonizers are of particular relevance in understanding the role of antimicrobial peptides in the establishment and survival of resident organisms and in host–microbe homeostasis.

Notwithstanding these difficulties, increasing numbers of studies are addressing questions related to the roles played by antimicrobial peptides in determining the outcome of microbial colonization of a host by use of animal and in vitro studies.

2.5.1. Animal Studies of the Roles of Antimicrobial Peptides in Host–Microbe Relationships

Some studies have used in vivo models to examine the influence of antimicrobial peptides on bacterial colonization and clearance or on susceptibility to disease. Blocking or inhibition of antimicrobial peptide production in amphibians and *Drosophila* resulted in increased or overwhelming infection (Meister, Lemaitre, and Hoffmann, 1997; Simmaco et al., 1998). Inhibition of antimicrobial peptide synthesis in frogs also led to increased colonization and growth of commensal populations (Simmaco et al., 1998). Studies of rodents (see following subsections) have supported a role for mammalian antimicrobial peptides in defense against infection, and they demonstrate antimicrobial peptide redundancy (or overlap of function of multiple antimicrobial peptides expressed at one site) in host innate defenses and suggest differences in the roles of the same peptide at different sites.

Skin

In a model of skin infections, mice in which the gene for cathelicidin CRAMP had been disrupted suffered more severe infections than wild-type mice following challenge with group A streptococci (Nizet et al., 2001).

Respiratory Tract

Overexpression of the human peptide LL-37 in a cystic fibrosis mouse model resulted not only in increased killing of *Pseudomonas aeruginosa* (Bals et al., 1999c), but also in reduced ability of *P. aeruginosa* to colonize the lung epithelium and in reduced inflammation and susceptibility to septic shock (Bals et al., 1999b). Mice deficient in mouse β-defensin 1 (mBD-1) expression did not show any overt signs of ill health or abnormality (Morrison et al., 2002a; Moser et al., 2002). In 24 h postinfection, *Haemophilus influenzae* was cleared less efficiently from lungs and airways of mutant mice compared with wild-type mice (Moser et al., 2002). However, mBD-1 appeared to be less important in defense against other pathogens, as mBD-1-deficient mice eliminated *Staphylococcus aureus* from their lungs as efficiently as wild-type mice (Morrison et al., 2002a), and mutant and wild-type mice were equally susceptible to infection, sepsis, and death following infection with *Streptococcus pneumoniae* (Moser et al., 2002).

Gastrointestinal Tract

Small intestinal crypts are normally devoid of colonizing microorganisms and experiments that used ablation of Paneth cells in mice indicated that antimicrobial factors of these cells, which include cryptdins, are not involved in preventing colonization of intestinal crypts (Garabedian et al., 1997). However, the authors did suggest that secretion of cryptdins may influence the composition of the flora at small intestinal sites adjacent to crypts (Garabedian et al., 1997), and the fact that exposure of Paneth cells to bacteria and bacterial cell products significantly increased cryptdin secretion further supports a role for cryptdins in modulating bacterial populations in the small intestine (Ayabe et al., 2000). Mice deficient in matrilysin expression were also unable to produce mature cryptdins, and these animals were less efficient at controlling colonization by orally administered bacteria (Wilson et al., 1999).

Urinary Tract

A number of studies have indicated an interesting relationship between antimicrobial peptides and defence of the bladder. The mBD-1-deficient mice were less able than wild-type mice to eliminate bacteria, particularly *S. aureus*, from the bladder (Morrison et al., 2002a), and oral administration of antimicrobial peptides protected mice from urinary tract infection (Håverson et al., 2000). Women with recurrent urinary tract infection were less able than other women to clear bacteria from the urethra (Kunin et al., 2002), although the levels of secretion of antimicrobial peptides by urethral tissues by these two groups were not assessed. Isoforms of β-defensins have been isolated from urine (Valore et al., 1998; Zucht et al., 1998; Hiratsuka et al., 2000), and their cytotoxic activities have led Zucht et al. (1998) to suggest they are involved in cell turnover/eradication at epithelia, which may be particularly important at a sparsely colonized site such as the bladder, in which bacterial adhesion to epithelial cells appears to be so important in the disease process. Hiratsuka et al. (2000) found low levels of expression of β-defensin 1 in diabetic mice and rats, leading Lehrer and Ganz (2002) to suggest this may be worthy of further study, as a possible explanation for the fact that people with diabetes suffer increased incidence of urinary tract infections.

Summary

The mBD-1-deficient mouse studies suggest redundancy in antimicrobial peptide expression and function, as deficiency in production of a single antimicrobial peptide did not have significant or overt consequences for the general health of animals. They also indicate that in the lung, one antimicrobial peptide may be more important in defence against one organism than another (*H. influenzae* compared with *S. aureus*), and a number of studies imply that an individual antimicrobial peptide may be more significant in defense of one site compared with another (e.g., bladder). It is possible that specialization of function of individual antimicrobial peptides is greater or more apparent at sites that are usually sparsely colonized or sterile (e.g., lung, bladder) compared with sites that are heavily colonized (e.g., mouth, intestine).

2.5.2. Studies of Induction of Antimicrobial Peptides by Colonizing Microorganisms

None of the preceding studies specifically examined the role of antimicrobial peptides in controlling resident populations, probably because there are considerable difficulties associated with the complex and labor-intensive microbiology required and the higher numbers of animals needed for gaining significant results. However, in vitro studies have also been undertaken to determine the nature of interactions between antimicrobial peptides and microbes, which may indicate their roles in vivo. Colonizing microorganisms, pathogenic or commensal, would benefit from surviving or evading the effects of antimicrobial peptides, and in vitro studies have explored this. Existing evidence indicates that common strategies are employed by both pathogens and nonpathogens. It is possible that colonization and survival may be facilitated by either a failure to induce antimicrobial peptide synthesis or by an ability to downregulate their expression, and examples can be found of each of these mechanisms.

Skin

Human skin is protected by constitutive production of HBD-1 and dermcidin and inducible expression of HBD2 and LL-37 (Chronnell et al., 2001; Schittek et al., 2001; Ong et al., 2002). Although low levels of induction of HBD-2 and LL-37 have been linked with susceptibility to skin infection by certain bacteria (Ong et al., 2002), no consistent relationship emerged between microbial pathogenicity and induction of skin antimicrobial peptide expression when in vitro models were used. Viable and nonviable *S. aureus* and *Candida albicans,* both opportunistic pathogens that are commonly carried asymptomatically, did not upregulate HBD2 production by cultured keratinocytes (Liu et al., 2002). However, Dinulos et al. (2003) found that keratinocyte HBD2 expression was upregulated by *S. aureus* as well as by *Staphylococcus epidermidis,* but not by *Streptococcus pyogenes.* Gram-negative colonization or infection of the skin is rare; *E. coli* did not induce keratinocytes to produce HBD2 (Liu et al., 2002 ; Dinulos et al., 2003), and, in a tissue-engineered human skin equivalent, HBD-2 was only weakly induced by a nonpathogenic laboratory K12 strain of *E. coli* (Schmid et al., 2001). In one study, viable cultures of *P. aeruginosa* induced keratinocyte HBD2 expression (Liu et al., 2002) but in another this organism did not (Dinulos et al., 2003), highlighting the importance of the experimental system used.

Oral Cavity

Krisanaprakornkit et al. (2000) proposed that oral commensal bacteria upregulate the synthesis of HBD-2 in gingival epithelial cells whereas pathogens do not, based on observations of differences in the abilities of cell walls from *Fusobacterium nucleatum* and *Porphyromonas gingivalis* to induce HBD-2. The microbiology of the periodontal pocket is complex, and defining the roles of individual organisms in the disease process is equally complex (Haffajee et al., 1999; Loesche and Grossman, 2001). Thus, although *P. gingivalis* is considered by most to be a periodontal pathogen, the role of *F. nucleatum* is equivocal. It is described by some as a putative periodontal pathogen, and it additionally demonstrates properties associated with pathogenicity (Han et al., 2000); on the other hand, it is widely distributed in dental plaque at both healthy and diseased sites. However, it is not a primary or early colonizer of the oral mucosa, and it is arguably these organisms that we should study to understand host–microbe homeostasis. *Candida* spp. are usually commensal colonizers of oral mucosa and cause disease only in immunocompromised or antibiotic-treated individuals or on unnatural inert surfaces, such as dentures. Jainkittivong, Johnson, and Yeh (1998) noted reduced salivary flow and lower salivary histatin concentrations in individuals whose oral mucosa were colonized by *Candida* compared with those whose oral mucosa were uncolonized. Whether this was a reflection of patient factors or the ability of *Candida* spp. to repress secretion of histatins was not explored. Kapas et al. (2001) found that oral keratinocytes were induced to upregulate adrenomedullin production by Gram-positive and Gram-negative bacteria, but not by *C. albicans*.

Respiratory Tract

Although animal models of lung infection are well developed, there is relatively little information regarding the abilities of respiratory organisms to induce expression of antimicrobial peptides by respiratory cells, although inducible and constitutive production of antimicrobial peptides is known to occur through the respiratory tract and increased levels of β-defensins and LL-37 have been recorded in infections and inflammation (Dauletbaev et al., 2002; Lee et al., 2002; Schaller-Bals, Schulze, and Bals, 2002). β-defensin synthesis was induced in vivo in mouse respiratory epithelia by *P. aeruginosa* (Bals et al., 1999a; Morrison, Davidson, and Dorin, 1999). Patients with cystic fibrosis, who are susceptible to infection by bacteria highly resistant to antimicrobial peptides, do not appear to upregulate β-defensins in response to inflammatory mediators (Dauletbaev et al., 2002).

Gastrointestinal Tract

Some studies have indicated that enteric pathogens upregulate antimicrobial peptide synthesis in gastrointestinal epithelial cells. Coculture of gastric epithelial cells with *Helicobacter pylori* resulted in induction of HBD2 production, and HBD1 was upregulated at high multiplicity of infection (O'Neil et al., 2000; Bajaj-Elliott et al., 2002). Examination of biopsies confirmed higher levels of expression of both β-defensins in *H. pylori* gastritis (Bajaj-Elliott et al., 2002). Intestinal cell lines were stimulated to produce HBD2 by exposure to *Salmonella enteritidis* (Takahashi et al., 2001), *S. enterica* serovar Dublin, and enteroinvasive *E. coli*, and there were higher levels of HBD2 in colorectal biopsies of inflamed tissues compared with noninflamed (O'Neil et al., 1999). However, although these pathogens regulated expression of β-defensins, the same species did not affect expression of the gene for LL-37 in human colonic epithelial cells, nor was LL-37 expression dependent on the presence of a normal microbiota (Hase et al., 2002). Another highly significant enteric pathogen, *Shigella dysenteriae*, downregulated synthesis of both LL-37 and HBD1 (Islam et al., 2001). Few studies have considered commensal or nonpathogenic organisms and antimicrobial peptide induction in these experimental systems, although an attenuated strain of *Salmonella typhi* did induce HBD2 expression in human small-intestinal xenografts (O'Neil et al., 1999).

2.5.3. The Roles of Resistance to Antimicrobial Peptides in Host–Microbe Interactions

A number of mechanisms have been described that increase microbial resistance to antimicrobial peptides (reviewed in Devine and Hancock, 2002, Peschel, 2002, and chapter 9 and 12 in this book), and these have been associated with virulence in certain intracellular or invasive pathogens. For example, *S. enterica* serovar Typhimurium pathogenicity is dependent on lipopolysaccharide (LPS) modifications that increase resistance to antimicrobial peptides (Ernst et al., 2001). Also, CRAMP-resistant mutants of a group A *Strep.* strain were more virulent in a mouse skin infection model, although both wild-type and mutant bacteria caused necrotic infection (Nizet et al., 2001). The innate resistance of *Burkholderia cepacia* and *S. aureus* to antimicrobial peptides has been linked to their ability to cause serious infection in patients with chronic granulomatous disease, whose neutrophils are deficient in oxidative killing mechanisms. These examples link bacterial resistance to antimicrobial peptides and virulence with intracellular survival or invasion. Before neutrophils are encountered or access to deeper tissues is gained, it would be anticipated that microbial colonization and persistence on mucosal tissues would be facilitated by an ability to resist killing by mucosal antimicrobial peptides.

Cell envelope modifications resulting in altered charge density or hydrophobicity have been associated with alterations in sensitivity to antimicrobial peptides (Ernst et al., 2001: Peschel, 2002). Unencapsulated nontypable *H. influenzae* (NTHi) is a common commensal of the upper respiratory tract and is sometimes associated with localized disease. In a manner analogous to lipid A modifications in *S. enterica* serovar Typhimurium, NTHi regulates acylation of lipid A. Underacylated mutants of *H. influenzae* were more susceptible to killing by HBD2, but killing by the more highly cationic HBD3 was unaffected (Starner et al., 2002). Swords et al. (2002) found that such mutants were also less able to colonize human airway xenografts and speculated that NTHi may differentially acylate lipid A during commensal and disease states.

Host mimicry by decoration of cell surfaces with phosphorylcholine (ChoP) occurs in a wide range of commensal and pathogenic organisms that colonize the oral cavity and upper respiratory tracts, for example, streptococci, pneumococci, *Neisseria* spp., *Haemophilus* spp., *Actinomyces* spp., *Fusobacterium* spp., mycoplasmas, and others (Lysenko et al., 2000; Schenkein et al., 2001). ChoP substitution of *H. influenzae* LPS resulted in a reduction in sensitivity to LL-37 (Lysenko et al., 2000), but did not affect sensitivity to HBD2 or HBD3 (Starner et al., 2002). Both *H. influenzae* and *S. pneumoniae* variants expressing ChoP were more efficient than ChoP$^-$ variants at colonization and persistence in animal models, and this molecule appears to be particularly important in the early stages of infection. Serino and Virji (2000) found ChoP on the surfaces of a large number of commensal and pathogenic *Neisseria* spp. However, in commensal strains, ChoP was present only on LPS, whereas in pathogenic strains it decorated pili. It is not known whether this link between the mode of ChoP attachment and commensalism/pathogenicity is related to interactions with antimicrobial peptides or to the fact that cell surface ChoP facilitates adhesion to the platelet-activating factor on host cells and also affects the host-signaling pathways initiated by bacterial colonization (Swords et al., 2002).

Gram-positive cell wall teichoic acids and lipoteichoic acids (LTAs) vary greatly with respect to modification of alditol groups by glycosyl residues or D-alanine. In commensal and pathogenic strains of *Staphylococcus* spp., *dlt* operon-mediated D-alanyl esterification of teichoic acids resulted in reduced sensitivity to antimicrobial peptides by decreasing cell wall negative charge, consequently reducing antimicrobial peptide binding (Peschel et al., 1999). A *dlt*-deficient mutant of *S. aureus* was less virulent in a mouse model than in the wild-type strain (Collins et al., 2002). Many Gram-positive species possess *dlt* operons; therefore this may be a common mechanism for resisting innate peptides, as well as peptides produced by other Gram-positive bacteria, such as lactococcin, nisin, and subtilin (Sahl, Jack, and Bierbaum, 1995; Nes et al., 1996).

Bacterial proteases are produced by a wide variety of pathogenic and nonpathogenic organisms, and these may help overcome or deregulate host defences and also cause direct damage to host tissues. Pathogenic or opportunistic organisms (*P. aeruginosa, Enterococcus faecalis, Proteus mirabilis,* and *S. pyogenes*) degraded LL-37 (Schmidtchen et al., 2002), whereas *P. gingivalis,* which is associated with periodontal disease, cleaved a range of nonhuman antimicrobial peptides (Devine et al., 1999). However, in *P. gingivalis* this ability did not necessarily correlate with sensitivity to killing by antimicrobial peptides, possibly because the protease-mediated peptide degradation occurred relatively slowly. Nor was it related to pathogenicity, as nonpathogenic oral anaerobes were also able to degrade antimicrobial peptides (Devine et al., 1999).

2.5.4. The Broader Relevance to Host–Microbe Interactions of Properties Affecting Resistance

The cellular properties and modifications described in the preceding subsection influence antimicrobial peptide resistance, but they often affect other activities, which are also highly significant in determining the outcomes of microbial colonization. Acylation of lipid A strongly influences interactions between LPS and host receptors, thereby helping to determine the signaling pathways initiated. It has been proposed that acylation affects the shape of lipid A, which in turn affects binding to host cell Toll-like receptors (Netea et al., 2002). PhoP-constitutive mutants of *S. enterica* serovar Typhimurium attenuated inflammatory responses of intestinal epithelial cells (Neish et al., 2000), and underacylated mutants of NTHi stimulated host cells less than wild-type bacteria did and stimulated less of the proinflammatory cytokines TNFα and IL-6 (Swords et al., 2002). ChoP on teichoic acids of *S. pneumoniae* and on LPSs of NTHi promotes adhesion to the platelet-activating factor on host cells, and the products of *dlt* provide other selective advantages to some bacteria, such as increased acid tolerance, greater intracellular polymer accumulation, and mediation of interbacterial aggregations involved in biofilm establishment (Clemans et al., 1999; Spatafora et al., 1999; Boyd et al., 2000). Finally, proteases can have many indirect effects beyond direct damage of host tissues and antimicrobial peptides, including deregulation of inflammatory processes and inactivation of host cell receptors. *P. gingivalis* proteases, and other virulence determinants, are environmentally controlled and appear to be downregulated during inflammation; this attenuation of virulence under certain conditions may contribute to the long-term survival of this organism within the hostile environment of the periodontal pocket (Percival et al., 1999).

2.6. CONCLUSIONS

There is not a clear-cut relationship between resistance to antimicrobial peptides or induction of antimicrobial peptide expression and pathogenicity or commensalism. As pathogens display a number of strategies for causing disease (Finlay and Falkow, 1997; Rhen, Eriksson, and Pettersson, 2000; Boldrick et al., 2002; Tato and Hunter, 2002), so do they employ varied strategies for overcoming or evading the actions of antimicrobial peptides. Nonpathogenic organisms also employ many of these strategies with respect to antimicrobial peptides. In some ways this is not surprising. Commensal and pathogenic organisms express many common surface molecules such as LPS or LTA, some of which are referred to as pathogen-associated molecular patterns (Medzhitov and Janeway, 2000). In fact, pathogens may differ only to a small extent from nonpathogens, and transition from commensal to pathogen may result from relatively small changes in gene expression. For example, this transition in a strain of *E. coli* resulted from genetic variation solely in a tissue-specific adhesion factor (Sokurenko et al., 1998). Many pathogenicity determinants are found clustered on mobile genetic elements and pathogenicity islands (Hentschel et al., 2000), which may be transmitted by horizontal gene transfer (Hacker et al., 1997). Pathogens and commensals often differ very little in terms of genetic composition, and it is not surprising that they adopt similar strategies for survival; pathogens in addition display properties that result in host damage (Hentschel et al., 2000) which may or may not involve antimicrobial peptides.

Current opinions concerning the interactions between host and commensal bacteria are undergoing some revision. It had been a widely assumed that climax communities of commensal bacteria are highly stable whereas pathogenic organisms are eliminated by the host defenses within a limited period of time. However, genetic studies of a range of organisms, including *Streptococcus mitis, E. coli, H. influenzae,* and *Moraxella catarrhalis,* have shown that mucosal populations of these species comprise clones that are in a constant state of change (Caugant et al., 1981; Trottier, Stenberg, and Svanborg-Eden, 1989; Klingman et al., 1995; Hohwy et al., 2001). Genotypes of *S. mitis* detected in the upper respiratory tract were established on the mucosal membranes for weeks but for less than 3 months. It appears that many commensal bacteria are no more stable than pathogenic bacteria that transiently colonize mucosal membranes (Hohwy et al., 2001). Thus it is perhaps no surprise that commensal and pathogenic bacteria are in some cases similar in their interactions with antimicrobial peptides. Hohwy et al. (2001) stated that the ecological pressure contributing to the turnover of some resident clones and not others was unknown; it is possible that antimicrobial peptides contribute to this selective pressure.

Most of the studies just discussed concern interactions between antimicrobial peptides or host cells and whole viable bacteria. This overlooks another function of antimicrobial peptides, which is that they bind avidly to many potentially proinflammatory molecules released from microorganisms, such as LPS, LTA, and DNA. In doing so, they inhibit responses of host cells and suppress an undesirable inflammatory response (Scott, Gold, and Hancock, 1999; Scott et al., 2000a, 2000b; Nagaoka et al., 2001), and this may be a key function of antimicrobial peptides. This suppression of inflammation and protection against sepsis may be through binding of proinflammatory effectors, such as LPS, or through direct action of the peptide on host cell gene expression (Scott et al., 2002). Epithelial cells are generally relatively nonresponsive to LPS (Svanborg, Godaly, and Hedlund, 1999; Neish et al., 2000), and antimicrobial peptides may contribute to this, thereby ensuring host–microbe homeostasis and toleration of resident populations. Some mucosal antimicrobial peptides are expressed constitutively whereas others are induced by microbes or inflammatory mediators. Perhaps constitutive "sentinel" antimicrobial peptides bind to and neutralize free microbial cellular components, thereby helping attenuate an immune response. Initiation of a host response may follow an increased microbial load or the presence of particular chemotypes of molecules such as LPS. Antimicrobial peptides may function in concert with other LPS-binding molecules such as lipopolysaccharide-binding proteins and bactericidal permeability-inducing proteins to minimize and regulate immune responses at mucosal sites assaulted by large numbers of bacteria and, in particular, their released cellular components. Although many more studies are needed to promote a full understanding of the roles of antimicrobial peptides in maintaining host–microbe homeostasis, it is becoming clear that these multifunctional molecules are essential in host defense against harmful microorganisms, in regulation of normal resident populations, and in regulating host responses to microorganisms and their products.

REFERENCES

Agnani, G., Tricot-Doleux, S., and Bonnaure-Mallet, M. (2000). Adherence of *Porphyromonas gingivalis* to gingival epithelial cells: Modulation of bacterial protein expression. *Oral Microbiology and Immunology*, 15, 48–52.

Allaker, R. P., Zihni, C., and Kapas, S. (1999). An investigation into the antimicrobial effects of adrenomedullin on members of the skin, oral, respiratory tract, and gut microbiota. *FEMS Immunology and Medical Microbiology*, 23, 289–93.

Ayabe, T., Satcheell, D. P., Wilson, C. L., Parks, W. C., Selsted, M. E., and Oullette, A. J. (2000). Secretion of microbicidal alpha-defensins by intestinal Paneth cells in response to bacteria. *Nature Immunology*, 1, 113–18.

Bajaj-Elliott, M., Fedeli, P., Smith, G. V., Domizio, P., Maher, L., Ali, R. S., Quinn, A. G., and Farthing, M. J. G. (2002). Modulation of host antimicrobial peptide (β-defensins 1 and 2) expression during gastritis. *Gut*, 51, 356–61.

Bals, R., Wang, X., Meegalla, R. L., Wattler, S., Weiner, D. J., Nehls, M. C., and Wilson, J. M. (1999a). Mouse β-defensin 3 is an inducible antimicrobial peptide expressed in the epithelia of multiple organs. *Infection and Immunity*, 67, 3542–7.

Bals, R., Weiner, D. J., Meegalla, R. L., and Wilson, J. M. (1999b). Transfer of a cathelicidin peptide antibiotic gene restores bacterial killing in a cystic fibrosis xenograft model. *Journal of Clinical Investigation*, 103, 1113–17.

Bals, R., Weiner, D. J., Moscioni, A. D., Meegalla, R. L., and Wilson, J. M. (1999c). Augmentation of innate host defence by expression of a cathelicidin antimicrobial peptide. *Infection and Immunity*, 67, 6084–9.

Bassler, B. L. (1999). How bacteria talk to each other: Regulation of gene expression by quorum sensing. *Current Opinion in Microbiology*, 2, 582–7.

Bauer, F., Schweimer, K., Klüver, E., Conejo-Garcia, J.-R., Forssmann, W.-G., Rösch, P., Adermann, K., and Sticht, H. (2001). Structure determination of human and murine β-defensins reveals structural conservation in the absence of significant sequence similarity. *Protein Science*, 10, 2470–9.

Berg, R. D. (1996). The indigenous gastrointestinal microflora. *Trends in Microbiology*, 4, 430–5.

Bobek, L. A. and Situ, H. (2003). MUC7 20-mer: Investigation of antimicrobial activity, secondary structure, and possible mechanism of antifungal action. *Antimicrobial Agents and Chemotherapy*, 47, 643–52.

Boldrick, J. C., Alizadeh, A. A., Diehn, M., Dudoit, S., Liu, C. L., Belcher, C. E., Botstein, D., Staudt, L. M., Brown, P. O., and Relman, D. A. (2002). Stereotypes and specific gene expression programs in human innate immune responses to bacteria. *Proceedings of the National Academy of Science, USA*, 99, 972–7.

Boman, H. G. (1996). Antimicrobial peptides: Key components needed in immunity. *Cell*, 65, 205–7.

Boman, H. G. (2000). Innate immunity and the normal microflora. *Immunological Reviews*, 173, 5–16.

Boyd, D. A., Cvitkovitch, D. G., Bleiweis, A. S., Kiruikhin, M. Y., Debabov, D. V., Neuhaus, F. C., and Hamilton, I. R. (2000). Defects in D-alanyl-lipoteichoic acid synthesis in *Streptococcus mutans* results in acid sensitivity. *Journal of Bacteriology*, 182, 6055–65.

Casadevall, A. and Pirofski, L.-A. (1999). Host–pathogen interactions: Redefining the basic concepts of virulence and pathogenicity. *Infection and Immunity*, 67, 3703–13.

Casadevall, A. and Pirofski, L.-A. (2000). Host–pathogen interactions: Basic concepts of microbial commensalisms, colonization, infection, and disease. *Infection and Immunity*, 68, 6511–18.

Caugant, D. A., Levin, B. R., and Selander, R. K. (1981). Genetic diversity and temporal variation in the *E. coli* population of a human host. *Genetics*, 98, 467–90.

Cebra, J. J. (1999). Influences of microbiota on intestinal immune system development. *Journal of Immunology*, 69, 1046S–51S.

Chronnell, C. M., Ghali, L. R., Ali, R. S., Quinn, A. G., Holland, D. B., Bull, J. J., Cunliffe, W. J., Mackay, I. A., Philpott, M. P., and Muller-Rover, S. (2001). Human beta defensin-1 and -2 expression in human pilosebaceous units: Upregulation in acne vulgaris lesions. *Journal of Investigative Dermatology*, 117, 1120–5.

Circo, R., Skerlavaj, B., Gennaro, R., Amoroso, A., and Zanetti, M. (2002). Structural and functional characterization of hBD-1(Ser35), a peptide deduced from a DEFB1 polymorphism. *Biochemical and Biophysical Research Communications*, 293, 586–92.

Clemans, D. L., Kolenbrander, P. E., Debabov, D. V., Zhang, Q., Lunsford, R. D., Sakone, H., Whitteker, C. J., Heaton, M. P., and Neuhaus, F. C. (1999). Insertional inactivation of genes responsible for the D-alanylation of lipoteichoic acid in *Streptococcus gordonii* DL1 (Challis) affects intrageneric coaggregations. *Infection and Immunity*, 67, 2464–74.

Cole, A. M., Ganz, T., Liese, A. M., Burdick, M. D., Liu, L., and Strieter, R. M. (2001). IFN-inducible ELR^{-} CXC chemokines display defensin-like antimicrobial activity. *Journal of Immunology*, 167, 623–7.

Collins, L. V., Kristian, S. A., Weidenmaeir, C., Faigle, M., van Kessel, K. P. M., van Strijp, J. A. G., Gotz, F., Neumeister, B., and Peschel, A. (2002). *Staphylococcus aureus* strains lacking D-alanine modifications of teichoic acids are highly susceptible to human neutrophil killing and are virulence attenuated in mice. *Journal of Infectious Diseases*, 186, 214–19.

Dauletbaev, N., Gropp, R., Frye, M., Loitsch, S., Wagner, T. O., and Bargon, J. (2002). Expression of human β-defensin (HBD-1 and HBD-2) mRNA in nasal epithelia of adult cystic fibrosis patients, healthy individuals and individuals with acute colds. *Respiration*, 69, 46–51.

de Kievit, T. R., and Iglewski, B. H. (2000). Bacterial quorum sensing in pathogenic relationships. *Infection and Immunity*, 68, 4839–49.

Del Pero, M., Boniotto, M., Zuccon, D., Cervella, P., Spanò, A., Amoroso, A., and Crovella, S. (2002). β-defensin 1 gene variability among non-human primates. *Immunogenetics*, 53, 907–13.

Devine, D. A., and Hancock, R. E. W. (2002). Cationic peptides: Distribution and mechanisms of resistance. *Current Pharmaceutical Design*, 8, 99–110.

Devine, D. A., Marsh, P. D., Percival, R. S., Rangarajan, M., and Curtis, M. A. (1999). Modulation of antibacterial peptide activity by products of *Porphyromonas gingivalis* and *Prevotella* spp. *Microbiology*, 145, 965–71.

Diamond, D. L., Kimball, J. R., Krisanaprakornkit, S., Ganz, T., and Dale, B. A. (2001). Detection of β-defensins secreted by human oral epithelial cells. *Journal of Immunological Methods*, 256, 65–76.

Dinulos, J. G., Mentele, L., Fredericks, L. P., Dale, B. A., and Darmstadt, G. L. (2003). Keratinocyte expression of human β defensin 2 following bacterial infection: Role in cutaneous host defence. *Clinical and Diagnostic Laboratory Immunology*, 10, 161–6.

Donlan, R. M. and Costerton, J. W. (2002). Biofilms: Survival mechanisms of clinically relevant microorganisms. *Clinical Microbiology Reviews*, 15, 167–93.

Dörk, T. and Sturhmann, M. (1998). Polymorphisms of the human *ß-defensin-1* gene. *Molecular and Cellular Probes*, 12, 171–3.

Dorschner, R. A., Pestonjamasp, V. K., Tamakuwala, S., Ohtake, T., Rudisill, J., Nizet, V., Agerberth, B., Gudmundssen, G. H., and Gallo, R. L. (2001). Cutaneous injury induces the release of cathelicidin antimicrobial peptides active against group A *Streptococcus*. *Journal of Investigative Dermatology*, 117, 91–7.

Ernst, R. K., Guina, T., and Miller, S. I. (2001). *Salmonella typhimurium* outer membrane remodelling: Role in resistance to host immunity. *Microbes and Infection*, 3, 1327–34.

Finlay, B. B. and Falkow, S. (1997). Common themes in microbial pathogenicity revisited. *Microbiology and Molecular Biology Reviews*, 61, 136–69.

Freitas, M., Axelsson, L.-G., Cayuela, C., Midvedt, T., and Trugnan, G. (2002). Microbial–host interactions specifically control the glycosylation pattern in intestinal mucosa. *Histochemistry and Cell Biology*, 118, 149–61.

Garabedian, E. M., Roberts, L. J. J., McNevin, M. S., and Gordon, J. I. (1997). Examining the role of Paneth cells in the small intestine by lineage ablation in transgenic mice. *Journal of Biological Chemistry*, 272, 23729–40.

Garcia, J.-R. C., Jaumann, F., Schulz, S., Krause, A., Rodriguez-Jiminez, J., Forssmann, U., Adermann, K., Klüver, E., Vogelmeier, C., Becker, D., Hedrich, R., Forssmann, W.-G., and Bals, R. (2001). Identification of a novel multifunctional β-defensin (human β-defensin 3) with specific antimicrobial activity. *Cell Tissue Research*, 306, 257–64.

Groisman, E. A. (2001). The pleiotropic two-component regulatory system PhoP-PhoQ. *Journal of Bacteriology*, 183, 1835–42.

Gudmundsson, G. H. and Agerberth, B. (1999). Neutrophil antibacterial peptides, multifunctional effector molecules in the mammalian immune system. *Journal of Immunological Methods*, 232, 45–54.

Gupta, S., and Maiden, M. C. J. (2001). Exploring the evolution of diversity in pathogen populations. *Trends in Microbiology*, 9, 181–5.

Hacker, J., Blum-Oehler, G., Mühldorfer, I., and Tschäpe, H. (1997). Pathogenicity islands of virulent bacteria: Structure, function, and impact on microbial evolution. *Molecular Microbiology*, 23, 1089–97.

Haffajee, A. D., Socransky, S. S., Ferres, M., and Ximenez-Fyvie, L. A. (1999). Plaque microbiology in health and disease. In *Dental Plaque Revisited*, ed. H. N. Newman and M. Wilson, pp. 255–82. Cardiff: Bioline.

Han, Y. W., Shi, W., Huang, G.T.-J., Haake, S. K., Park, N-H., Kuramitsu, H., and Genco, R. J. (2000). Interactions between periodontal bacteria and human oral epithelial cells: *Fusobacterium nucleatum* adheres to and invades epithelial cells. *Infection and Immunity*, 68, 3140–6.

Hancock, R. E. W. and Diamond, G. (2000). The role of cationic antimicrobial peptides in innate host defences. *Trends in Microbiology*, 8, 402–10.

Hase, K., Eckmann, L., Leopard, J. D., Varki, N., and Kagnoff, M. F. (2002). Cell differentiation is a key determinant of cathelicidin LL-37/human cationic antimicrobial protein 18 expression by human colon epithelium. *Infection and Immunity*, 70, 953–63.

Håverson, L. A., Engberg, I., Baltzer, L., Dolphin, G., Hanson, L. Å., and Mattsby-Baltzer, I. (2000). Human lactoferrin and peptides derived from a surface-exposed helical region reduce experimental *Escherichia coli* urinary tract infection in mice. *Infection and Immunity*, 68, 5816–23.

Hentschel, U., Steinert, M., and Hacker, J. (2000). Common molecular mechanisms of symbiosis and pathogenesis. *Trends in Microbiology*, 8, 226–31.

Hiratsuka, T., Nakazato, M., Minematsu, T., Chino, N., Nakanishi, T., Shimizu, A., Kangawa, K., and Matsukuru, S. (2000). Structural analysis of human β-defensin-1 and its significance in urinary tract infection. *Nephron*, 85, 34–40.

Hohwy, J., Reinholdt, J., and Kilian, M. (2001). Population dynamics of *Streptococcus mitis* in its natural habitat. *Infection and Immunity*, 69, 6055–63.

Hong, R. W., Shchepetov, M., Weiser, J. N., and Axelsen, P. H. (2003). Transcriptional profile of the *Escherichia coli* response to the antimicrobial insect peptide cecropin A. *Antimicrobial Agents and Chemotherapy*, 47, 1–6.

Hooper, L., Falk, P. G., and Gordon, J. I. (2000). Analyzing the molecular foundations of commensalisms in the mouse intestine. *Current Opinion in Microbiology*, 3, 79–85.

Hughes, A. L. (1999). Evolutionary diversification of the mammalian defensins. *Cellular and Molecular Life Sciences*, 56, 94–103.

Huttner, K. M. and Bevins, C. L. (1999). Antimicrobial peptides as mediators of epithelial host defense. *Pediatric Research*, 45, 785–94.

Islam, D., Bandholtz, L., Nilsson, J., Wigzell, H., Christensson, B., Agerberth, B., and Gudmundsson, G. H. (2001). Downregulation of bactericidal peptides in enteric infections: A novel immune escape mechanism with bacterial DNA as a potential regulator. *Nature Medicine*, 7, 180–85.

Jainkittivong, A., Johnson, D. A., and Yeh, C. K. (1998). The relationship between salivary histatin levels and oral yeast carriage. *Oral Microbiology and Immunology*, 13, 181–7.

Jia, H. P., Schutte, B. C., Schudy, A., Linzmeier, R., Guthmiller, J. M., Johnson, G. K., Tack, B. F., Mitros, J. P., Rosenthal, A., Ganz, T., and McCray, P. B. Jr. (2001). Discovery of new β-defensins using a genomics based approach. *Gene*, 263, 211–18.

Jia, H. P., Wowk, S. A., Schutte, B. C., Lee, S. K., Vivado, A., Tack, B. F., Bevins, C. L., and McCray, P. B. (2000). A novel murine β-defensin expressed in tongue, esophagus, and trachea. *Journal of Biological Chemistry*, 275, 33314–20.

Jolley, K. A., Kalmusova, J., Feil, E. J., Gupta, S., Musilek, M., Kriz, P., and Maiden, M. C. J. (2000). Carried meningococci in the Czech Republic: A diverse recombining population. *Journal of Clinical Microbiology*, 38, 4492–8.

Kapas, S., Bansal, A., Bhargava, V., Maher, R., Malli, D., Hagi-Pavli, E., and Allaker, R. P. (2001). Adrenomedullin expression in pathogen-challenged oral epithelial cells. *Peptides*, 22, 1485–9.

Klingman, K. L., Pye, A., Murphy, T. F., and Hill, S. (1995). Dynamics of respiratory tract colonization by *Branhamella catarrhalis* in bronchiectasis. *American Journal of Respiratory Critical Care Medicine*, 152, 1072–8.

Kohashi, O., Ono, T., Ohki, K., Soejima, T., Moriya, T., Umeda, A., Meno, Y., Amako, K., Funakosi, S., Masuda, M., and Fujii, N. (1992). Bactericidal activities of rat defensins and synthetic rabbit defensins on staphylococci, *Klebsiella pneumoniae* (Chedid, 277, and 8N3), *Pseudomonas aeruginosa* (mucoid and nonmucoid strains), *Salmonella typhimurium* (Ra, Rc, Rd, and Re of LPS mutants) and *Escherichia coli*. *Microbiology and Immunology*, 36, 369–80. [Erratum appears in *Microbiology and Immunology* 1992, 36, following p. 909.]

Kolenbrander, P. E., Andersen, R. N., Blehert, D. S., Egland, P. G., Foster, J. S., and Palmer, R. J. (2002). Communication among oral bacteria. *Microbiology and Molecular Biology Reviews*, 66, 486–505.

Könönen, E., Jousimies-Somer, H., Bryk, A., Kilpi, T., and Kilian, M. (2002). Establishment of streptococci in the upper respiratory tract: Longitudinal changes in the mouth and nasopharynx up to 2 years of age. *Journal of Medical Microbiology*, 51, 723–30.

Krisanaprakornkit, S., Kimball, J. R., Weinberg, A., Darveau, R. P., Bainbridge, B. W., and Dale, B. A. (2000). Inducible expression of human β-defensin 2 by *Fusobacterium nucleatum* in oral epithelial cells: Multiple signalling pathways and role of commensal bacteria in innate immunity and the epithelial barrier. *Infection and Immunity*, 68, 2907–15.

Kunin, C. M., Evans, C., Bartholomew, D., and Bates, D. G. (2002). The antimicrobial defense mechanism of the female urethra: a reassessment. *Journal of Urology*, 168, 413–9.

Lawrence, I. C., Fiocchi, C., and Chakravarti, S. (2001). Ulcerative colitis and Crohn's disease: Distinctive gene expression and novel susceptibility candidate genes. *Human Molecular Genetics*, 10, 445–56.

Lee, S. H., Kim, J. E., Lim, H. H., and Choi, J. O. (2002). Antimicrobial defensin peptides of the human nasal mucosa. *Annals of Otology, Rhinology and Laryngology*, 111, 135–41.

Lehrer, R. I. and Ganz, T. (2002). Defensins of vertebrate animals. *Current Opinion in Immunology*, 14, 96–102.

Levy, O. (2000). A neutrophil-derived anti-infective molecule: Bactericidal/permeability-increasing protein. *Antimicrobial Agents and Chemotherapy*, 44, 2925–31.

Li, P., Chan, H. P., He, B., So, S. C., Chung, Y. W., Shang, Q., Zhang, Y.-D., and Zhang, Y.-L. (2001). An antimicrobial peptide gene found in the male reproductive system of rats. *Science*, 291, 1783–5.

Liu, A. Y., Destoumieux, D., Wong, A. V., Park, C. H., Valore, E. V., Lui, L., and Ganz, T. (2002). Human β-defensin-2 production in keratinocytes is regulated by interleukin-1, bacteria, and the state of differentiation. *Journal of Investigative Dermatology*, 118, 275–81.

Loesche, W. J. and Grossman, N. S. (2001). Periodontal disease as a specific, albeit chronic, infection: Diagnosis and treatment. *Clinical Microbiology Reviews*, 14, 727–52.

Lysenko, E. S., Gould, J., Bals, R., Wilson, J. M., and Weiser, J. N. (2000). Bacterial phosphorylcholine decreases susceptibility to the antimicrobial peptide LL-37/hCAP18 expressed in the upper respiratory tract. *Infection and Immunity*, 68, 1664–71.

Mak, P., Szewczyk, A., Mickowska, B., Kicinska, A., and Dubin, A. (2001). Effect of antimicrobial apomyoglobin 56–131 peptide on liposomes and planar lipid bilayer membrane. *International Journal of Antimicrobial Agents*, 17, 137–42.

Marsh, P. D. and Bradshaw, D. J. (1999). Microbial community aspects of dental plaque. In *Dental Plaque Revisited*, ed. H. N. Newman and M. Wilson, pp. 237–53. Cardiff: Bioline.

Martin, A. (2002). Phylogenetic approaches for describing and comparing the diversity of microbial communities. *Applied and Environmental Microbiology*, 68, 3673–82.

Matsushita, I., Hasegawa, K., Nakata, K., Yasuda, K., Tokunaga, K., and Keicho, N. (2002). Genetic variants of human β-defensin-1 and chronic obstructive pulmonary disease. *Biochemical and Biophysical Research Communications*, 291, 17–22.

Mead, G. C. and Barrow, P. A. (1990). *Salmonella* control in poultry by "competitive exclusion" or immunization. *Letters in Applied Microbiology*, 10, 221–7.

Medzhitov, R. and Janeway, C. (2000). Innate immune recognition: Mechanisms and pathways. *Immunological Reviews*, 173, 89–97.

Meister, M., Lemaitre, B., and Hoffmann, J. A. (1997). Antimicrobial peptide defense in Drosophila. *Bioessays*, 19, 1019–26.

Morrison, G. M., Davidson, D. J., and Dorin, J. R. (1999). A novel mouse β-defensin DefB2, which is upregulated in the airways by lipopolysaccharide. *FEBS Letters*, 442, 112–16.

Morrison, G. M., Kilanowski, F., Davidson, D., and Dorin, J. (2002a). Characterization of the mouse beta defensin 1, *Defb1*, mutant mouse model. *Infection and Immunity*, 70, 3053–60.

Morrison, G. M., Rolfe, M., Kilanowski, F. M., Cross, S. H., and Dorin, J. R. (2002b). Identification and characterization of a novel murine beta-defensin-related gene. *Mammalian Genome*, 13, 445–51.

Moser, C., Weiner, D. J., Lysenko, E., Bals, R., Weiser, J. N., and Wilson, J. M. (2002). β-defensin 1 contributes to pulmonary innate immunity in mice. *Infection and Immunity*, 70, 3068–72.

Miyasaki, K. T., Bodeau, A. L., Ganz, T., Selsted, M. E., and Lehrer, R. I. (1990). *In vitro* sensitivity of oral, gram-negative, facultative bacteria to the bactericidal activity of human neutrophil defensins. *Infection and Immunity*, 58, 3934–40.

Nagaoka, I., Hirota, S., Niyonsaba, F., Hirata, M., Adachi, Y., Tamura, H., and Heumann, D. (2001). Cathelicidin family of antibacterial peptides CAP18 and CAP11 inhibit the expression of TNF-α by blocking the binding of LPS to CD14$^+$ cells. *Journal of Immunology*, 167, 3329–38.

Neish, A. S., Gewirtz, A. T., Zeng, A. T., Young, A. N., Hobert, M. E., Karmali, V., Rao, A. S., and Madara, J. L. (2000). Prokaryotic regulation of epithelial responses by inhibition of IκB-α ubiquitination. *Science*, 289, 1560–3.

Nes, I. F., Diep, D. B., Havarstein, L. S., Brurberg, M. B., Eijsink, V., and Holo, H. (1996). Biosynthesis of bacteriocins in lactic acid bacteria. *Antonie van Leeuwenhoek*, 70, 113–28.

Netea, M. G., van Deuren, M., Kullberg, B. J., Cavaillon, J.-M., and Van der Meer, J. W. M. (2002). Does the shape of lipid A determine the interaction of LPS with Toll-like receptors? *Trends in Immunology*, 23, 135–9.

Nizet, V., Ohtake, T., Trowbridge, J., Rudisill, J., Dorschner, R. A., Pestonjamasp, V., Piraino, J., Huttner, K., and Gallo, R. L. (2001). Innate antimicrobial peptide protects the skin from invasive bacterial infection. *Nature* (*London*), 414, 454–7.

Odell, E. W., Sarra, R., Foxworthy, D. S., Chapple, D. S., and Evans, R. W. (1996). Antibacterial activity of peptides homologous to a loop region in human lactoferrin. *FEBS Letters*, 382, 175–8.

O'Neil, D. A., Porter, E. M., Elewaut, D., Anderson, G. M., Eckmann, L., Ganz, T., and Kagnoff, M. F. (1999). Expression and regulation of the human β-defensins hBD-1 and hBD-2 in intestinal epithelium. *Journal of Immunology*, 163, 6718–24.

Ong, P. Y., Ahtake, T., Brandt, C., Strickland, I., Boguniewicz, M., Ganz, T., Gallo, R. L., and Leung, D. Y. M. (2002). Endogenous antimicrobial peptides and skin infections in atopic dermatitis. *New England Journal of Medicine*, 347, 1151–60.

Ouellette, A. J., Satchell, D. P., Hsieh, M. M., Hagen, S. J., and Selsted, M. E. (2000). Characterization of luminal paneth cell alpha-defensins in mouse small intestine. Attenuated antimicrobial activities of peptides with truncated amino termini. *Journal of Biological Chemistry*, 275, 33969–73.

Park, C. H., Valore, E. V., Waring, A. J., and Ganz, T. (2001). Hepcidin, a urinary antimicrobial peptide synthesized in the liver. *Journal of Biological Chemistry*, 276, 7806–10.

Paster, B. J., Boches, S. K., Galvin, J. L., Ericson, R. E., Lau, C. N., Levanos, V. A., Sahasrabudhe, A., and Dewhirst, F. E. (2001). Bacterial diversity in human subgingival plaque. *Journal of Bacteriology*, 183, 3770–83.

Percival, R. S., Marsh, P. D., Devine, D. A., Rangarajan, M., Aduse-Opoku, J., and Curtis, M. A. (1999). The effect of temperature on growth and protease activity of *Porphyromonas gingivalis* W50. *Infection and Immunity*, 67, 1917–21.

Peschel, A. (2002). How do bacteria resist human antimicrobial peptides? *Trends in Microbiology*, 10, 179–86.

Peschel, A., Otto, M., Jack, R. W., Kalbacher, H., Jung, G., and Götz, F. (1999). Inactivation of the *dlt* operon in *Staphylococcus aureus* confers sensitivity to defensins, protegrins and other antimicrobial peptides. *Journal of Biological Chemistry*, 274, 8405–10.

Raj, P. A., Antonyraj, K. J., and Karunakaran, T. (2000). Large-scale synthesis and functional elements for the antimicrobial activity of defensins. *Biochemical Journal*, 347, 633–41.

Rasmussen, T. T., Kirkeby, L. P., Poulsen, K., Reinholdt, J., and Kilian, M. (2000). Resident aerobic microflora of the adult human nasal cavity. *Acta Pathologica Microbiologica et Immunologica Scandinavica (APMIS)*, 108, 663–75.

Rhen, M., Eriksson, S., and Pettersson, S. (2000). Bacterial adaptation to host innate immunity. *Current Opinion in Microbiology*, 3, 60–64.

Roos, K., Håkansson, E. G., and Holm, S. (2000). Effect of recolonisation with "interfering" α streptococci on recurrences of acute and secretory otitis media in children: randomised placebo controlled trial. *British Medical Journal*, 322, 210–12.

Sahl, H.-G., Jack, R. W., and Bierbaum, G. (1995). Biosynthesis and biological activities of lantibiotics with unique post-translational modifications. *European Journal of Biochemistry*, 230, 827–53.

Savage, D. (1977). Microbial ecology of the gastrointestinal tract. *Annual Review of Microbiology*, 31, 107–33.

Schaller-Bals, S., Schulze, A., and Bals, R. (2002). Increased levels of antimicrobial peptides in tracheal aspirates of newborn infants during infection. *American Journal of Respiratory and Critical Care Medicine*, 165, 992–5.

Schenkein, H. A., Berry, C. R., Purkall, D., Burmeister, J. A., Brooks, C. N., and Tew, J. G. (2001). Phosphorylcholine-dependent cross-reactivity between dental plaque bacteria and oxidised low-density lipoproteins. *Infection and Immunity*, 69, 6612–17.

Schittek, B., Hipfel, R., Sauer, B., Bauer, J., Kalbacher, H., Stevanovic, S., Schirle, M., Schroeder, K., Blin, N., Meier, F., Rassner, G., and Garbe, C. (2001). Dermcidin, a novel human antibiotic peptide secreted by sweat glands. *Nature Immunology*, 2, 1133–7.

Schmid, P., Grenet, O., Medina, J., Chibout, S., Osborne, C., and Cox, D. A. (2001). An intrinsic antibiotic mechanism in wounds and tissue-engineered skin. *Journal of Investigative Dermatology*, 116, 471–2.

Schmidtchen, A., Frick, I., Andersson, E., Tapper, H., and Björck, L. (2002). Proteinases of common pathogenic bacteria degrade and inactivate the antibacterial peptide LL-37. *Molecular Microbiology*, 46, 157–68.

Schutte, B. C. Mitros, J. P., Bartlett, J. A., Walters, J. D., Jia, H. P., Welsh, M. J., Casavant, T. L., and McCray, Jr. P. B. (2002). Discovery of five conserved beta-defensin gene clusters using a computational search strategy. *Proceedings of the National Academy of Sciences USA*, 99, 2129–33.

Scott, M. G., Davidson, D. J., Gold, M. R., Bowdish, D., and Hancock, R. E. W. (2002). The human antimicrobial peptide LL-37 is a multifunctional modulator of innate immune response. *Journal of Immunology*, 169, 3883–91.

Scott, M. G., Gold, M. R., and Hancock, R. E. W. (1999). Interaction of cationic peptides with lipoteichoic acid and Gram-positive bacteria. *Infection and Immunity*, 67, 6445–53.

Scott, M. G., Rosenberger, C. M., Gold, M. R., Finlay, B. B., and Hancock, R. E. W. (2000a). An α-helical cationic antimicrobial peptide selectively modulates macrophage responses to lipopolysaccharide and directly alters macrophage gene expression. *Journal of Immunology*, 165, 3358–65.

Scott, M. G., Vreugdenhil, A. C. E., Buurman, W. A., Hancock, R. E. W., and Gold, M. R. (2000b). Cutting edge: Cationic antimicrobial peptides block the binding of lipopolysaccharide (LPS) to LPS binding protein. *Journal of Immunology*. 164, 549–53.

Serino, L. and Virji, M. (2000). Phosphorylcholine decoration of lipopolysaccharide differentiates commensal *Neisseria* from pathogenic strains: Identification of *licA*-type genes in commensal *Neisseria*. *Molecular Microbiology*, 35, 1550–9.

Simmaco, M., Mangoni, M. L., Boman, A., Barra, D., and Boman, H. G. (1998). Experimental infections of *Rana esculenta* with *Aeromonas hydrophila*: A molecular mechanism for the control of the normal flora. *Scandinavian Journal of Immunology*, 48, 357–63.

Sokurenko, E. V., Chesnokova, V., Dykhuizen, D., Ofek, I., Wu, X., Krogfelt, K. A., Struve, C., Schrembri, M. A., and Hasty, D. L. (1998). Pathogenic adaptation of *Escherichia coli* by natural variation of the FimH adhesion. *Proceedings of the National Academy of Sciences, USA*, 95, 8922–26.

Spatafora, G. A., Sheets, M., June, R., Luyimbazi, D., Howard, K., Hulbert, R., Barnard, D., el Janne, M., and Hudson, M. C. (1999). Regulated expression of the *Streptococcus mutans dlt* genes correlates with intracellular polysaccharide accumulation. *Journal of Bacteriology*, 181, 2363–73.

Starner, T. D., Swords, W. E., Apicella, M. A., and McCray, P. B. (2002). Susceptibility of nontypeable *Haemophilus influenzae* to human β-defensins is influenced by lipooligosaccharide acylation. *Infection and Immunity*, 70, 5287–9.

Svanborg, C., Godaly, G., and Hedlund, M. (1999). Cytokine responses during mucosal infections: Role in disease pathogenesis and host defence. *Current Opinion in Microbiology*, 2, 99–105.

Swords, W. E., Chance, D. L., Cohn, L. A., Shao, J., Apicella, M. A., and Smith, A. L. (2002). Acylation of the lipooligosaccharide of *Haemophilus influenzae* and colonization: An *htrB* mutation diminishes the colonization of human airway epithelial cells. *Infection and Immunity*, 70, 4661–8.

Takahashi, A., Wade, A., Ogushi, K., Maeda, K., Kawahara, T., Mawatari, K., Kurazono, H., Moss, J., Hirayama, T., and Nakaya, Y. (2001). Production of β-defensin-2 by human colonic epithelial cells induced by *Salmonella enteritidis* flagella filament structural protein. *FEBS Letters*, 508, 484–8.

Tam, J. P., Lu, Y. A., Yang, J. L., and Chui, K. W. (1999). An unusual structural motif of antimicrobial peptides containing end-to-end macrocycle and cystine-knot disulfides. *Proceedings of the National Academy of Sciences, USA*, 96, 8913–18.

Tannock, G. W. (1999). *Medical Importance of Normal Microflora*. Dordrecht, The Netherlands: Kluwer Academic.

Tato, C. M. and Hunter, C. A. (2002). Host pathogen interactions: Subversion and utilization of the NF-κB pathway during infection. *Infection and Immunity*, 70, 3311–17.

Trottier, S., Stenberg, K., and Svanborg-Eden, C. (1989). Turnover of non-typable *Haemophilus influenzae* in the nasopharynx of healthy children *Journal of Clinical Microbiology*, 27, 2175–9.

Valore, E. V., Park, C. H., Quayle, A. J., Wiles, K. R., McCray, P. B., and Ganz, T. (1998). Human β-defensin-1: An antimicrobial peptide of urogenital tissues. *Journal of Clinical Investigation*, 101, 1633–42.

Vatta, S., Boniotto, M., Bevilacqua, E., Belgrano, A., Pirulli, D., Crovella, S., and Amoroso, A. (2000). Human beta defensin 1 gene: Six new variants. *Human Mutation*, 15, 582–3.

von Horsten, H. H., Derr, P., and Kirchhoff, C. (2002). Novel antimicrobial peptide of human epididymal duct origin. *Biology of Reproduction*, 67, 804–13.

Wehkamp, J., Schwind, B., Herrlinger, K. R., Baxmann, S., Schmidt, K., Duchrow, M., Wohlschlager, C., Feller, A. C., Stange, E. F., and Fellermann, K. (2002). Innate immunity and colonic inflammation: Enhanced expression of epithelial α-defensins. *Digestive Diseases and Sciences*, 47, 1349–55.

Wilson, C. L., Ouellette, A. J., Satchell, D. P., Ayabe, T., Lopez-Boado, Y. S., Stratman, J. L., Hultgren, S. J., Matrisian, L. M., and Parks, W. C. (1999). Regulation of intestinal alpha-defensin activation by the metalloproteinase matrilysin in innate host defence. *Science*, 286, 113–17.

Wilson, M. J., Weightman, A. J., and Wade, W. G. (1997). Applications of molecular ecology in the characterization of uncultured microorganisms associated with human disease. *Reviews of Medical Microbiology*, 8, 91–101.

Yamaguchi, Y., Nagase, T., Makita, R., Fukuhara, S., Tomita, T., Tominaga, T.,

Kurihara, H., and Ouchi, Y. (2002). Identification of multiple novel epididymis-specific β-defensin isoforms in humans and mice. *Journal of Immunology*, 169, 2516–23.

Yang, D., Biragyn, A., Kwak, L. W., and Oppenheim, J. J. (2002). Mammalian defensins in immunity: More than just microbicidal. *Trends in Immunology*, 23, 291–6.

Zucht, H. D., Bragowsky, J., Schrader, M., Liepke, C., Jurgens, M., Schulz-Knappe, P., and Forssmann, W. G. (1998). Human β-defensin-1: A urinary peptide present in variant molecular forms and its putative functional implication. *European Journal of Medical Research*, 3, 315–23.

CHAPTER 3

Multiple functions of antimicrobial peptides in host immunity

De Yang and Joost J. Oppenheim

3.1. INTRODUCTION

Mammals live in an environment full of potentially pathogenic microorganisms. However, individuals rarely become infected because of the barrier function of the skin and epithelia that prevent microbial entry by mechanical separation and generation of antimicrobial substances. Even when the barrier is breached and microorganisms do enter the host, the pathogens, under most circumstances, are contained and ultimately eliminated by the host immune system. The mammalian immune system comprises both innate and adaptive components. Innate immunity represents the first line of constitutively preexisting host defense that is rapidly mobilized following the detection of microbial invasion (Hoffmann et al., 1999; Medzhitov and Janeway, 1997). The effector branch of innate immunity consists of two major components. One involves the release, activation, or both, of a variety of extracellular humoral mediators such as complement, cytokines, and antimicrobial substances. The other is based on recruitment to the sites of microbial invasion and activation of cells, such as phagocytic granulocytes, monocytes/macrophages, and, in some cases, natural-killer (NK) cells to combat the invading pathogens. Adaptive immunity is induced when lymphocytes are activated in response to antigens presented by antigen-presenting cells (APCs), in particular dendritic cells (DCs) (Banchereau and Steinman, 1998; Hoffmann et al., 1999). The T-cell antigen receptors (TCRs) recognize antigenic epitopes bound to the major histocompatibility complex (MHC) on the surface of APCs. $CD8^+$ T cells, once activated after recognition of antigenic epitopes bound to the MHC class I, differentiate into cytotoxic T cells that directly kill cells infected by intracellular pathogens. The complex of MHC class II and antigenic epitope triggers the activation of $CD4^+$, generating

T-helper cells that, by producing various cytokines, promote B-cell activation and enhance the efficiency of phagocytes to eliminate pathogens. B-cell activation leads to the production of antigen-specific antibodies that neutralize pathogen-derived toxins, block the infectivity of invading pathogens, and promote their opsonization and elimination by phagocytes. Thus adaptive immunity, in addition to generating antibodies and cytotoxic T cells capable of combating particular infectious organisms, also further activates and promotes innate host defenses.

Antimicrobial substances comprise microbicidal chemicals (e.g., hydrogen peroxide, nitric oxide) and a wide variety of host gene-encoded antimicrobial proteins. Those antimicrobial proteins with fewer than 100 amino acid residues are often collectively referred to as antimicrobial peptides (Boman, 1998; Ganz and Lehrer, 1998; Lehrer and Ganz, 1999; Scott and Hancock, 2002; Zasloff, 2002). Antimicrobial peptides are diverse in size and structure and include defensins, cathelicidins, histatins, lactoferrins, and many others. Mammalian antimicrobial peptides are predominantly produced by phagocytes (including neutrophils and monocytes/macrophages) and/or epithelial cells that line the skin, digestive, respiratory, and genitourinary tracts. They are constitutively expressed and/or induced by direct microbial contact, microbial products [e.g., lipopolysaccharides (LPSs)], or proinflammatory cytokines (IL-1β, TNFα, etc.). In the course of infection, they are released or produced either systemically or locally at the site of pathogen entry. Although the antibiotic activities of mammalian antimicrobial peptides have been extensively studied for over twenty years, their effects on cells engaged in innate and adaptive immunity have become evident only over the past several years (Gudmundsson and Agerberth, 1999; Scott and Hancock, 2002; Yang, Chertov, and Oppenheim, 2001b).

The classification, structures, gene expression, and processing of mammalian antimicrobial peptides, their antimicrobial spectra and mechanisms, and potential involvement in human diseases are covered in detail in other chapters of this book. In this chapter, we focus on the multiple receptor-mediated effects of mammalian antimicrobial peptides on host cells (Table 3.1) and their resultant roles in host innate and adaptive antimicrobial immunity.

3.2. CONTRIBUTION OF ANTIMICROBIAL PEPTIDES TO INNATE ANTIMICROBIAL IMMUNITY

3.2.1. Direct Antimicrobial Activity

All antimicrobial peptides identified so far share the capacity to kill a broad spectrum of both Gram-positive and Gram-negative bacteria, fungi,

Table 3.1. *Functions of mammalian antimicrobial peptides*[1]

Mammalian antimicrobial peptides		Antimicrobial activity	Chemotactic activity	Cell activation	Cell differentiation	Cytokine induction	Chemokine induction	Complement regulation	Dendritic-cell maturation	Adjuvant activity	Neutralizing endotoxin	Anti-inflammatory	Apoptotic effect	Mitogenic effect	Suppressing glucocorticoid production
One disulfide bond	Lactoferricin	+										+	+		
	Bactenecin	+													
Two disulfide bonds	Protegrin	+													
Three disulfide bonds	α-defensin	+	+	+		+	+	+		+		+		+	+
	β-defensin	+	+	+	+	+	+		+	+					
	θ-defensin	+													
Four disulfide bonds	Hepcidin	+													
Linear, α-helical	LL-37	+	+	+		+	+				+				
	CRAMP[2]	+									+				
Linear, not α-helical	Histatin	+		+								+			
	PR-39	+	+	+								+			
	Bac 5, 7	+	+								+				
	Indolicidin	+									+				

[1]The + sign indicates that such an effect has been reported for a given antimicrobial peptide.

[2]CRAMP, cathelin-related antimicrobial peptide.

and some parasites (Arnold, Cole, and McGhee, 1977; Bensch et al., 1995; Boman, 1998; Ganz and Lehrer, 1998; Ganz et al., 1985; Harder et al., 1997; Larrick et al., 1995; Lehrer and Ganz, 1999; Oppenheim et al., 1988; Scott and Hancock, 2002; Selsted et al., 1993; Selsted, Szklarek, and Lehrer, 1984; Tang et al., 1999; Zanetti, Gennaro, and Romeo, 1995; Zasloff, 2002). With the exception of θ-defensins (Tang et al., 1999) and human β-defensin 3 (HBD3) (Harder et al., 2001), the in vitro antibacterial and antifungal activities of many defensins and cathelicidins are ablated in the presence of physiological concentration of salt (i.e., 150 mM of NaCl) or serum (Bals et al., 1998; Ganz and Lehrer, 1998; Johansson et al., 1998; Larrick et al., 1995; Lehrer and Ganz, 1999). Therefore the microbicidal activities of most antimicrobial peptides in vivo may be manifested only at sites with a low concentration of NaCl and serum, such as in phagocytic vacuoles of phagocytes and on the surfaces of skin and mucosal epithelium (Ganz and Lehrer, 1998; Lehrer and Ganz, 1999; Schild and Kellenberger, 2001). Certain antimicrobial peptides also demonstrate, in addition to antibacterial and antifungal activities, in vitro antiviral activity. Two rabbit phagocyte-derived α-defensins are able to neutralize the infectivity of herpes simplex and influenza viruses, but are not active on cytomegalovirus, echovirus, and reovirus (Lehrer et al., 1985). The α-defensins derived from neutrophils of guinea pigs, rabbits, and rats have been shown to have modest anti-HIV-1 activities (Nakashima et al., 1993). Recently, both θ-defensins and human neutrophil-derived α-defensins (HNP1–HNP3) have also been shown to inhibit the replication of various viruses, including HIV (Bastian and Schäfer, 2001; Zhang et al., 2002).

3.2.2. Antimicrobial Peptides Act as Chemoattractants for Cells Engaged in Innate Host Defense

Many antimicrobial peptides can act as chemoattractants for cells participating in innate host defenses (Table 3.2). For phagocytes (e.g., neutrophils, monocytes, and macrophages) to combat invading bacteria, they have to be recruited to sites of bacterial entry. Leukocyte recruitment is guided by chemotactic factors including chemokines and other chemoattractants (Baggiolini, 1998; Murphy, 1994). HNP1–HNP3 were reported to be chemotactic for monocytes in 1989 (Territo et al., 1989). PR-39, a porcine cathelicidin, chemoattracts neutrophils, but not mononuclear cells (Huang, Ross, and Blecha, 1997). LL-37, the human cathelicidin, is chemotactic for both neutrophils and monocytes (Agerberth et al., 2000; Yang, Chen, Schmidt, et al., 2000). HBD3, HBD4, and proBac7, a bovine cathelicidin, are also chemotactic for monocytes/macrophages (Garcia, Jaumann, et al., 2001; Garcia,

Table 3.2. *The chemotactic cell targets of mammalian antimicrobial peptides*[1]

Mammalian antimicrobial peptide	Neutrophil	Monocyte/ macrophage	Mast cell	T lymphocyte			Dendritic cell	
				CD4		CD8		
				Naive	Memory		Immature	Mature
HNP1–HNP3	–	+	?	+	–	+	+	–
HBD1	–	–	?	–	+	+	+	–
HBD2	–	–	+	–	+	+	+	–
HBD3	–	+	?	–	+	+	+	–
HBD4	?	+	?		?		?	
mBD2, 3	–	–	?		?		?	
LL-37	+	+	+		+		–	
Histatin 5	?	+	?		?		?	
PR-39	+	–	?		–		?	
proBac7	?	+	?		?		?	

[1]HNP, human neutrophil-derived α-defensin; HBD, human β-defensin; mBD, murine β-defensin; ?, not yet tested.

Table 3.3. *Key features of immature and mature dendritic cells*[1]

Characteristic mediators		iDC	mDC	Function
Secretory factor	Type 1 interferon	+	–	Antiviral
	Microbicidal peptide	+	–	Antimicrobial
	CXCL1, CCL2, 3, 5	+	–	Inflammatory chemokines
	CCL17, 18, 19, 21, 22	–	+	T-cell-attracting chemokines
	IL-2 and IL-12	–	+	T-cell activation
Receptor	CCR1, 2, 3, 5, 6, 8	+	–	iDC recruitment
	CXCR5, CCR7	–	+	mDC trafficking
	Toll-like receptors	+	?	Pattern recognition
	Fc opsonin receptors	High	Low	
	Complement receptors	High	Low	
	C-type lectin receptors	High	Low	Antigen uptake
	Scavenger receptors	High	Low	
Endosomal component	Intracellular MHC class II	High	Low	Antigen processing
	HLA-DM or H-2DM	+	–	
Surface molecule	Surface MHC class II	Low	High	
	CD40, CD83	Low	High	Antigen presentation
	CD54, 58, CD80, 86	Low	High	

[1]IL, interleukin; CXCL and CXCR, CXC chemokine and receptor, respectively; CCL and CCR, CC chemokine and receptor, respectively; ?, unknown.

Krause, et al., 2001; Verbanac, Zanetti, and Romeo, 1993; Yang et al., 2002). In addition, both human LL-37 and HBD2 have also been shown to chemoattract mast cells (Niyonsaba et al., 2002a, 2002b). Attracting mast cells to sites of microbial entry also indirectly promotes phagocyte recruitment (see next subsection for details). The capacity of antimicrobial peptides to chemoattract phagocytes and mast cells suggests that they can potentially contribute either directly or indirectly to the recruitment of phagocytes to sites of bacterial entry.

DCs go through two stages of differentiation, immature DCs (iDCs) and mature DCs (mDCs), each with distinct properties and functions (Table 3.3). Aside from their roles in the induction and regulation of adaptive immunity (Banchereau and Steinman, 1998; Cella, Sallusto, and Lanzavecchia, 1997;

Liu, 2001), as will be discussed later in this chapter, DCs also participate in innate antimicrobial immunity (Liu, 2001) in at least two ways. First, DC precursor cells and iDCs can directly phagocytize and kill pathogens (Liu, 2001). Second, iDCs, particularly upon activation, produce numerous mediators including cytokines, chemokines, and antimicrobial peptides (Duits et al., 2002) that participate in innate antimicrobial immunity. Interestingly, several mammalian defensins can also function as chemoattractants for DCs (Table 3.2). For example, HNPs are selectively chemotactic for human iDCs generated from human $CD34^+$ DC precursor cells (Yang, Chen, Chertov, et al., 2000). HBD1 and HBD2 have also been recently found to be selectively chemotactic for human iDCs, but not for mDCs (Yang, Chertov, et al., 1999). Similar to HBDs, mouse β-defensin 2 (mBD2) and mBD3 are also selectively chemotactic for iDCs generated from mouse bone marrow progenitor cells (Biragyn et al., 2001). Thus defensins may enhance the recruitment of iDCs to the sites of infection based on their chemotactic activities, thereby contributing to innate antimicrobial immunity.

3.2.3. Antimicrobial Peptides Induce Mediators of Inflammation

Human, rabbit, and guinea pig α-defensins induce mast cell degranulation and histamine release (Befus et al., 1999; Yamashita and Saito, 1989). Recently, HBD2, LL-37, and histatins have also been shown to activate rat mast cells, resulting in the release of histamine and prostaglandin D_2 (Niyonsaba et al., 2001; Yoshida et al., 2001). HNP1–HNP3 can augment IL-8 gene transcription and IL-8 protein production by bronchial epithelial cells (van Wetering et al., 1997a, 1997b). Because mast cell granule products increase neutrophil influx (Echtenacher, Mannel, and Hultner, 1996; Malaviya et al., 1996) and IL-8 is a potent neutrophil chemotactic factor (Baggiolini, 1998; Murphy, 1994), these antimicrobial peptides may therefore indirectly promote the recruitment and accumulation of phagocytic neutrophils at inflammatory sites. Degranulation of the recruited neutrophils releases more defensins (Chertov et al., 1996; Ganz, 1987; Taub et al., 1996) and consequently generates more IL-8 (Baggiolini, Dewalk, and Moser, 1994; Oppenheim et al., 1991), both of which result in a positive feedback loop that promotes the recruitment of inflammatory cells.

More recently, LL-37 has also been shown to enhance the expression of a variety of genes by macrophages (Scott et al., 2002). Of particular interest is the upregulation of the chemokines IL-8 and monocyte chemoattractant protein (MCP)-1 and their corresponding receptors, CXCR2 and CCR2, by LL-37 (Scott et al., 2002). CXCR2 is expressed by neutrophils, monocytes, and

T cells whereas CCR2 is expressed by monocytes and iDCs (Baggiolini, 1998; Murphy, 1996; Sozzani et al., 1999). Therefore induction of IL-8 and MCP-1 and upregulation of their corresponding receptors potentially also promote the recruitment of phagocytes to inflammatory sites. HNP1–HNP3, in contrast to LL-37, have been reported to increase the production of TNFα and IL-1 while decreasing the production of IL-10 by monocytes (Chaly et al., 2000). Increased levels of proinflammatory factors (IL-1, TNFα, and histamine) and suppressed levels of IL-10 at the site of microbial infection are likely to amplify local inflammatory responses.

3.2.4. Antimicrobial Peptides Affect the Functions of Phagocytes and Complement System

Once recruited to the site of infection, phagocytes work in concert with the complement system to destroy microbial invaders. Guinea pig α-defensins are reported to induce neutrophil aggregation (a sign of activation) as well as to enhance the expression of adhesion molecules, including ICAM-1, CD11b, and CD11c by neutrophils (Yomogida et al., 1996). A synthetic α-helical antimicrobial peptide, CEMA, is reported to increase ICAM-1 expression by macrophages (Scott and Hancock, 2002; Scott, Rosenberger, et al., 2000). Upregulation of adhesion molecules on phagocytes not only facilitates their recruitment, but also promotes their activation, leading to enhanced microbicidal activity (Baggiolini, 1998; Hayflick, Kilgannon, and Gallatin, 1998; Patarroyo, 1994). In addition, human α-defensins can directly activate phagocytes to enhance phagocytosis (Ichinose et al., 1996) and to induce the production of reactive oxygen intermediates, potent bactericidal molecules (Porro et al., 2001). Moreover, human α-defensins can bind to complement C1q to either enhance (Prohaszka et al., 1997) or suppress (van den Berg et al., 1998) the activation of the classical pathway of complement in vitro, depending on the experimental conditions. Presumably α-defensins have similar effects in vivo and therefore participate in regulating the complement system.

Certain α-defensins have also been reported to inhibit the production of immunosuppressive adrenal steroid hormones (Tominaga et al., 1990; Zhu et al., 1988). During systemic infections, α-defensin levels in plasma can reach up to 100 μg/ml, a concentration sufficient to interfere with the production of adrenal glucocorticoids (Panyutich et al., 1993; Shiomi et al., 1993). Because glucocorticoids are potent immunosuppressive mediators, α-defensins may thus also enhance systemic antimicrobial immunity in vivo by inhibiting the production of glucocorticoids. Overall, the capabilities of antimicrobial peptides to directly kill or inactivate microorganisms, to enhance

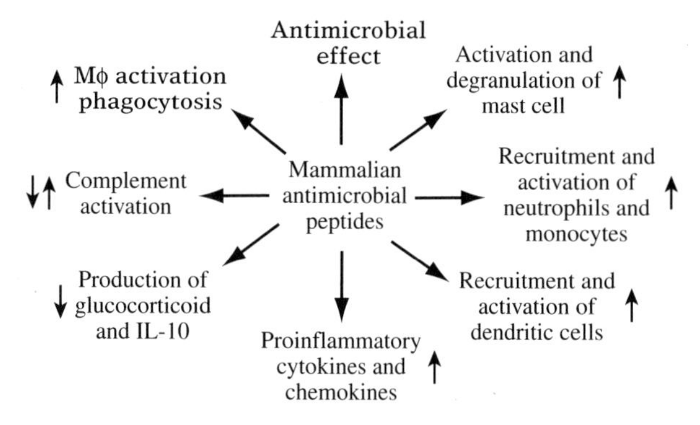

Figure 3.1. Contribution of mammalian antimicrobial peptides to host innate antimicrobial defense. ↑, promotion; ↓, suppression.

phagocytosis, to promote phagocyte and mast cell recruitment, to enhance production of proinflammatory cytokines, to suppress anti-inflammatory or immunosuppressive mediators, and to regulate complement activation suggest that they are vital contributors to innate host immunity against microbial infection (Fig. 3.1). This has been demonstrated in vivo for some cathelicidins and defensins. For example, overexpression of human cathelicidin/LL-37 in mouse airway increases the resistance of the aminals to *Pseudomonas aeruginosa* challenge (Bals et al., 1999a, 1999b). Conversely, knockout of *Cnlp*, the mouse gene coding for cathelicidin/cathelicidin-related antimicrobial peptide (CRAMP), results in increased susceptibility of the mice to infection by Group A *Streptococcus* (Nizet et al., 2001). Furthermore, inactivation of the gene for matrilysin, an enzyme required for the generation of mature Paneth cell α-defensins, leads to the deficiency of functional α-defensin and an increase in susceptibility of mice to oral bacterial challenge (Wilson et al., 1999). Even knockout of a single defensin, mBD1, results in delayed clearance of *Haemophilus influenzae* from the lung (Moser et al., 2002) and increased colonization by *Staphylococcus* species in the bladder (Morrison et al., 2002).

3.3. CONTRIBUTION OF ANTIMICROBIAL PEPTIDES TO ADAPTIVE ANTIMICROBIAL IMMUNITY

3.3.1. Antimicrobial Peptides Promote the Recuitment of Immature Dendritic Cells to Infected Tissue

The induction of the adaptive antimicrobial immune response begins at sites of infection where iDCs take up and process microbial antigens and display antigenic epitopes in the antigen-binding troughs of MHC class I and MHC class II on the cell surface. Immature DCs express an array of surface and intracellular molecules that function to engulf and process antigens (Table 3.3). For efficient antigen uptake and processing, iDCs need to migrate to infected tissue where microbial antigens are abundant. A number of defensins and cathelicidins including HNP1–HNP3 (Territo et al., 1989), HBD3 (Garcia, Jaumann, et al., 2001; Yang et al., 2002), HBD4 (Garcia, Krause, et al., 2001), LL-37 (Agerberth et al., 2000; Yang, Chen, Schmidt, et al., 2000), and proBac7 (Verbanac et al., 1993), have been shown to be chemotactic for monocytes. Because monocytes are DC precursors and can differentiate into myeloid iDCs (Randolph et al., 1998, 1999; Yang, Howard, et al., 1999), enhanced recruitment of monocytes to infected tissue increases the local number of iDCs. In addition, α- and β-defensins are chemotactic for iDCs (Biragyn et al., 2001; Yang, Chen, Chertov, et al., 2000; Yang, Chertov, et al., 1999) and thus potentially have the capacity to directly attract iDCs to infected tissue. Furthermore, the recruitment of iDCs to sites of infection may also be enhanced indirectly by the induction of the production of chemokines that attract iDCs such as MCP-1 (as discussed in Subsection 3.2.3). Thus antimicrobial peptides are likely to promote antigen uptake and processing by augmenting the recruitment of iDCs to infected tissues.

3.3.2. Antimicrobial Peptides Activate and Induce the Maturation of Dendritic Cells

After antigen uptake and processing, iDCs undergo the process of maturation to become mDCs that acquire the capacity to migrate to regional lymph nodes for antigen presentation to naïve T cells. Mature DCs express CCR7, a chemokine receptor responsible for directing mDC trafficking from the inflammatory site by means of lymphatics to lymph nodes. This is associated with upregulation of surface costimulatory molecules including CD40, CD80, CD83, CD86, and surface MHC class II (Table 3.3) that are required for efficient stimulation of T-cell activation and expansion. Very recently, mBD2 has been shown to activate murine bone-marrow-derived iDCs, resulting in the induction of a variety of cytokines including IL-12 (Biragyn et al., 2002). mBD2 also induces the maturation of iDCs into phenotypically characteristic mDCs, as demonstrated by the upregulation of CD86, MHC class II, as well as CCR7 (Biragyn et al., 2002). Furthermore, many other antimicrobial peptides may stimulate the maturation of DCs indirectly by inducing the production of IL-1β and TNFα, proinflammatory cytokines that can induce DCs to mature (Banchereau and Steinman, 1998; Cella et al.,

1997). Thus antimicrobial peptides can promote the maturation of DCs either directly, as in the case of mBD2, or indirectly, by inducing inflammatory cytokines.

3.3.3. Antimicrobial Peptides Function as T-Cell Chemoattractants

Successful presentation of microbial antigens to lymphocytes in the secondary lymphoid organs results in the activation and clonal proliferative expansion of both T and B lymphocytes, ultimately leading to the generation of antigen-specific effector T cells and antibodies (Banchereau and Steinman, 1998; Hoffmann et al., 1999; Thompson, 1995). CD4 effector T cells can produce cytokines (e.g., IFNγ) capable of activating phagocytes to eliminate pathogens more efficiently, and CD8 effector T cells can directly kill cells infected by intracellular pathogens. To carry out their effector functions, both CD4 and CD8 effector T cells also have to be recruited to sites of infection, a process often guided by chemotactic factors (Baggiolini, 1998; Murphy, 1996; Murphy, 1994; Zlotnik and Yoshie, 2000). Interestingly, several antimicrobial peptides are also T-cell chemoattractants. For example, HNP1–HNP3 are chemotactic for CD45RA naïve $CD4^+$ T cells and $CD8^+$ T cells (Chertov et al., 1996; Yang, Chen, Chertov, et al., 2000). In contrast, HBD2 is chemotactic for CD45RO (memory or effector) T cells (Yang, Chertov, et al., 1999). In addition, LL-37 also chemoattracts peripheral blood T cells (Agerberth et al., 2000; Yang, Chen, Schmidt, et al., 2000). Although the consequence of α-defensin's effect on CD45RA naïve $CD4^+$ T cells is unclear, β-defensins and LL-37 can potentially facilitate the recruitment of memory and effector T cells to sites of infection.

3.3.4. Immunoenhancing Activities of Antimicrobial Peptides In Vivo

The effects of mammalian antimicrobial peptides on DCs and T cells suggest that they should be able to promote in vivo antigen-specific immune responses. This hypothesis has been supported by evidence obtained with certain defensins. The contribution of α-defensins to in vivo T-cell recruitment is based on data that show that subcutaneous injection of HNP1 and HNP2 into severe combined immunodeficiency mice reconstituted with human peripheral blood lymphocytes results in the recruitment of human T cells to the site of injection (Chertov et al., 1996). In addition, simultaneous administration of ovalbumin (OVA) and a mixture of HNP1–HNP3 intranasally into C57BL/6 mice enhances the production of OVA-specific serum IgG antibody and the ex vivo generation of IFNγ, IL-5, IL-6, and IL-10 by OVA-specific $CD4^+$ T cells, providing the first clear evidence that α-defensins are capable of promoting both humoral and cellular adaptive immune responses (Lillard et al., 1999). Furthermore, intraperitoneal injection of HNP1–HNP3 together with keyhole limpet hemocyanin or B-cell lymphoma idiotype antigen into mice not only augmented the serum levels of antigen-specific IgG, but also enhanced the resistance of immunized mice to tumor challenge (Tani et al., 2000). The immunoenhancing activities of murine β-defensins have also been investigated with a DNA vaccine approach in which mBD2 or mBD3 is fused with tumor antigen consisting of the V_H and V_L fragments of B-cell lymphoma-derived immunoglobulin. Immunization of mice with lymphoma antigens fused with mBDs induced not only humoral immune responses against otherwise nonimmunogenic B-cell lymphoma antigens, but also, in the case of mBD2, protected the immunized mice from subsequent tumor challenge (Biragyn et al., 2001, 2002). Because induction of antitumor protection relies on the generation of tumor-specific cellular immunity, it appears that β-defensins are capable of promoting both humoral and cellular adaptive immune responses (Biragyn et al., 2001, 2002).

3.3.5. Mechanistic Basis for the Effects of Antimicrobial Peptides on Host Immune Cells

The mechanisms by which mammalian antimicrobial peptides regulate the migration, proliferation, differentiation, activation, gene expression, and mediator production of host immune cells have, in most cases, not been elucidated as yet. To transduce signals in host cells, antimicrobial peptides must either bind to cell surface receptors or enter the cells to interact with intracellular molecular targets. The capacity of mammalian antimicrobial peptides, in particular defensins and cathelicidins, to induce chemotaxis of immune cells is relatively well characterized (Table 3.2). Clearly the chemotactic activity of an individual antimicrobial peptide is selective for a particular spectrum of target cells (Table 3.2). Additionally, chemotaxis induced by either defensins or LL-37 could be inhibited by pretreatment of the target cells with pertussis toxin (Niyonsaba et al., 2002a, 2002b; Yang, Chen, Chertov, et al., 2000; Yang, Chen, Schmidt, et al., 2000; Yang, Chertov, et al., 1999), a toxin capable of inhibiting Gi-protein-coupled receptor (GPCR) signaling by adenosine diphosphate-ribosylating Giα subunit (Kaslow and Burns, 1992), indicating the involvement of GPCRs. These characteristics, together with the fact that leukocyte chemotaxis in response to all chemokines and chemoattractants is mediated by GPCRs (Baggiolini, 1998; Murphy, 1994, 1996) prompted us to test the possibility that defensins and LL-37 use GPCRs. Because HBD1 and HBD2 selectively chemoattracted human iDCs, but not by mDCs, we focused on GPCRs selectively expressed by iDCs, but not by mDCs. Although iDCs express CCR1, 2, 5, 6, and 8 (Table 3.3), only cells expressing CCR6 could migrate in response to HBD1 and HBD2. In addition, HBD2-induced chemotaxis of $CCR6^{+}$ cells was cross desensitized by CCL20, the chemokine ligand for CCR6 (Baba et al., 1997; Liao et al., 1997; Power et al., 1997; Yang, Howard, et al., 1999), and was also inhibited by the anti-CCR6 antibody. Therefore we established that HBD1–HBD2-induced chemotaxis of iDCs and memory T cells was mediated by CCR6 (Yang, Chertov, et al., 1999; Yang et al., 2001a). Furthermore, HBD3 (Yang et al., 2002), mBD2, and mDB3 (Biragyn et al., 2001) also utilize CCR6 to chemoattract iDCs. Additional evidence that β-defensins share CCR6 with CCL20 came from structural studies. Recently the x-ray crystallography and nuclear magnetic resonance structures of HBD2 as well as the nuclear magnetic resonance structures of mouse and human CCL20/MIP-3α have been solved (Hoover et al., 2000, 2002; Pérez-Cañadillas et al., 2001; Sawai et al., 2001). Comparison of their solution structures revealed that, although HBD2 and CCL20 have no apparent similarity at the primary structural level, both have similar topological motifs. An Asp^{4}-Leu^{9} motif in HBD2 that resembles the Asp^{5}-Leu^{8} motif of CCL20 is considered to be responsible for specific interaction with CCR6, providing a structural basis for the capacity of both β-defensins and CCL20 to interact with the same receptor (Pérez-Cañadillas et al., 2001).

The chemotactic activity of LL-37 for phagocytes appears to be mediated by another chemotactic receptor, formyl peptide-receptor-like 1 (Yang, Chen, Schmidt, et al., 2000; Yang et al., 2001a). An individual antimicrobial peptide may use more than one GPCR. One such example is HBD3 that is known to be chemotactic for both iDCs and monocytes (Table 3.2). Although HBD3-induced migration of iDCs appears to be mediated by CCR6 (Yang et al., 2002), its effect on monocyte migration (Garcia et al., 2001; Yang et al., 2002) must be mediated by another GPCR because monocytes do not express CCR6 (Yang, Howard, et al., 1999). Another example may be α-defensins. HNP induces chemotaxis of iDCs and subsets of T cells in a pertussis toxin-sensitive manner, suggesting the utilization of a GPCR (Yang, Chen, Chertov, 2000). Interestingly, the suppressive effect of α-defensins on glucocorticoid production is achieved by blocking the adrenocorticotropin receptor (Solomon et al., 1991; Zhu and Solomon, 1992), which, not surprisingly, is also a GPCR.

Furthermore, as has been discussed, mBD2 utilizes CCR6 to induce iDC migration (Biragyn et al., 2001), but its capacity to stimulate DC activation and maturation appears to be mediated by Toll-like receptor 4 (Biragyn et al., 2002).

There is a suggestion that some antimicrobial peptides may directly enter the host cells. For instance, HNP1–HNP3 have been reported to be associated with the lymphocyte nuclear fraction (Blomqvist et al., 1999). PR-39 has been demonstrated to enter human microvascular endothelial cells to bind to several SH3 domain-containing proteins (Chan and Gallo, 1998). However, peptide mediators in general cannot traverse the cytoplasmic membrane of host cells. Because PR-39 also binds to NIH 3T3 fibroblasts in a saturable manner (suggestive of a class of binding sites or receptors) (Chan and Gallo, 1998), its entry into cells may also be receptor mediated. The observed nuclear localization of HNP1–HNP3 might simply be due to GPCR-mediated internalization followed by nuclear translocation as a result of their high cationicity. Consequently it is unlikely that most mammalian antimicrobial peptides can enter host cells directly in the absence of receptors.

In summary, antimicrobial peptides are abundant at sites of infection based on their production by epithelial cells and keratinocytes and their release by infiltrating phagocytes. How do they contribute to adaptive antimicrobial immunity? The induction of adaptive antimicrobial immunity potentially begins in infected tissues when iDCs take up microbial antigens. Subsequently, iDCs differentiate into mature antigen-presenting DCs that display processed antigenic peptides in the form of MHC class II–peptide complexes on their surface and develop the capacity to traffic to secondary lymphoid organs for antigen presentation to naïve T cells bearing the antigen-specific T-cell receptors. Based on the data available so far, several in vivo scenarios can be envisaged (Fig 3.2). First, mammalian antimicrobial peptides can themselves promote the migration of DC precursors (monocytes) and iDCs to infected tissues directly or indirectly, through the induction of chemokines. Additionally, antimicrobial peptides can augment the presentation of microbial antigens by promoting maturation of DCs directly, such as mBD2, or indirectly, through inducing the production of TNFα and IL-1 [both are known inducers of DC maturation (Banchereau and Steinman, 1998; Yang, Howard, et al., 1999)]. Promoting the uptake, processing, and presentation of microbial antigens can certainly enhance the induction phase of adaptive antimicrobial immunity. Furthermore, antimicrobial peptides can also augment the effector phase of adaptive antimicrobial immune response by promoting effector T cells to infiltrate the infected tissue.

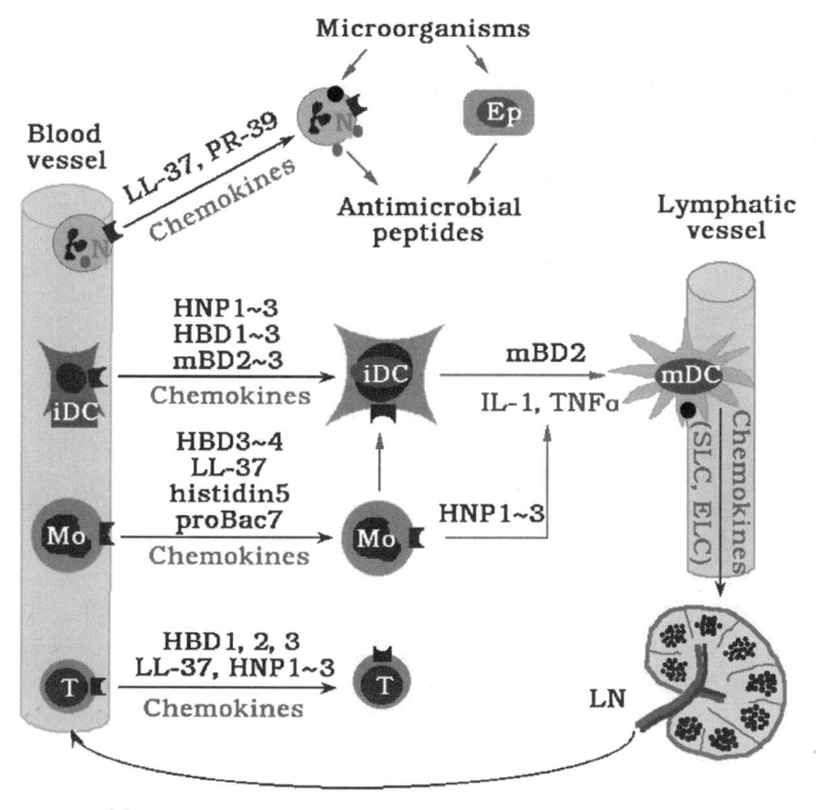

Figure 3.2. Schematic illustration of the potential mechanisms by which mammalian antimicrobial peptides enhance host adaptive antimicrobial immunity. In infected tissue, antimicrobial peptides are abundant because of the production predominantly by epithelial cells (Ep, including keratinocytes) and infiltrating neutrophils (N). Antimicrobial peptides form chemotactic gradients and potentially participate in the recruitment of both DC precursors (Mo) and iDCs to sites of infection. In addition, antimicrobial peptides promote the maturation of iDCs to mDCs directly or indirectly through inducing the production of TNFα and IL-1. Participation of antimicrobial peptides in the recruitment and maturation of DCs certainly contributes to promote antigen capture and presentation, thereby enhancing the induction of adaptive antimicrobial immune response. Furthermore, antimicrobial peptides may also contribute to the effector phase of adaptive antimicrobial immunity by facilitating the recruitment of effector T cells to infected tissues. Chemokines have similar roles, particularly in the recruitment of leukocytes to infected tissue. SLC, secondary lymphoid tissue chemokine; ELC, EBI1-ligand chemokine; LN, lymph node. (See color section.)

3.4. NEGATIVE REGULATION OF HOST INFLAMMATORY AND IMMUNE RESPONSES BY MAMMALIAN ANTIMICROBIAL PEPTIDES

3.4.1. Neutralization of Microbial Components

As previously discussed, it is obvious that mammalian antimicrobial peptides are multifunctional promoters of host innate and adaptive antimicrobial immunity. However, they also have the potential to downregulate destructive host inflammatory responses. The killing of bacteria by antimicrobial peptides, phagocytes, and complement system results in the release of bacterial components, such as LPSs from Gram-negative bacteria and lipoteichoic acids (LTA) from Gram-positive bacteria. These bacterial products, if allowed to enter the circulation, have a detrimental outcome, such as septic shock, by inducing the production of high levels of systemic proinflammatory cytokines including TNFα, IL-1β, and IL-6. Many antimicrobial peptides have the capability to bind to bacterial LPS or LTA and thus to reduce their ability to induce the production of proimflammatory cytokines. Many of the cathelicidins including LL-37, sheep myeloid antimicrobial peptide (SMAP)-29, Bac 2A-NH2 (a linear form of Bac 2A), and indolicidin have the capacity to neutralize LPS both in vitro and in vivo, reducing LPS-induced production of proinflammatory cytokines (Hirata et al., 1994; Larrick et al., 1994, 1995; Scott et al., 2002; Scott and Hancock, 2002; Scott, Rosenberger, et al., 2000; Scott, Vreugdenhil, et al., 2000; Tack et al., 2002). Defensins (HNP1 and HBD2), although less potent, have a similar effect on LPS (Scott and Hancock, 2002; Scott, Rosenberger, et al., 2000). In addition to LPS, LTA can also be neutralized by certain antimicrobial peptides (Scott, Gold, and Hancock, 1999; Scott and Hancock, 2002). In agreement with this property, administration of human cathelicidin/LL-37 to mice has been shown to prevent otherwise lethal endotoxemia (Kirikae et al., 1998; Scott et al., 2002). This protection has also been demonstrated by overexpression of LL-37 in mice through adenovirus-mediated gene transfer (Bals et al., 1999b). Thus antimicrobial peptides can serve as buffers for LPS and LTA, reducing the severity of host inflammatory responses to bacteria that can lead to lethal conditions such as septic shock.

In addition to acting as a buffering factor for microbial products, LL-37 has also been shown to suppress the macrophage production of proinflammatory cytokines in response to LPS and to simultaneously upregulate IL-10 production (Scott et al., 2002; Scott, Rosenberger, et al., 2000). Therefore LL-37 may contribute to the negative regulation of inflammation by functioning as more than just a buffer for microbial products.

3.4.2. Reducing the Production of Proinflammatory Mediators

In addition to neutralizing microbial products, some antimicrobial peptides have the capacity to inhibit the production of proinflammatory cytokines. Lactoferrin has been shown to inhibit neutrophil TNFα production in response to LPS in vitro (Appelmelk et al., 1994; Crouch, Slater, and Fletcher, 1992) and in vivo (Zhang, Mann, and Tsai, 1999). Histatin 5 suppresses proinflammatory cytokine production by human fibroblasts in response to *Porphyromonas gingivalis* (Imatani et al., 2000). LL-37 can inhibit the production of proinflammatory cytokines such as IL-1, TNFα, and MIP-1β by a macrophage cell line (Scott et al., 2002). In addition, some antimicrobial peptides can inhibit certain proteases that promote the generation of proinflammatory mediators from extracellular matrix. In this regard, histatin 3 and histatin 5 have been shown to inhibit the serine endoproteases (Basak et al., 1997) whereas ProBac5 can inhibit cathepsin L, a cysteine protease (Verbanac et al., 1993). HNP, on the other hand, can suppress serine protease-induced epithelial cell detachment (van Wetering et al., 1997a). Inhibition of the production or activity of proinflammatory mediators may help limit tissue injury in host inflammatory responses against infection.

3.4.3. Inhibition of Antimicrobial Effectors

Reactive oxygen intermediates generated by nicotinamide–adenine dinucleotide phosphate (NADPH) oxidase in phagocytes are important antimicrobial effectors. HNPs have been shown to suppress the activation of phagocyte NADPH oxidase (Tal and Irit, 1993; Tal, Michaela, and Irit, 1998) and to downregulate the generation of superoxide by neutrophils (Kaplan, Heine, and Simmons, 1999). PR-39 can not only inhibit phagocyte NADPH oxidase (Shi et al., 1996) but also suppress the generation of reactive oxygen by cultured endothelial cells, perfused rat lung (Al Mehdi et al., 1998), and by rat mesenteric venules subjected to ischemia reperfusion (Korthuis et al., 1999). Not only do they inhibit the generation of reactive oxygen, but some antimicrobial peptides can also induce the apoptosis of immune cells. Lactoferricin, the antimicrobial portion of lactoferrin, induces apoptosis of THP-1 cells, a human monocytic leukemia cell line (Yoo et al., 1997). BMAP-27 and BMAP-28, members of bovine cathelicidins, are also capable of inducing apoptosis of activated T cells (Risso, Zanetti, and Gennaro, 1998).

The capacity of antimicrobial peptides to limit both innate and adaptive immune responses seems to be counterproductive at a first glance. However, after the invaded pathogens or pathogen-infected cells are eliminated, the inflammatory and immune responses have to be silenced to restore homeostasis. By and large, the concentrations required for antimicrobial peptides to downregulate immune responses are higher than those needed to promote antimicrobial immunity. Therefore the functions of antimicrobial peptides in the earlier stages of infection may be predominantly immunoenhancing. In the late stage, especially after pathogens are destroyed, they may shift to downregulate innate and adaptive immune responses.

It is conceivable that some antimicrobial peptides may even promote wound healing. HNP1–HNP3 have been shown to be mitogenic for epithelial cells and fibroblasts at certain concentrations (Aarbiou et al., 2002; Bateman et al., 1991; Murphy et al., 1992). In addition, HBD1 can induce keratinocyte differentiation when the cells are transfected to express this antimicrobial peptide (Frye, Bargon, and Gropp, 2001). Furthermore, PR-39 has been shown to be able to induce the production of syndecan-1 and syndecan-4, major components of cell coat, in cultured fibroblasts and epithelial cells as well as in mouse skin wounds. Thus these antimicrobial peptides potentially promote tissue repair.

3.5. CONCLUSION

It has become evident that mammalian antimicrobial peptides have, in addition to their well-recognized direct antimicrobial activities, multiple roles in host innate and adaptive immune responses against invasion by microorganism. A number of the antimicrobial peptides have potent receptor-dependent effects on humoral and cellular components of the host immune system. However, new antimicrobial peptides with unidentified host effects keep emerging (Cutuli et al., 2000; Hobta et al., 2001; Park et al., 2001; Yamaguchi et al., 2001, 2002). Many HBD and mBD genes are being discovered through use of the available gene databases with a computational search tool based on hidden Markov models (Schutte et al., 2002). The immunomodulatory activities of these antimicrobial peptides have not been evaluated as yet. Most of the known immunomodulatory effects overlap with other humoral components, such as cytokines and chemotactic factors of the immune system (Yang et al., 2002). Therefore additional studies are needed to distinguish their relative contribution to host antimicrobial immunity from that of other mediators. How antimicrobial peptides signal gene activation in target cells awaits further studies. Based on their immunomodulatory activities, antimicrobial peptides or their modified forms may have therapeutic potential as immunoadjuvants or even as topical antimicrobial agents. However, it should also be noted that the current effort to develop some antimicrobial peptides into systemic antibiotics (Hancock, 1999) may be complicated by generating undesirable immunomodulatory effects.

ACKNOWLEDGMENT

This project has been funded in part with Federal funds from the National Cancer Institute, National Institutes of Health, under contract number NO1-CO-12400. The content of this chapter does not necessarily reflect the views or policies of the Department of Health and Human Services, nor does mention of trade names, commercial products, or organizations imply endorsement by the U.S. Government. The publisher or recipient acknowledges right of the U.S. Government to retain a nonexclusive, royalty-free license in and to any copyright covering the chapter.

REFERENCES

Aarbiou, J., Ertmann, M., van Wetering, S., van Noort, P., Rook, D., Rabe, K. F., Litvinov, S. V., van Krieken, J. H., de Boer, W. I., and Hiemstra, P. S. (2002). Human neutrophil defensins induce lung epithelial cell proliferation in vitro. *Journal of Leukocyte Biology*, 72, 167–74.

Agerberth, B., Charo, J., Werr, J., Olsson, B., Idali, F., Lindbom, L., Kiessling, R., Jomvall, H., Wigzell, H., and Gudmundsson, G. H. (2000). The human antimicrobial and chemotactic peptides LL-37 and α-defensins are expressed by specific lymphocyte and monocyte populations. *Blood*, 96, 3086–93.

Al Mehdi, A. B., Zhao, G., Dodia, C., Tozawa, K., Costa, K., Muzykantov, V., Ross, C., Blecha, F., Dinauer, M., and Fisher, A. B. (1998). Endothelial NADPH oxidase as the source of oxidants in lungs exposed to ischemia or high K+. *Circulation Research*, 83, 730–7.

Appelmelk, B. J., An, Y. Q., Geerts, M., Thijs, B. J., de Boer, H. A., MacLaren, D. M., de Graaff, J., and Nuijens, J. H. (1994). Lactoferrin is a lipid A-binding protein. *Infection and Immunity*, 62, 2628–32.

Arnold, R. R., Cole, M. F., and McGhee, A. (1977). Bactericidal effect for human lactoferrin. *Science*, 197, 263–5.

Baba, M., Imai, T., Nishimura, M., Kakizaki, M., Takagi, S., Hieshima, K., Nomiyama, H., and Yoshie, O. (1997). Identification of CCR6, the specific receptor for a novel lymphocyte-directed CC chemokine LARC. *Journal of Biological Chemistry*, 272, 14893–8.

Baggiolini, M. (1998). Chemokines and leukocyte traffic. *Nature (London)*, 392, 565–8.

Baggiolini, M. B., Dewalk, D., and Moser, B. (1994). Interleukin-8 and related chemotactic cytokines: CXC and CC chemokines. *Advances in Immunology*, 55, 97–179.

Bals, R., Wang, X., Wu, Z., Bafna, V., Zasloff, M., and Wilson, J. M. (1998). Human β-defensin 2 is a salt-sensitive peptide antibiotic expressed in human lung. *Journal of Clinical Investigation*, 102, 874–80.

Bals, R., Weiner, D. J., Meegalla, R. L., and Wilson, J. M. (1999a). Transfer of a cathelicidin peptide antibiotic gene restores bacterial killing in a cystic fibrosis xenograft model. *Journal of Clinical Investigation*, 103, 1113–17.

Bals, R., Weiner, D. J., Moscioni, A. D., Meegalla, R. L., and Wilson, J. M. (1999b). Augmentation of innate host defense by expression of a cathelicidin antimicrobial peptide. *Infection and Immunity*, 67, 6084–9.

Banchereau, J. and Steinman, R. M. (1998). Dendritic cells and the control of immunity. *Nature (London)*, 392, 245–51.

Basak, A., Ernst, B., Brewer, D., Seidah, N. G., Munzer, J. S., Lazure, C., and Lajoie, G. A. (1997). Histidine-rich human salivary peptides are inhibitors of proprotein convertases furin and PC7 but act as substrates for PC1. *Journal of Peptide Research*, 49, 596–603.

Bastian, A. and Schäfer, H. (2001). Human α-defensin 1 (HNP-1) inhibits adenoviral infection in vitro. *Regulatory Peptides*, 101, 157–61.

Bateman, A., Singh, A., Congote, L. F., and Solomon, S. (1991). The effect of HNP1 and related neutrophil granule peptides on DNA synthesis in HL60 cells. *Regulatory Peptides*, 35, 135–43.

Befus, A. D., Mowat, C., Gilchrist, M., Hu, J., Solomon, S., and Bateman, A. (1999). Neutrophil defensins induce histamine secretion from mast cells: Mechanisms of action. *Journal of Immunology*, 163, 947–53.

Bensch, K. W., Raida, M., Magert, H.-J., Schulz-Knappe, P., and Forssmann, W.-G. (1995). hBD-1: A novel β-defensin from human plasma. *FEBS Letters*, 368, 331–5.

Biragyn, A., Ruffini, P. A., Leifer, C. A., Klyushnenkova, E., Shakhov, A., Chertov, O., Shirakawa, A. K., Farber, J. M., Segal, D. M., Oppenheim, J. J., and Kwak, L. W. (2002). Toll-like receptor 4-dependent activation of dendritic cells by β-defensin 2. *Science*, 298, 1025–9.

Biragyn, A., Surenhu, M., Yang, D., Ruffini, P. A., Haines, B. A., Klyushnenkova, E., Oppenheim, J. J., and Kwak, L. W. (2001). Mediators of innate immunity that target immature, but not mature, dendritic cells induce antitumor immunity when genetically fused with nonimmunogenic tumor antigens. *Journal of Immunology*, 167, 6644–53.

Blomqvist, M., Bergquist, J., Westman, A., Hakansson, K., Hakansson, P., Fredman, P., and Ekman, R. (1999). Identification of defensins in human lymphocyte nuclei. *European Journal of Biochemistry*, 263, 312–18.

Boman, H. G. (1998). Gene-encoded peptide antibiotics and the concept of innate immunity: An update review. *Scandinavian Journal of Immunology*, 48, 15–25.
Cella, M., Sallusto, F., and Lanzavecchia, A. (1997). Origin, maturation and antigen presenting function of dendritic cells. *Current Opinion in Immunology*, 9, 10–16.
Chaly, Y. V., Paleolog, E. M., Kolesnikova, T. S., Tikhonov, I. I., Petratchenko, E. V., and Voitenok, N. N. (2000). Neutrophil α-defensin human neutrophil peptide modulates cytokine production in human monocytes and adhesion molecule expression in endothelial cells. *European Cytokine Network*, 11, 257–60.
Chan, Y. R. and Gallo, R. L. (1998). PR-39, a syndecan-inducing antimicrobial peptide, binds and affects p130(Cas). *Journal of Biological Chemistry*, 273, 28978–85.
Chertov, O., Michiel, D. F., Xu, L., Wang, J. M., Tani, K., Murphy, W. J., Longo, D. L., Taub, D. D., and Oppenheim, J. J. (1996). Identification of defensin-1, defensin-2, and CAP37/azurocidin as T-cell chemoattractant proteins released from interleukin-8-stimulated neutrophils. *Journal of Biological Chemistry*, 271, 2935–40.
Crouch, S. P., Slater, K. J., and Fletcher, J. (1992). Regulation of cytokine release from mononuclear cells by the iron-binding protein lactoferrin. *Blood*, 80, 235–40.
Cutuli, M., Cristiani, S., Lipton, J. M., and Catania, A. (2000). Antimicrobial effects of α-MSH peptides. *Journal of Leukocyte Biology*, 67, 233–9.
Duits, L. A., Ravensbergen, B., Rademaker, M., Hiemstra, P. S., and Nibbering, P. H. (2002). Expression of β-defensin 1 and 2 mRNA by human monocytes, macrophages and dendritic cells. *Immunology*, 106, 517–25.
Echtenacher, B., Mannel, D. N., and Hultner, L. (1996). Critical protective role of mast cells in a model of acute septic peritonitis. *Nature (London)*, 381, 75–7.
Frye, M., Bargon, J., and Gropp, R. (2001). Expression of human β-defensin-1 promotes differentiation of keratinocytes. *Journal of Molecular Medicine*, 79, 275–82.
Ganz, T. (1987). Extracellular release of antimicrobial defensins by human polymorphonuclear leukocytes. *Infection and Immunity*, 55, 568–71.
Ganz, T. and Lehrer, R. I. (1998). Antimicrobial peptides of vertebrates. *Current Opinion in Immunology*, 10, 41–4.
Ganz, T., Selsted, M. E., Szklarek, D., Harwig, S. S. L., Daher, K., Bainton, D. F., and Lehrer, R. I. (1985). Defensins: Natural peptide antibiotics of human neutrophils. *Journal of Clinical Investigation*, 76, 1427–35.
Garcia, J. R., Jaumann, F., Schulz, S., Krause, A., Rodriguez-Jimenez, J., Forssmann, U., Adermann, K., Kluver, E., Vogelmeier, C., Becker, D., Hedrich, R., Forssmann, W. G., and Bals, R. (2001). Identification of a novel,

multifunctional β-defensin (human β-defensin 3) with specific antimicrobial activity: Its interaction with plasma membranes of Xenopus oocytes and the induction of macrophage chemoattraction. *Cell and Tissue Research*, 306, 257–64.

Garcia, J. R., Krause, A., Schulz, S., Rodriguez-Jimenez, F. J., Kluver, E., Adermann, K., Forssmann, U., Frimpong-Boateng, A., Bals, R., and Forssmann, W. G. (2001). Human β-defensin 4: A novel inducible peptide with a specific salt-sensitive spectrum of antimicrobial activity. *FASEB Journal*, 15, 1819–21.

Gudmundsson, G. H. and Agerberth, B. (1999). Neutrophil antibacterial peptides, multifunctional effector molecules in the mammalian immune systen. *Journal of Immunological Methods*, 232, 45–54.

Hancock, R. E. (1999). Host defence (cationic) peptides: What is their future clinical potential? *Drugs*, 57, 469–73.

Harder, J., Bartels, J., Christophers, E., and Schröder, J. M. (2001). Isolation and characterization of human β-defensin-3, a novel human inducible peptide antibiotics. *Journal of Biological Chemistry*, 276, 5707–13.

Harder, J., Bartels, J., Christophers, E., and Schröder, J. M. (1997). A peptide antibiotic from human skin. *Nature (London)*, 387, 861.

Hayflick, J. S., Kilgannon, P., and Gallatin, W. M. (1998). The intercellular adhesion molecule (ICAM) family of proteins. New members and novel functions. *Immunologic Research*, 17, 313–27.

Hirata, M., Shimomura, Y., Yoshida, M., Morgan, J. G., Palings, I., Wilson, D., Yen, M. H., Wright, S. C., and Larrick, J. W. (1994). Characterization of a rabbit cationic protein (CAP18) with lipopolysaccharide-inhibitory activity. *Infection and Immunity*, 62, 1421–6.

Hobta, A., Lisovskiy, I., Mikhalap, S., Kolybo, D., Romanyuk, S., Soldatkina, M., Markeyeva, N., Garmanchouk, L., Sidorenko, S. P., and Pogrebnoy, P. V. (2001). Epidermoid carcinoma-derived antimicrobial peptide (ECAP) inhibits phosphorylation by protein kinases in vitro. *Cell Biochemistry and Function*, 19, 291–8.

Hoffmann, J. A., Kafatos, F. C., Janeway Jr., C. A., and Ezekowitz, R. A. B. (1999). Phylogenetic perspectives in innate immunity. *Science*, 284, 1313–18.

Hoover, D. M., Boulegue, C., Yang, D., Oppenheim, J. J., Tucker, K. D., Lu, W., and Lubkowski, J. (2002). The structure of human MIP-3α/CCL20: Linking antimicrobial and CCR6 receptor binding activities with human β-defensins. *Journal of Biological Chemistry*, 277, 37647–54.

Hoover, D. M., Rajashankar, K. R., Blumenthal, R., Puri, A., Oppenheim, J. J., Chertov, O., and Lubkowski, J. (2000). The structure of human β-defensin-2 shows evidence of higher order oligomerization. *Journal of Biological Chemistry*, 275, 32911–18.

Huang, H. J., Ross, C. R., and Blecha, F. (1997). Chemoattractant properties of PR-39, a neutrophil antibacterial peptide. *Journal of Leukocyte Biology*, 61, 624–9.

Ichinose, M., Asai, M., Imai, K., and Sawada, M. (1996). Enhancement of phagocytosis by corticostatin I (CSI) in cultured mouse peritoneal macrophages. *Immunopharmacology*, 35, 103–9.

Imatani, T., Kato, T., Minaguchi, K., and Okuda, K. (2000). Histatin 5 inhibits inflammatory cytokine induction from human gingival fibroblasts by *Porphyromonas gingivalis*. *Oral Microbiology and Immunology*, 15, 378–82.

Johansson, J., Gudmundsson, G. H., Rottenberg, M. E., Berndt, K. D., and Agerberth, B. (1998). Conformation-dependent antibacterial activity of naturally occurring human peptide LL-37. *Journal of Biological Chemistry*, 273, 3718–24.

Kaplan, S. S., Heine, R. P., and Simmons, R. L. (1999). Defensins impair phagocytic killing by neutrophils in biomaterial-related infection. *Infection and Immunity*, 67, 1640–5.

Kaslow, H. R. and Burns, D. L. (1992). Pertussis toxin and target eukaryotic cells: binding, entry, and activation. *FASEB Journal*, 6, 2684–90.

Kirikae, T., Hirata, M., Yamasu, H., Kirikae, F., Tamura, H., Kayama, F., Nakatsuka, K., Yokochi, T., and Nakano, M. (1998). Protective effects of a human 18-kilodalton cationic antimicrobial protein (CAP18)-derived peptide against murine endotoxemia. *Infection and Immunity*, 66, 1861–8.

Korthuis, R. J., Gute, D. C., Blecha, F., and Ross, C. R. (1999). PR-39, a proline/arginine-rich antimicrobial peptide, prevents postischemic microvascular dysfunction. *American Journal of Physiology*, 277(3, Pt. 2), H1007–13.

Larrick, J. W., Hirata, M., Balint, R. F., Lee, J., Zhong, J., and Wright, S. C. (1995). Human CAP18: A novel antimicrobial lipopolysaccharide-binding protein. *Infection and Immunity*, 63, 1291–7.

Larrick, J. W., Hirata, M., Zheng, H., Zhong, J., Bolin, D., Cavaillon, J.-M., Warren, H. S., and Wright, S. C. (1994). A novel granulocyte-derived peptide with lipopolysaccharide-neutralizing activity. *Journal of Immunology*, 152, 231–40.

Lehrer, R. I., Daher, K., Ganz, T., and Selsted, M. E. (1985). Direct inactivation of viruses by MCP-1 and MCP-2, natural peptide antibiotics from rabbit leukocytes. *Journal of Virology*, 54, 467–72.

Lehrer, R. I. and Ganz, T. (1999). Antimicrobial peptides in mammalian and insect host defense. *Current Opinion in Immunology*, 11, 23–7.

Liao, F., Alderson, R., Su, J., Ullrich, S. J., Kreider, B. L., and Farber, J. M. (1997). STRL22 is a receptor for the CC chemokine MIP-3α. *Biochemical and Biophysical Research Communications*, 236, 212–17.

Lillard Jr., J. W., Boyaka, P. N., Chertov, O., Oppenheim, J. J., and McGhee, J. R. (1999). Mechanisms for induction of acquired host immunity by neutrophil peptide defensins. *Proceedings of the National Academy of Sciences, USA*, 96, 651–6.

Liu, Y. J. (2001). Dendritic cell subsets and lineages, and their functions in innate and adaptive immunity. *Cell*, 106, 259–62.

Malaviya, R., Ikeda, T., Ross, E., and Abraham, S. N. (1996). Mast cell modulation of neutrophil influx and bacterial clearance at sites of infection through TNFα. *Nature (London)*, 381, 77–80.

Medzhitov, R. and Janeway Jr., C. A. (1997). Innate immunity: The virtues of a nonclonal system of recognition. *Cell*, 91, 295–8.

Morrison, G., Kilanowski, F., Davidson, D., and Dorin, J. (2002). Characterization of the mouse β defensin 1, Defb1, mutant mouse model. *Infection and Immunity*, 70, 3053–60.

Moser, C., Weiner, D. J., Lysenko, E., Bals, R., Weiser, J. N., and Wilson, J. M. (2002). β-defensin 1 contributes to pulmonary innate immunity in mice. *Infection and Immunity*, 70, 3068–72.

Murphy, C. J., Foster, B. A., Mannis, M. J., Selsted, M. E., and Reid, T. W. (1992). Defensins are mitogenic for epithelial cells and fibroblasts. *Journal of Cellular Physiology*, 155, 408–13.

Murphy, P. M. (1996). Chemokine receptors: Structure, function and role in microbial pathogenesis. *Cytokine and Growth Factor Reviews*, 7, 47–64.

Murphy, P. M. (1994). The molecular biology of leukocyte chemoattractant receptors. *Annual Review of Immunology*, 12, 593–633.

Nakashima, H., Yamamoto, N., Masuda, M., and Fujii, N. (1993). Defensins inhibit HIV replication in vitro. *AIDS*, 7, 1129–1129.

Niyonsaba, F., Iwabuchi, K., Matsuda, H., Ogawa, H., and Nagaoka, I. (2002a). Epithelial cell-derived human beta-defensin-2 acts as a chemotaxin for mast cells through a pertussis toxin-sensitive and phospholipase C-dependent pathway. *International Immunology*, 14, 421–6.

Niyonsaba, F., Iwabuchi, K., Someya, A., Hirata, M., Matsuda, H., Ogawa, H., and Nagaoka, I. (2002b). A cathelicidin family of human antibacterial peptide LL-37 induces mast cell chemotaxis. *Immunology*, 106, 20–6.

Niyonsaba, F., Someya, A., Hirata, M., Ogawa, H., and Nagaoka, I. (2001). Evaluation of the effects of peptide antibiotics human β-defensin-1/2 and LL-37 on histamine release and prostaglandin D_2 production from mast cells. *European Journal of Immunology*, 31, 1066–75.

Nizet, V., Ohtake, T., Lauth, X., Trowbridge, J., Rudisill, J., Dorschner, R. A., Pestonjamasp, V., Piraino, J., Huttner, K., and Gallo, R. L. (2001). Innate antimicrobial peptide protects the skin from invasive bacterial infection. *Nature (London)*, 414, 454–7.

Oppenheim, F. G., Xu, T., McMillian, F. M., Levitz, S. M., Diamond, R. D., Offner, G. D., and Troxler, R. F. (1988). Histatins, a novel family of histidine-rich proteins in human parotid secretion. Isolation, characterization, primary structure, and fungistatic effects on Candida albicans. *Journal of Biological Chemistry*, 263, 7472–77.

Oppenheim, J. J., Zachariae, C. O., Mukaida, N., and Matsushima, K. (1991). Properties of the novel proinflammatory supergene "intercrine" cytokine family. *Annual Review of Immunology*, 9, 617–48.

Panyutich, A. V., Panyutich, E. A., Krapivin, V. A., Baturevich, E. A., and Ganz, T. (1993). Plasma defensin concentrations are elevated in patients with septicemia or bacterial meningitis. *Journal of Laboratory and Clinical Medicine*, 122, 202–7.

Park, C. H., V. Valore, E., Waring, A. J., and Ganz, T. (2001). Hepcidin, a urinary antimicrobial peptide synthesized in the liver. *Journal of Biological Chemistry*, 276, 7806–10.

Patarroyo, M. (1994). Adhesion molecules mediating recruitment of monocytes to inflamed tissue. *Immunobiology*, 191, 474–7.

Pérez-Cañadillas, J. M., Zaballos, A., Gutiérrez, J., Varona, R., Roncal, F., Albar, J. P., Márquez, G., and Bruix, M. (2001). NMR solution structure of murine CCL20/MIP-3α, a chemokine that specifically chemoattracts immature dendritic cells and lymphocytes through its highly specific interaction with the β-chemokine receptor CCR6. *Journal of Biological Chemistry*, 276, 28372–9.

Porro, G. A., Lee, J. H., de Azavedo, J., Crandall, I., Whitehead, T., Tullis, E., Ganz, T., Liu, M., Slutsky, A. S., and Zhang, H. (2001). Direct and indirect bacterial killing functions of neutrophil defensins in lung explants. *American Journal of Physiology*, 281, L1240–7.

Power, C. A., Church, D. J., Meyer, A., Alouani, S., Proudfoot, A. E. I., Clark-Lewis, I., Sozzani, S., Mantovani, A., and Wells, T. N. C. (1997). Cloning and characterization of a specific receptor for the novel CC chemokine MIP-3α from lung dendritic cells. *Journal of Experimental Medicine*, 186, 825–35.

Prohaszka, Z., Nemet, K., Csermely, P., Hudecs, F., Mezo, G., and Fust, G. (1997). Defensins purified from human granulocytes bind C1q and activate the classical complement pathway like the transmembrane glycoprotein gp41 of HIV-1. *Molecular Immunology*, 34, 809–16.

Randolph, G. J., Beaulieu, S., Lebecque, S., Steinman, R. M., and Muller, W. A. (1998). Differentiation of monocytes into dendritic cells in a model of transendothelial trafficking. *Science*, 282, 480–3.

Randolph, G. J., Inaba, K., Robbiani, D. F., Steinman, R. M., and Muller, W. A. (1999). Differentiation of phagocytic monocytes into lymph node dendritic cells in vivo. *Immunity*, 11, 753–61.

Risso, A., Zanetti, M., and Gennaro, R. (1998). Cytotoxicity and apoptosis mediated by two peptides of innate immunity. *Cellular Immunology*, 189, 107–15.

Sawai, M. V., Jia, H. P., Liu, L., Aseyev, V., Wiencek, J. M., McCray Jr., P. B., Ganz, T., R. Kearney, W., and Tack, B. F. (2001). The NMR structure of human β-defensin-2 reveals a novel α-helical segment. *Biochemistry*, 40, 3810–16.

Schild, L. and Kellenberger, S. (2001). Structure function relationships of ENaC and its role in sodium handling. *Advances in Experimental Medicine and Biology*, 502, 305–14.

Schutte, B. C., Mitros, J. P., Bartlett, J. A., Walters, J. D., Jia, H. P., Welsh, M. J., Casavant, T. L., and McCray Jr., P. B. (2002). Discovery of five conserved beta-defensin gene clusters using a computational search strategy. *Proceedings of the National Academy of Sciences, USA*, 99, 2129–33.

Scott, M. G., Davidson, D. J., Gold, M. R., Bowdish, D., and Hancock, R. E. W. (2002). The human antimicrobial peptide LL-37 is a multifunctional modulator of innate immune response. *Journal of Immunology*, 169, 3883–91.

Scott, M. G., Gold, M. R., and Hancock, R. E. W. (1999). Interaction of cationic peptides with lipoteichoic acid and Gram-positive bacteria. *Infection and Immunity*, 67, 6445–53.

Scott, M. G. and Hancock, R. E. W. (2002). Cationic antimicrobial peptides and their multifunctional role in the immune system. *Critical Reviews in Immunology*, 20, 407–31.

Scott, M. G., Rosenberger, C. M., Gold, M. R., Finlay, B. B., and Hancock, R. E. W. (2000). An α-helical cationic antimicrobial peptide selectively modulates macrophage responses to lipopolysaccharide and directly alters macrophage gene expression. *Journal of Immunology*, 165, 3358–65.

Scott, M. G., Vreugdenhil, A. C., Buurman, W. A., and Hancock, R. E. W. (2000). Cutting edge: Cationic antimicrobial peptides block the binding of lipopolysaccharide (LPS) to LPS binding protein. *Journal of Immunology*, 164, 549–53.

Selsted, M. E., Szklarek, D., and Lehrer, R. I. (1984). Purification and antibacterial activities of antimicrobial peptides of rabbit granulocytes. *Infection and Immunity*, 45, 150–4.

Selsted, M. E., Tang, Y.-Q., Morris, W. L., McGuire, P. A., Novotny, M. J., Smith, W., Henschen, A. H., and Cullor, J. S. (1993). Purification, primary structures, and antimicrobial activities of β-defensins, a new family of antimicrobial peptides from bovine neutrophils. *Journal of Biological Chemistry*, 268, 6641–8.

Shi, J., Ross, C. R., Leto, T. L., and Blecha, F. (1996). PR-39, a proline-rich antibacterial peptide that inhibits phagocyte NADPH oxidase activity by binding to Src homology 3 domains of p47 phox. *Proceedings of the National Academy of Sciences, USA*, 93, 6014–18.

Shiomi, K., Nakazato, M., Ihi, T., Kangawa, K., Matsuo, H., and Matsukura, S. (1993). Establishment of radioimmunoassay for human neutrophil peptides and their increases in plasma and neutrophil in infection. *Biochemical and Biophysical Research Communications*, 195, 1336–44.

Solomon, S., Hu, J., Zhu, Q., Belcourt, D., Bennett, H. P. J., Bateman, A., and Antakly, T. (1991). Corticostatic peptides. *Journal of Steroid Biochemistry and Molecular Biology*, 40, 391–8.

Sozzani, S., Allavena, P., Vecchi, A., and Mantovani, A. (1999). The role of chemokines in the regulation of dendritic cell trafficking. *Journal of Leukocyte Biology*, 66, 1–9.

Tack, B. F., Sawai, M. V., Kearney, W. R., Robertson, A. D., Sherman, M. A., Wang, W., Hong, T., Boo, L. M., Wu, H., Waring, A. J., and Lehrer, R. I. (2002). SMAP-29 has two LPS-binding sites and a central hinge. *European Journal of Biochemistry*, 269, 1181–9.

Tal, T. and Irit, A. (1993). Defensin interferes with the activation of NADPH oxidase in a cell-free system. *Biochemical and Biophysical Research Communications*, 196, 630–41.

Tal, T., Michaela, S., and Irit, A. (1998). Cationic proteins of neutrophil azurophilic granules: protein-protein interaction and blockade of NAPDH oxidase activation. *Journal of Leukocyte Biology*, 63, 305–11.

Tang, Y.-Q., Yuan, J., Osapay, G., Osapay, K., Tran, D., Miller, C. J., Ouellette, A. J., and Selsted, M. E. (1999). A cyclic antimicrobial peptide produced in primate leukocytes by the ligation of two truncated α-defensins. *Science*, 286, 498–502.

Tani, K., Murphy, W. J., Chertov, O., Salcedo, R., Koh, C. Y., Utsunomiya, I., Funakoshi, S., Asai, O., Herrmann, S. H., Wang, J. M., Kwak, L. W., and Oppenheim, J. J. (2000). Defensins act as potent adjuvants that promote cellular and humoral immune responses in mice to a lymphoma idiotype and carrier antigens. *International Immunology*, 12, 691–700.

Taub, D. D., Anver, M., Oppenheim, J. J., Longo, D. L., and Murphy, W. J. (1996). T lymphocyte recruitment by interleukin-8 (IL-8). IL-8-induced degranulation of neutrophils releases potent chemoattractants for human T lymphocytes both in vitro and in vivo. *Journal of Clinical Investigation*, 97, 1931–41.

Territo, M. C., Ganz, T., Selsted, M. E., and Lehrer, R. (1989). Monocyte-chemotactic activity of defensins from human neutrophils. *Journal of Clinical Investigation*, 84, 2017–20.

Thompson, C. B. (1995). New insights into V(D)J recombination and its role in the evolution of the immune system. *Immunity*, 3, 531–9.

Tominaga, T., Fukata, J., Naito, Y., Funakoshi, S., Fujii, N., and Imura, H. (1990). Effects of corticostatin-I on rat adrenal cells in vitro. *Journal of Endocrinology*, 125, 287–92.

van den Berg, R. H., Faber-Krol, M. C., van Wetering, S., Hiemstra, P. S., and Daha, M. R. (1998). Inhibition of activation of classical pathway of complement by human neutrophil defensins. *Blood*, 92, 3898–903.

van Wetering, S., Mannesse-Lazeroms, S. P. G., Dijkman, J. H., and Hiemstra, P. S. (1997a). Effect of neutrophil serine proteinases and defensins on lung epithelial cells: modulation of cytotoxicity and IL-8 production. *Journal of Leukocyte Biology*, 62, 217–26.

van Wetering, S., Mannesse-Lazeroms, S. P. G., van Sterkenburg, M. A. J. A., Daha, M. R., Dijkman, J. H., and Hiemstra, P. S. (1997b). Effect of defensins on interleukin-8 synthesis in airway epithelial cells. *American Journal of Physiology*, 272, L888–96.

Verbanac, D., Zanetti, M., and Romeo, D. (1993). Chemotactic and protease-inhibiting activities of antibiotic peptide precursors. *FEBS Letters*, 371, 255–8.

Wilson, C. L., Ouellette, A. J., Satchell, D. P., Ayabe, T., Lopez-Boado, Y. S., Stratman, J. L., Hultgren, S. J., Matrisian, L. M., and Parks, W. C. (1999). Regulation of intestinal α-defensin activation by the metalloproteinase matrilysin in innate host defense. *Science*, 286, 113–17.

Yamaguchi, Y., Fukuhara, S., Nagase, T., Tomita, T., Hitomi, S., Kimura, S., Kurihara, H., and Ouchi, Y. (2001). A novel mouse β-defensin, mBD-6, predominantly expressed in skeletal muscle. *Journal of Biological Chemistry*, 276, 31510–14.

Yamaguchi, Y., Nagase, T., Mikita, R., Fukuhara, S., Tomita, T., Tominaga, T., Kurihara, H., and Ouchi, Y. (2002). Identification of multiple novel epididymis-specific β-defensin isoforms in humans and mice. *Journal of Immunology*, 169, 2516–23.

Yamashita, T. and Saito, K. (1989). Purification, primary structure, and biological activity of guinea pig neutrophil cationic peptides. *Infection and Immunity*, 57, 2405–9.

Yang, D., Biragyn, A., Kwak, L. W., and Oppenheim, J. J. (2002). Mammalian defensins in immunity: More than just microbicidal. *Trends in Immunology*, 23, 291–6.

Yang, D., Chen, Q., Chertov, O., and Oppenheim, J. J. (2000). Human neutrophil defensins selectively chemoattract naïve T and immature dendritic cells. *Journal of Leukocyte Biology*, 68, 9–14.

Yang, D., Chen, Q., Schmidt, A. P., Anderson, G. M., Wang, J. M., Wooters, J., Oppenheim, J. J., and Chertov, O. (2000). LL-37, the neutrophil granule- and epithelial cell-derived cathelicidin, utilizes formyl peptide receptor-like 1 (FPRL1) as a receptor to chemoattract human peripheral blood neutrophils, monocytes, and T cells. *Journal of Experimental Medicine*, 192, 1069–74.

Yang, D., Chertov, O., Bykovskaia, S. N., Chen, Q., Buffo, M. J., Shogan, J., Anderson, M., Schroder, J. M., Wang, J. M., Howard, O. M. Z., and Oppenheim, J. J. (1999). β-defensins: Linking innate and adaptive immunity through dendritic and T cell CCR6. *Science*, 286, 525–8.

Yang, D., Chertov, O., and Oppenheim, J. J. (2001a). Participation of mammalian defensins and cathelicidins in antimicrobial immunity: Receptors and activities of human defensins and cathelicidin (LL37). *Journal of Leukocyte Biology*, 69, 691–7.

Yang, D., Chertov, O., and Oppenheim, J. J. (2001b). The role of mammalian antimicrobial peptides and proteins in awakening of innate host defenses and adaptive immunity. *Cellular and Molecular Life Sciences*, 58, 978–89.

Yang, D., Howard, O. M. Z., Chen, Q., and Oppenheim, J. J. (1999). Cutting edge: Immature dendritic cells generated from monocytes in the presence of TGF-β1 express functional C-C chemokine receptor 6. *Journal of Immunology*, 163, 1737–41.

Yomogida, S., Nagaoka, I., Saito, K., and Yamashita, E. (1996). Evaluation of the effects of defensins on neutrophil functions. *Inflammation Research*, 45, 62–7.

Yoo, Y. C., Watanabe, R., Koike, Y., Mitobe, M., Shimazaki, K., Watanabe, S., and Azuma, I. (1997). Apoptosis in human leukemic cells induced by lactoferricin, a bovine milk protein-derived peptide: involvement of reactive oxygen species. *Biochemical and Biophysical Research Communications*, 237, 624–8.

Yoshida, M., Kimura, T., Kitaichi, K., Suzuki, R., Baba, K., Matsushima, M., Tatsumi, Y., Shibata, E., Takagi, K., Hasegawa, T., and Takagi, K. (2001). Induction of histamine release from rat peritoneal mast cells by histatins. *Biological and Pharmaceutical Bulletin*, 24, 1267–70.

Zanetti, M., Gennaro, R., and Romeo, D. (1995). Cathelicidins: a novel protein family with a common proregion and a variable C-terminal antimicrobial domain. *FEBS Letters*, 374, 1–5.

Zasloff, M. (2002). Antimicrobial peptides of multicellular organisms. *Nature (London)*, 415, 389–95.

Zhang, G. H., Mann, D. M., and Tsai, C. M. (1999). Neutralization of endotoxin in vitro and in vivo by a human lactoferrin-derived peptide. *Infection and Immunity*, 67, 1353–8.

Zhang, L., Yu, W., He, T., Yu, J., Caffrey, R. E., Dalmasso, E. A., Fu, S., Pham, T., Mei, J., Ho, J. J., Zhang, W., Lopez, P., and Ho, D. D. (2002). Contribution of human α-defensin 1, 2, and 3 to the anti-HIV-1 activity of CD8 antiviral factor. *Science*, 298, 995–1000.

Zhu, Q., Hu, J., Mulay, S., Esch, F., Shimasaki, S., and Solomon, S. (1988). Isolation and structure of corticostatin peptides from rabbit fetal and adult lung. *Proceedings of the National Academy of Sciences, USA*, 85, 592–6.

Zhu, Q. and Solomon, S. (1992). Isolation and mode of action of rabbit corticostatic (antiadrenocorticotropin) peptides. *Endocrinology*, 130, 1413–23.

Zlotnik, A. and Yoshie, O. (2000). Chemokines: A new classification system and their role in immunity. *Immunity*, 12, 121–7.

CHAPTER 4

Therapeutic potential and applications of innate immunity peptides

Timothy J. Falla and Lijuan Zhang

4.1. INTRODUCTION

The use of peptides of the innate immune system as anti-infective therapeutic agents has attracted attention and investment for almost 20 years. However, such peptides have failed to gain "large pharma" acceptance and failed to achieve approval of the U.S. Food and Drug Administration (FDA) in a number of topical indications such as impetigo, diabetic foot ulcers, and oral mucositis. However, from this experience and from advances in the field there is now renewed effort to develop one of nature's most prolific defense mechanisms into a valuable preventative and therapeutic agent. The driving force for such an initiative is an ever-decreasing resource of novel bioactive molecules and an ever-increasing expense of discovering such molecules by complex molecular and genetic screening. Companies that are involved in the development of such peptides have focused on cost of goods, delivery, and stability and on addressing such issues, and they have begun to produce realistic product profiles. At the same time, further properties of these peptides based on their ability to modulate the immune system have provided a wealth of applications including wound healing, anti-inflammation, and innate immunity stimulation to combat disease. The decrease in peptide manufacturing costs combined with a concerted effort by researchers to create more druglike peptides is likely to create a new generation of topical therapeutics aimed at disease prevention and cure. In this chapter we define the rationale for the clinical use of peptides based on those found in the innate immune system. This rationale is centered on the natural role of these bioactive peptides in the mammalian immune system. In addition, we outline the path taken to achieve this goal by the leading peptide-development companies around the world.

4.2. INNATE IMMUNITY PEPTIDES: MUCH-NEEDED NOVEL THERAPEUTICS

The innate immunity of vertebrates to microbial invasion is arbitrated by a network of host–defense mechanisms involving both the long-lasting highly specific responses of the cell-mediated immune system and a nonspecific chemical defense system based on a series of broad-spectrum antimicrobial peptides that are analogous to those found in insects (Nicolas and Mor, 1995; Hoffmann and Reichhart, 2002). Vertebrate antibiotic peptides secreted by nonlymphoid cells of the mucosal surfaces of the respiratory and gastrointestinal tracts as well as by the granular glands of the skin reportedly cause the lysis of numerous pathogenic microorganisms, including viruses, Gram-positive and Gram-negative bacteria, protozoa, yeasts, and fungi, as well as of cancer cells (Boman, 1996; Ganz and Lehrer, 1997; Ganz and Weiss, 1997; Hancock and Scott, 2000).

Antimicrobial peptides isolated from vertebrates have three characteristic properties: They are relatively small (20–46 amino acid residues), basic (lysine rich or arginine rich), and amphipathic. Although these peptides differ widely in length and amino acid sequence, they may be grouped into four broad families based on characteristic structural features (Hancock and Lehrer, 1998; Hancock, 2001). Although the precise mechanism of action of these peptides remains to be clearly defined, it is likely to involve multiple targets within the cell, as indicated by both model membrane studies and gene array analysis (Wu and Hancock, 1999; Friedrich et al., 2001; Zhang, Rozek, and Hancock, 2001; Hong et al., 2003). In addition, such peptides have the capacity to form channels or pores within the microbial membrane in order to permeate the cell and impair its ability to carry out anabolic processes (Ludtke et al., 1996; Oren and Shai, 1998; Park, Kim, and Kim, 1998; Shai, 1999). This secondary, innate immune system provides vertebrates with a repertoire of small peptides that are promptly synthesized upon induction, easily stored in large amounts, and readily available for antimicrobial warfare (Sorensen et al., 2001; Bulow et al., 2002).

There is a growing need for new antimicrobials, especially because drug discovery has been limited to modifications of traditional antimicrobial compounds. Until the year 2000, with the introduction of linezolid (Durrant, 2001; Ford, Zurenko, and Barbachyn, 2001), it had been nearly 30 years since a truly new class of antibiotics was introduced. This gap between the ability of microbes to become resistant to drugs and the scientific community's ability to combat them has many in the medical profession fearing the worst. The appearance of methicillin-resistant *Staphylococcus aureus* has left doctors with only one drug to combat this strain, vancomycin (Michel and Gutmann, 1997). However, strains with intermediate levels of vancomycin resistance have been isolated (Michel and Gutmann, 1997; Tenover, 1999). This means that the last line of defense against these deadly bacteria may soon be ineffective.

The world pharmaceutical market is estimated to be over $290 billion a year (all dollar amounts are in U.S. dollar; 1 billion = 10^9), of which approximately 12% is anti-infective agents. This $37 billion market is made up of about 65% antibacterials, 15% antivirals, 11% antifungals, and 9% antiprotozoals and vaccines and by 2010 is likely to reach $44 billion. This growth in anti-infectives in part is driven by the emergence of new pathogens and new strains resistant to currently employed drugs and by a growing immunosuppressed patient population susceptible to hospital-acquired infections. However, in part because of a dwindling supply of new chemical entities with appropriate bioactivities, and economics, many large pharmaceutical companies have left the field of antibiotic research and development (e.g., Eli Lilly, Proctor & Gamble, and Bristol Myers Squibb between 2001 and 2003). In addition, there are many new and emerging therapeutic areas for which the return for large pharmaceutical companies is greater than those from anti-infectives. The role of advancing novel antimicrobial agents into and through the clinic is now falling to the smaller companies (e.g., Cubist Pharmaceuticals, Lexington, Massachusetts; Versicor, Fremont, California; and Genome Therapeutics, Waltham, Massachusetts), and the search for new chemical diversity is ever widening and increasing in complexity. Examples of this intensive search include the sampling of oceans by Nereus Pharmaceuticals (San Diego, California) and sampling environmental nucleic acid by companies such as Diversa Corp. (La Jolla, California), and Cubist Pharmaceuticals (Lexington, Massachusetts).

Antimicrobial peptides certainly fill many of the requirements for development as useful therapeutics. However, the failure to get FDA approval to date has made future attempts more difficult and has not improved the view of the field held by the larger pharmaceutical companies. This is of significance as both Magainin Pharmaceuticals and IntraBiotics Inc. demonstrated, with GlaxoSmithKline (Research Triangle Park, North Carolina) and Pharmacia Corp. (Peapack, New Jersey), respectively, that large pharma assistance is very valuable in the drug development process. Micrologix Inc. has now taken this approach a stage further by licensing a specific peptide program to Fujisawa Healthcare Inc. (Fujisawa Pharmaceuticals, Osaka, Japan) to help it progress through the final stages of clinical development.

In this chapter we review the potential of peptides of the innate immune system used on their natural role in mammals. From this we outline how this activity may be applied to potential therapeutics and the progress of small biotechnology companies in attempting to exploit this potential.

4.3. ANTIMICROBIAL PEPTIDES OF THE MAMMALIAN INNATE IMMUNE SYSTEM IN NATURE

4.3.1. Mammalian Antimicrobial Peptides

Defensins

Defensins are cysteine-rich cationic peptides that function in antimicrobial defense in both invertebrates and vertebrates (Ganz, Selsted, and Lehrer, 1990; Lehrer and Ganz, 1992, 2002b). Mammalian defensins contain three subclasses, named α- and β-, and circular θ-defensins (Trabi, Schirra, and Craik, 2001). All three subclasses contain six cysteines and all have largely β-sheet structures that are stabilized by three intramolecular disulfide bonds (Ganz, Oren, and Lehrer, 1992; Lehrer and Ganz, 1992). The α- and β-defensins differ only in the spacing and connectivity of their cysteines.

In humans there are six α-defensins, four of which are sequestered in secretory granules within neutrophils and are termed human neutrophil peptides (HNP1–HNP4), whereas the remaining two (human α-defensins 5 and 6) are secreted from Paneth cells in the gastrointestinal tract (Jones and Bevins, 1992; Wimley, Selsted, and White, 1994; Yang, Chertov, and Oppenheim, 2001b). Mice express numerous Paneth cell α-defensins isoforms, termed cryptdins (crypt defensins). Six cryptdins have been purified (Yamaguchi et al., 2001; Ayabe et al., 2002a, 2002b).

The α-defensins (HNP1–HNP4) have broad antimicrobial activity against Gram-negative and Gram-positive bacteria, fungi, and enveloped viruses (Lehrer and Ganz, 1996; Ganz and Lehrer, 1998; Chaly et al., 2000). The human β-defensins (HBDs) are also called epithelial defensins, and four of them, HBD1–HBD4, have been isolated and characterized (Harder et al., 1997, 2001). Recently, two more β-defensin isoforms, HBD5 and HBD6, with the six conserved cysteine motifs identical to HBD4 were characterized from the human genomic sequences, and their mouse homologs have also been identified (Jia et al., 2001; Schutte et al., 2002). The cyclic θ-defensins, RTD1, RTD2, and RTD3, were originally isolated in bone marrow from the rhesus monkey, *Rhesus macaque,* granulocytes (Trabi et al., 2001). They have 18 residues and are formed by the fusion of two truncated α-defensin precursors. Recently the human counterpart, retrocyclin, has been identified as derived from a pseudogene in the same locus where α- and β-defensin genes cluster on the chromosome (Cole et al., 2002).

Seven mouse β-defensins (mBD1–mBD6, Bin1b) and two rat β-defensins have been described (Jia et al., 1999, 2000; Com et al., 2003). Most of them are expressed in the epithelial cells of various organs, such as lung, trachea, and kidney, except that mBD6 is predominately expressed in skeletal muscles in addition to the esophagus, tongue, and trachea (Yamaguchi et al., 2001), and Bin1b appears to be a natural epididymis-specific antimicrobial peptide that plays a role in reproductive tract host defense and male fertility in mouse (Li et al., 2001).

Cathelicidins

Cathelicidins are a novel family of antimicrobial peptide precursors from the cytoplasmic granules of mammalian neutrophils (Lehrer and Ganz, 2002a; Ramanathan et al., 2002; Zaiou and Gallo, 2002). Unlike defensins that are fully processed before storage, cathelicidins are stored as precursors that require additional processing. They are characterized by a conserved N-terminal region that shares homology in the proregion with cathelin, a 96-amino-acid oligopeptide originally isolated from porcine neutrophils (Gennaro and Zanetti, 2000), whereas the C-terminal antimicrobial domain can vary considerably in both primary sequence and length, varying from 13 to 30 residues, and has no homology among different members of cathelicidins (Zanetti, Gennaro, and Romeo, 1995). Members of this family of antimicrobial peptides have been found in several mammalian species including humans, mice, pigs, cattle, sheep, rabbits, and monkeys. Proteolytic processing at the cathelin and peptide junction site is generally required for microbicidal activity with one exception, in that rabbit cathelicidins, p15a and p15b, lack this consensus site, their cathelin domains are less anionic, and they are microbicidal without proteolytic processing (Zarember et al., 2002). Cathelicidins are located in the types of neutrophil granules that are readily released into the extracellular fluid, that is, the specific (secondary) granules in human and murine neutrophils and the large granules in bovine neutrophils, suggesting that the peptides may function primarily in extracellular spaces (Ganz and Weiss, 1997; Sorensen et al., 2001). The structurally diverse cathelicidin-derived antimicrobial peptides of animals provide interesting models for pharmaceutical development.

Humans apparently have only one cathelicidin gene. Its product, hCAP-18, is present in the secondary (specific) granules of neutrophils (Sorensen et al., 2001). The C-terminal antimicrobial peptide, LL-37, is liberated by proteinase 3 cleavage coincident with degranulation and secretion (De et al.,

2000). Many nonmyeloid tissues also express hCAP-18, including epididymis, spermatids, keratinocytes, epithelial cells, and various lymphocytes (Sorensen et al., 1997). Rhesus monkey bone marrow expresses a cathelicidin with a C-terminal domain comprising a 37-residue α-helical peptide (RL-37) that resembles human LL-37 (Zhao et al., 2001). Whereas human LL-37 contains five acidic residues and has a net charge of +6, rhesus RL-37 has only two acidic residues and a net charge of +8 (Zhao et al., 2001). Ruminants' neutrophils contain many cathelicidins. Bovine cathelicidins include molecules with antimicrobial domains encoding a cyclic dodecapeptide termed bactenecin (Romeo et al., 1988), a tryptophan-rich tridecapeptide, indolicidin (Del Sal et al., 1992; Selsted et al., 1992), proline- and arginine-rich Bac5 and Bac7 peptides (Frank et al., 1990), and several α-helical peptides (Tossi et al., 1995; Skerlavaj et al., 1996). These structural analogs were also identified in sheep and goat (Skerlavaj et al., 1996). Indolicidin is a cationic 13-amino-acid peptide with 39% tryptophan, 23% proline, and an amidated carboxyl terminus produced in the granules of bovine neutrophils (Selsted et al., 1992). The structure of indolicidin was determined by two-dimensional nonmagnetic resonance scanning and shown to form an extended boat-shaped conformation when bound to dodecylphosphocholine (DPC) (Falla, Karunaratne, and Hancock, 1996; Friedrich et al., 2001). Its short size and the fact that it has a broad spectrum of antimicrobial activity have made indolicidin and closely related analogs potential therapeutic candidates. Tritrypticin is also a 13-amino-acid tryptophan-rich cationic antimicrobial peptide and is found in porcine bone marrow (Nagpal et al., 1999). It has a broad spectrum of antibacterial and antifungal activity and has hemolytic activity comparable with that of indolicidin.

PR-39 and protegrin are two cathelicidin peptides produced in pigs. PR-39 is a proline–arginine-rich antibacterial peptide that was isolated from pig intestine in 1991 (Gudmundsson et al., 1995). It belongs to a group of linear peptides of innate immunity isolated from mammals and invertebrates and characterized by a high content of proline residues (up to 50%). Members of this group are predominantly active against Gram-negative bacterial species that, unlike most membrane-active peptides, kill by a nonlytic mechanism. Evidence is accumulating that the Proline-rich peptides enter the cells without membrane disruption and, once in the cytoplasm, bind to and inhibit the activity of specific molecular processes (Gao et al., 2000). PR-39 is a multifunctional peptide with activities related to wound healing, inflammation, and neutrophil activity. These activities are discussed in detail elsewhere in this chapter.

Protegrins are five analogous antimicrobial peptides that contain 16 to 18 amino acids that were originally purified from porcine leukocytes (Kokryakov et al., 1993). They have two intromolecular disulfide bonds folded into a rigid β-sheet structure and stored as inactive proforms in porcine neutrophil granules activated extracellularly by neutrophil elastase.

Other Antimicrobial Peptides of Mammalian Origin

Recently an interesting group of multicysteine-containing peptides, named hepcidins, was identified in humans (Park et al., 2001). The first of these was isolated from human blood ultrafiltrate and from urine. It exhibited antifungal activity against *Candida albicans, Aspergillus fumigatus,* and *Aspergillus niger* and antibacterial activity against *Escherichia coli, S. aureus, Staphylococcus epidermidis,* and group B *Streptococcus* (Park et al., 2001; Shike et al., 2002). Hepcidins have eight cysteine residues that form four intramolecular disulfide bonds. They are synthesized predominantly in the liver as 84-amino-acid precursors that are subsequently processed and secreted in a 25-amino-acid peptide form. Three forms of hepcidin differing by amino-terminal truncation have been characterized. Reverse translation and database search identified homologous liver cDNAs in species from fish to human (chromosome 19) (Park et al., 2001). In addition to their antimicrobial activity, hepcidins are also suggested to play an important role in iron homeostasis under various pathophysiological conditions, which may support the pharmaceutical use of hepcidin agonists and antagonists in various iron homeostasis disorders (Nicolas et al., 2002a, 2002b, 2002c).

Histatins are histidine-rich antimicrobial peptides secreted by parotid, submandibular, and sublingual glands from human. Three peptides, histatin 1, 3, and 5, have been characterized and contain 38-, 32-, and 24-amino-acid residues, respectively, and all contain seven histidine residues (Oppenheim et al., 1988). The first 22 amino acid residues of all three histatins appear to be identical, and the carboxyl-terminal 7 residues of histatins 1 and 3 are also identical. The complete sequence of histatin 5 is contained within the amino-terminal 24 residues of histatin 3. Structural data suggest that histatins 1 and 3 are derived from different structural genes, whereas histatin 5 is a proteolytic product of histatin 3. All three histatins exhibit the ability to kill the pathogenic yeast *C. albicans.* Unlike most lytic antimicrobial peptides, histatin 5 has been shown to translocate across the yeast membrane and the targets to the mitochondria, suggesting an unusual antifungal mechanism of action (Helmerhorst et al., 1999, 2001).

Dermcidin, a 47-amino-acid human peptide that, after proteolytic processing of a precursor protein that is constitutively expressed in the sweat glands, is secreted into the sweat and transported to the epidermal surface (Schittek et al., 2001; Flad et al., 2002). This peptide has a broad spectrum of activity in response to a variety of pathogenic microorganisms and no homology to other known antimicrobial peptides.

4.3.2. Mammalian Antimicrobial Peptides In Innate Defenses

The role of antimicrobial peptides in innate defenses was initially suggested from the observation that they can be rapidly induced in insects and amphibian skins by microbial infection and that they are the major protein species in neutrophils of mammals. After more than a decade of extensive studies, it became more evident that peptides are expressed, both constitutively and inducible, in every niche of the human body susceptible to invasion and colonization by pathogenic microbes. The skin and mucosal membranes line body cavities exposed to the exterior, such as the respiratory tract, gastrointestinal tract, and the genitourinary tract, and all are defended by antimicrobial peptides to prevent the entry and colonization by pathogenic microbes.

Healthy human skin is partly protected against bacterial infection by the secretion of a peptide called dermcidin from the sweat glands (Schittek et al., 2001; Flad et al., 2002). The propeptide is specifically and constitutively expressed in the sweat glands, secreted into the sweat, and transported to the epidermal surface. In sweat, a proteolytically processed 47-amino-acid peptide is generated and has demonstrated antimicrobial activity in response to a variety of pathogenic microorganisms (Schittek et al., 2001; Flad et al., 2002). The activity of the peptide is maintained over a broad pH range and in high salt concentrations that resemble the human sweat environment. In addition to dermcidin, a number of dermcidin-related peptides are also present in the skin in significant quantities (Flad et al., 2002). This indicates that sweat plays a role in the regulation of human skin flora through the presence of antimicrobial peptide(s) and may help limit infection by potential pathogens in the first few hours following bacterial colonization.

Defensins and LL-37/CAP-18 are also an essential part of cutaneous innate immunity. In normal skin these peptides are barely detectable, but they accumulate in skin affected by inflammatory processes. A large increase in the expression of cathelicidins was observed in human and murine skin after sterile incision and in the mouse following infection (Dorschner et al., 2001). Immunohistochemical analysis showed the presence of abundant LL-37 and HBD2 in the superficial epidermis of all patients with psoriasis (Ong et al., 2002). This may explain why bacterial skin infection is rare among patients with psoriasis. In contrast, the level of these peptides was significantly decreased in acute and chronic lesions from patients with atopic dermatitis

(Ong et al., 2002). The fact that *S. aureus* often colonizes the skin of such patients provides a compelling link between innate antimicrobial peptide deficiency and infection.

HBD1 is consistently expressed in skin samples from various body sites, whereas expression of HBD2 is more variable and is more readily detectable in facial skin and foreskin compared with skin from abdomen and breast (Ali et al., 2001). Studies have shown that HBD2 is absent in the full-thickness burn wound and burn blister fluid, suggesting a possible therapeutic role for antimicrobial peptides in the management of burn wounds (Ortega, Ganz, and Milner, 2000).

The airway surface is an important host defense against pulmonary infection. Secretion of proteins with antimicrobial activity from epithelial cells onto the airway surface represents an important component of this innate immune system. Nasal and lung secretions are rich in lysozyme, lactoferrin, secretory leukoprotease inhibitor, and defensins (Travis et al., 2001; Frye et al., 2000; Dauletbaev et al., 2002). Defensins are the best characterized epithelial-derived peptide antibiotics. The tracheal antimicrobial peptide (TAP) and the lingual antimicrobial peptide (LAP) were among the first β-defensins isolated from cow trachea and tongue and were found to be expressed in multiple epithelia, including the lung (Diamond et al., 1991). In humans, HBD1 was initially purified from hemodialysate and subsequently shown to be expressed in the surface epithelia and submucosal glands of the human airways. Although HBD2 and HBD3 are most abundant in inflamed skin, they have also been detected in the respiratory tract (Singh et al., 1998; Harder et al., 2001). Both HBD2 and HBD3 are inducible by bacterial components and infection, further suggesting their primary role in host defenses. The human cathelicidin peptide, LL-37/hCAP-18, is expressed in myeloid cells and was also shown to be expressed in humans diffusely throughout the epithelia in many organs, including surface airway epithelia and serous and mucous cells of the proximal airway submucosal glands (Frohm Nilsson et al., 1999). This has led to the speculation that this innate immune component is probably intrinsic to most, if not all, epithelia, providing a barrier function. The fact that both β-defensin 1 and 2 and LL-37/hCAP-18 are salt sensitive has led to speculation that impairment of this innate defense may have contributed to the pathogenesis of cystic fibrosis (CF) (Bals et al., 1998).

Neutrophil defensins (α-defensins) also play important roles in mucosal defenses. They are prominently expressed in rabbit kidney (Wu, Daniel, and Bateman, 1998), and in female reproductive epithelium to prevent initial colonization by pathogenic microorganisms (Svinarich et al., 1997). Women infected with *Neisseria gonorrhoeae, Trichomonas vaginalis,* or *Chlamydia trachomatis* had higher median levels of α-defensins (hNP1–hNP3) in the vagina than did uninfected women (Wiesenfeld et al., 2002). In gastric biopsies, HBD2 was undetectable in normal gastric antrum but a marked increase was observed in *Helicobacter pylori*-positive gastritis compared with control tissue (Uehara et al., 2003). Constitutive expression of HBD1 was observed in normal gastric mucosa, and there was a significant increase in gastritis. Modulation of β-defensin expression by pathogenic and/or inflammatory stimuli, as well as their cellular localization, places these antimicrobial peptides in the front line of innate host defense in the human stomach. Both HBD1 and HBD2 also have been implicated as important defense factors in the endometrium in protecting against infection in the uterus (Svinarich et al., 1997).

In the intestine, there are marked differences in the regional distribution of antimicrobial peptides. The α-defensins, HD5 and HD6, as well as the antimicrobial protein lysozyme, are expressed strictly by Paneth cells in the crypt region of the small intestine (Jones and Bevins, 1992; Ouellette, 1997, 1999) and can be aberrantly produced in metaplastic Paneth cells in the colons of patients with ulcerative colitis (Cunliffe et al., 2001). In contrast, HBDs are ubiquitously expressed by intestinal epithelium throughout the large and small intestine, with HBD1 constitutively expressed and HBD2 induced in response to proinflammatory stimuli that activate the transcription factor NF-κB (O'Neil et al., 2000). LL-37 and hCAP-18 are shown to be expressed within epithelial cells located at the surface and upper crypts of a normal human colon, but little or no expression is seen within the deeper colon crypts or within epithelial cells of the small intestine. Differentiated human colon epithelium seems to express LL-37/hCAP-18 as part of its repertoire of innate defense molecules (Hase et al., 2002).

Direct evidence for the role of antimicrobial peptides in innate defense has been developed in the mouse. MBD1-deficient mice displayed delayed clearance of *Haemophilus influenzae* from lung (Moser et al., 2002). Recently a link between peptide deficiency and human disease may also have been demonstrated. Morbus Kostmann is a disease resulting in severe congenital neutropenia. Neutrophils from patients with morbus Kostmann were deficient in the cathelin LL-37 and had reduced concentrations of HNP1–HNP3 (Putsep et al., 2002). In contrast, LL-37 could not be detected in either plasma or saliva from such patients, although lactoferrin concentrations and oxidative burst were normal in all patients (Putsep et al., 2002). Patients suffering from morbus Kostmann experienced severe periodontal disease. However, one patient with morbus Kostmann, following a bone marrow transplantation from a healthy individual, showed almost normal concentrations of LL-37 and normal dental status.

Of note is that certain antimicrobial peptides may play a role in the inflammatory response. Neutrophils are thought to be involved in the pathogenesis of various inflammatory lung disorders, including chronic bronchitis and chronic obstructive pulmonary disease (Hiemstra, van Wetering, and Stolk, 1998). One of the targets of the neutrophil is the lung epithelium, and in vitro studies have revealed that both serine proteinases and neutrophil defensins markedly affect the integrity of the epithelial layer, decreasing the frequency of ciliary beat, increasing the secretion of mucus, and inducing the synthesis of epithelium-derived mediators that may influence the amplification and resolution of neutrophil-dominated inflammation (Hiemstra et al., 1998). In vitro results suggest that neutrophil defensins and serine proteinases cause injury and stimulate epithelial cells to produce chemokines that attract more neutrophils to the site of inflammation, a situation also described in idiopathic pulmonary fibrosis (Okrent, Lichtenstein, and Ganz, 1990). Furthermore, increased numbers of neutrophils are also present in the airways of patients with asthma (Nadel and Stockley, 1998). Because defensins are able to induce histamine release by mast cells and increase airway hyperresponsiveness to histamine, it is tempting to speculate that defensins may also contribute to the inflammatory processes in asthma (Nadel and Stockley, 1998).

4.3.3. Role of Mammalian Antimicrobial Peptides Beyond Innate Defenses

Innate immunity is nonspecific and includes, in addition to antimicrobial peptides, pattern-recognition molecules/receptors, complement, inflammatory mediators, and cytokines. There is a growing body of evidence indicating that antimicrobial peptides confer an impressive variety of activities that have an impact on the quality and effectiveness of the innate immune responses and inflammation by directly acting on cells of the immune system.

Lipopolysaccharide (LPS) is associated with Gram-negative sepsis, a syndrome that accounts for more than 2% of all hospital admissions and more than 400,000 deaths each year in the United States. Conventional antibiotics, while killing bacteria, also facilitate bacterial release of LPS, accelerating the inflammation process. A good number of peptides such as indolicidin (Falla et al., 1996), sheep peptide SMAP-29 (Tack et al., 2002), and human CAP-18 (Nagaoka et al., 2002) are capable of binding to the surface LPS of intact bacterial cells as well as LPS released from phagocytosed bacteria, thus suppressing LPS-stimulated production of proinflammatory cytokines (TNF-α) by means of Toll-like receptors (TLRs). LL-37 is a potent antisepsis agent with the ability to inhibit macrophage stimulation by LPS, lipoteichoic acid, and noncapped lipoarabinomannan, and protects mice against lethal endotoxemia (Scott et al., 2002). It is known that LPS binds to a LPS-binding protein (LBP) in the plasma, activating CD14 by binding to TLRs on macrophages and other cells. Mammalian TLRs function as sensors of infection and induce the activation of innate and adaptive immune responses. This process results in the release of large quantities of inflammatory cytokines (Schroder et al., 2000; Triantafilou and Triantafilou, 2002). The expression of some antimicrobial peptides apparently is upregulated by means of the same pathway in response to bacterial components. Upregulation of HBD2 was observed with cells transfected with TLR-2, but not wild-type cells, following stimulation with bacterial lipoprotein (Birchler et al., 2001). The gene for defensins HBD2 and TAP includes three NF-κB concensus sequences upstream of the transcriptional initiation site, and possible NF-κB regulation of β-defensin induction has been described (Liu et al., 1998). More evidence suggests that LPS-triggered induction of hBD2 mRNA in human tracheobronchial epithelial cells through CD14 and TLRs apparently involves activation of NF-κB (Becker et al., 2000). The fact that the NF-κB family of transcription factors plays an important role as a central regulator of innate and adaptive immune functions suggests that peptides such as HBD2 could be directly involved in connecting innate and adaptive immune responses.

Innate immunity peptides link the innate immune system to adaptive immunity also through the chemotaxis of specialized lymphocytes (Koczulla and Bals, 2003). Inflammatory responses are multistep processes involving the production of various chemotactic factors, resulting in an orchestrated recruitment of neutrophils, mast cells, monocytes, and T cells (Proost, Wuyts, and van Damme, 1996). During the acute phase of inflammatory diseases, the predominantly migrating cells are neutrophils and mast cells, and, in the subsequent chronic phase, monocytes and lymphocytes take over as the primary cell types. The mast cell is one of the major effector cells in inflammatory reactions and can be found in most tissues throughout the body. During inflammation, an increase in the number of mast cells in the local milieu occurs, and such accumulation requires directed migration of this cell population.

Studies suggest that LL-37 not only blocks activation of the LPS-stimulated CD14-TLRs signal pathway but also directly upregulates genes that encode chemokines and chemokine receptors, as indicated by a gene array profiling study (Scott et al., 2002). In addition, LL-37 showed direct chemotactic activity for several cell types. It stimulates the degranulation of mast cells and induces mast cell chemotaxis through a Gi protein–phospholipase C signaling pathway (Niyonsaba et al., 2002b). The peptide is also chemotactic to freshly isolated human peripheral blood neutrophils, monocytes, and T cells that utilize formyl peptide receptor-like 1 (FPRL1) as a receptor to mediate its chemotactic and Ca^{2+}-mobilizing effects (Yang, Chertov, and Oppenheim, 2001a). Mobilization of intracellular calcium is one of the factors induced by inflammation. This mechanism is potentially important in vivo because the chemotactic activity of LL-37, unlike its antimicrobial activity, is not significantly inhibited by the presence of 10% human serum (Johansson et al., 1998). Therefore LL-37 may link innate and adaptive immunity not only by limiting the damage caused by bacterial products but also by recruiting immune cells to the site of infection. The local modulation of the innate immune system offers a wealth of potential clinical applications not simply limited to infectious diseases but also including inflammatory disorders.

Defensins are good chemoattractants. Neutrophil granule-derived cathepsin G, azurocidin/CAP-37, and α-defensins have been shown to be chemotactic for mononuclear cells and neutrophils (Yang et al., 2000). Analysis of the chemotactic activity of α-defensins shows that they induce both $CD45RA^{+}$ and CD8 T-lymphocyte-cell migration at concentrations 10- to 100-fold below those required for direct bactericidal activity (Agerberth et al., 2000). Additionally, α- and β-defensins form chemotactic gradients for the migration of immature dendritic cells (DCs) (Yang et al., 2000). Recruiting immature DCs to sites of infection is one way for neutrophil granule proteins to initiate adaptive immune responses, because DCs are central to the control of adaptive immunity, and their ability to activate antigen-specific T cells depends on their maturation state (Chertov et al., 2000; Duits et al., 2002). Human β-defensins (HBD1 and HBD2) are chemoattractants for DCs and memory CD4 T cells that seem to be mediated through the chemokine receptor CCR6 (Niyonsaba et al., 2002a). In addition, monocytes, monocyte-derived macrophages (MDMs), and monocyte-derived DCs all have upregulated levels of HBD1 (Duits et al., 2002). Apparently these peptides participate in alerting, mobilizing, and amplifying innate and adaptive antimicrobial immunity of the host.

Neutrophil defensins have been shown to enhance proliferation of lung epithelial cells in vitro at physiological concentrations (Aarbiou et al., 2002). Defensin HBD1 levels correlate with differentiation and proliferation of HaCa T cells. The HBD1 mRNA levels remained low during proliferation but were highly induced on differentiation, whereas hBD2 expression was unaffected under these conditions (Aarbiou et al., 2002). Consistent with this, HBD1 also was overexpressed in keratinocytes. Promoting cell proliferation and differentiation processes clearly demonstrates a potential role for defensins in wound repair.

Interestingly, unlike defensins and LL-37, PR-39 (the porcine cathelicidin peptide) has been shown to prevent neutrophil recruitment following ischemia and reperfusion, possibly by downregulation of endothelial cell adhesion molecules (Hoffmeyer et al., 2000). PR-39 prevents platelet-activating factor-induced neutrophil chemotaxis as well as phorbol-myristate-acetate-(PMA-) stimulated intercellular adhesion molecule-1 expression by cultured endothelial cells (Hoffmeyer et al., 2000). PR-39 has also been shown to be a powerful inhibitor of neutrophilic nicotinamide-adenine dinucleotide phosphate (NADPH) oxidase activity, a contributing factor to the pathogenesis of ischemia-reperfusion injury in a variety of organs, including the brain (Korthuis et al., 1999), large intestine (Panes and Granger, 1998), and heart (Shi et al., 1996). The fact that PR-39 exerts a protective effect in various animal models of ischemia-reperfusion injury, preventing postischemic oxidant production, suggests that it may be therapeutically useful for prevention of neutrophil adhesion and activation during the postischemic inflammatory response. In addition, PR-39 exerts other potentially exploitable biological activities, such as induction of syndecan expression in mesenchymal cells, suggesting a role of this peptide in wound repair and inflammation (Chan and Gallo, 1998). This peptide is also a potent inducer of angiogenesis both in vitro and in vivo. Coronary flow studies demonstrated that PR-39-induced angiogenesis results in the production of functional blood vessels.

4.4. POTENTIAL CLINICAL DEVELOPMENT FOR INNATE IMMUNITY PEPTIDES

Innate immunity peptides demonstrate their greatest clinical promise in the area of topical application for specific local disease. This rationale is based on the role of such peptides in nature and their ease of delivery, stability, half-life, and potential to be efficacious. This section describes of number of these areas and data supporting the application of innate immunity peptides.

Eczema, also called atopic dermatitis, is an itchy allergic condition that often causes scratching that leaves the skin inflamed (Ong et al., 2002). It affects about 15 million people in the United States. About 90% of sufferers experience long-term staphylococcal skin infections. As previously mentioned, patients with eczema may suffer from a deficiency in antimicrobial peptides such as β-defensins and cathelicidins in their skin (Ong et al., 2002). The restoration of normal peptide levels by use of β-defensins and cathelicidins, or peptides that exhibit improved relevant activities, provides a very interesting therapeutic potential.

Although still in debate, CF has been speculated to be associated with inefficient antimicrobial activity of endogenous peptides in the high-salt airway environment of the CF lung. To supplement this defense mechanism, a number of researchers have designed salt-insensitive antimicrobial peptides that have demonstrated efficacy in the chronic rat lung infection model (Desai, Hancock, and Finlay, 2003). As a result, IntraBiotics Pharmaceuticals, Inc. (http://www.intrabiotics.com) has taken its protegrin peptide through Phase I clinical trials as an aerosolized therapeutic. Other biotechnology companies, including Helix BioMedix Inc. (http://www.helixbiomedix.com), have advanced antimicrobial peptides to the preclinical stage of development, demonstrating efficacy in the infected lung model. Aerosol-delivered traditional antimicrobials such as aztreonam and azithromycin remain the only other agents currently in development for microbial complications of the CF lung. However, the significant factors in CF lung disease are both infection and inflammation. There may well be a role for an innate immunity peptide to reduce both bioburden and inflammation, producing an extremely valuable treatment for improving the lung function of patients with CF.

Acne is the most common inflammatory skin disease of adolescence and early adulthood, with nearly 20% of all visits to dermatologists related to its evaluation and treatment. Data from the 1996 U.S. census indicate that approximately 45 million Americans are affected by acne. Current therapy for acne includes topical antibiotics and the use of systemic antibiotics approved for other indications, yet increasing resistance to these antibiotics is severely limiting their use. Micrologix Biotech Inc (www.mbiotech.com) announced in November 2001 the results of a Phase IIa clinical trial for MBI 594AN, an indolicidin variant, for acne treatment. The efficacy of MBI 594AN was significant when compared with that of a placebo control, and it had no apparent side effects. A Phase IIb clinical trial was initiated in February 2003 in collaboration with Fujisawa Healthcare Inc. More applications are anticipated that involve topical applications of antimicrobial peptides for acne treatment.

Colonization of central venous catheters (CVCs) and subsequent sepsis often originate from colonization of the patient's skin around the catheter insertion side by bacteria and fungi. Such organisms migrate down the catheter tract to colonize the implanted portion of the device (Verghese, Padmaja, and

Koshi, 1998; Eastman et al., 2001). The microbes then break away from the colonized catheter, seeding into the blood and causing subsequent bloodstream infections. Once in the blood, they multiply rapidly and begin to infect and kill healthy cells. In many cases the microorganisms that cause these infections have developed resistance to conventional antimicrobial drugs. In the United States, more than 5 million CVCs are sold each year. The Centers for Disease Control and Prevention estimates that in the United States a CVC-related bloodstream infection develops in 250,000 to 400,000 patients each year, resulting in approximately 70,000 deaths. In 2000, Micrologix Biotech Inc. completed safety and efficacy tests of MBI 226, another indolicidin variant, and entered Phase III clinical trials in the United States for prevention of CVC-associated infections. This application has received fast-track status from the U.S. FDA.

Burn wound is another promising area for topical use of peptides because patients are at high risk to septic complications that are due to the emergence of drug-resistant microbes. In a rat infection model (animals received a 20% total body surface area partial-thickness burn by immersion in 60 °C water for 20 s followed by wound seeding with 10^6 colony-forming units of Silvadene-resistant *Pseudomonas aeruginosa*), a topical application or intradermal injection of protegrin-1 was administered, followed by harvesting of wound tissues aseptically at different time points for quantitative bacterial counts (Steinstraesser et al., 2001). In vivo and in vitro experiments revealed rapid and significant decreases in bacterial counts for protegrin-1-treated groups compared with controls. Peptides may potentially be used as an alternative or adjunct therapy to standard agents used to treat wound infections. Demegen Inc. (http://www.demegen.com) has developed such a therapeutic (Demegel or D2A21-gel) and demonstrated efficacy in a *P. aeruginosa*-infected burn wound (Chalekson, Neumeister, and Jaynes, 2002). To truly exploit the potential of such peptides, the combination of antimicrobial and wound-healing properties seems an attainable goal. However, this may only be possible with a combination of two distinct peptides.

Dental hygiene is another area for peptide application. As mentioned earlier, histatins, histidine-rich peptides found within parotid and submandibular secretions, are a novel class of endogenous peptides with antimicrobial properties. The early preclinical studies on three topical histatin variants (histatin 5, P-113, and P-113D) showed efficacy in reducing the development of plaque and gingivitis in beagle dogs (Paquette et al., 1997, 2002).

One promising approach to the control of transmission of sexually transmitted diseases (STDs) is the use of topically applied microbicides that inactivate the relevant pathogens. *N. gonorrhoeae* causes more than 400,000 infections annually in the United States (Workowski, Levine, and Wasserheit, 2002). Although most of these infections respond to appropriate antibiotics, the emergence of strains resistant to benzylpenicillin and/or tetracycline requires more costly treatments. More often patients with Chlamydial disease are coinfected with *N. gonorrhoeae. Haemophilus ducreyi* causes chancroid, a genital ulcer that is painful and can persist for several months. Inguinal lymphoadenopathy occurs in up to 50% of patients with chancroids (Morse, 1989). Bacteria-related STDs have been associated with increased heterosexual transmission of human immunodeficiency virus (HIV) (Cameron et al., 1989; Plummer et al., 1991; Wasserheit, 1992). Vulvovaginal candidiasos (VVC) is a mucosal infection caused by *Candida* species (Sobel, 1988a, 1988b). *C. albicans* is the causative agent of VVC in approximately 85%–90% of patients with positive vaginal fungal cultures (Fidel and Sobel, 1996). The remainder of the cases are due to nonalbicans species, the most common of which are *C. glabrata* and *C. tropicalis.* In the United States alone, approximately 13 million cases of VVC occur annually, accounting for 10 million gynecologic office visits.

Antimicrobial peptides have attracted an increasing amount of interest in the development of microbicides because of their broad spectrum of activity against bacteria, fungi, enveloped viruses, and parasites. In addition, the lack of cross resistance with other commonly used antibacterial (e.g., β-lactams) and antifungal (e.g., azoles) agents makes them attractive potential therapeutics for STDs and vaginitis. Although this is a relatively new area for peptides, with few published results available, studies have already shown efficacy of peptides against pathogens, especially those associated with STDs. The porcine leukocyte protegrin has been shown to be active against *H. ducreyi, N. gonorrhoeae,* and *C. trachomatis* (Qu et al., 1996; Yasin et al., 1996; Fortney et al., 1998). In addition, certain peptides such as indolicidin have been shown to be active against HIV-1 in vitro (Robinson et al., 1998). The role of innate immunity peptides in defense against viral attack is an emerging area with therapeutic potential.

As described in earlier sections, the role of innate immunity peptides is not limited to infection prevention but also includes local immune modulation. It should be noted that Inimex Pharmaceuticals Inc. (www.inimexpharma.com), a Vancouver, British Columbia, based biopharmaceutical company, is pursuing the use of peptides to boost innate immunity, rather than as antimicrobials, to prevent or treat human disease. Preclinical data indicate that the company's lead compounds, although lacking antimicrobial activity in vitro, can protect against disease caused by bacterial infections in animal models.

4.5. COMMERCIALIZATION OF INNATE IMMUNITY PEPTIDES

The use of peptides of the innate immune system as anti-infective therapeutic agents has attracted a great deal of attention from academic groups and companies since the early 1980s. Despite the fact that such peptides failed to achieve FDA approval, there is now renewed effort to develop one of nature's most prolific defense mechanisms into valuable preventative and therapeutic agents. The driving force for such an initiative is an ever-decreasing resource of novel bioactive molecules and an ever-increasing expense of discovering such molecules by complex molecular and genetic screening. Companies involved in the development of such peptides have focused on cost of goods, delivery, and stability and on addressing such issues, and they have begun to produce realistic product profiles. At the same time, further properties of these peptides based on their ability to modulate the immune system have provided a wealth of applications, including wound healing and anti-inflammation. The decrease in peptide manufacturing costs combined with a concerted effort by researchers to create more druglike peptides is likely to create a new generation of topical therapeutics aimed at disease prevention and cure.

The current statuses of development programs being pursued by specific companies are outlined in the following subsections and are based on public disclosures and press releases.

4.5.1. IntraBiotics Inc.

IntraBiotics (Mountain View, California: www.intrabiotics.com) has concentrated its efforts on the topical antimicrobial market, building on technology initially licensed from the Lehrer laboratory at UCLA. Iseganan (IB-367), developed by IntraBiotics, was based on the initial discovery of protegrins in porcine leukocytes (Kokryakov et al., 1993). Early development, including an extensive structure–activity relationship (SAR) process, for the initial indication of prophylaxis of oral mucositis in radiotherapy and chemotherapy patients, was previously reviewed in detail (Chen et al., 2000). However, it is interesting to note that at the end of the SAR study the lead molecule (IB-367) varied little from the natural porcine peptide PG-1:

PG-1, RGGRLCYCRRRFCVCVGR;
IB-367, RGGLCYCRGRFCVCVGR.

Iseganan demonstrated an excellent in vitro profile, exhibiting broad-spectrum antimicrobial activity, low resistance emergence, and the maintenance of activity in saliva (Mosca et al., 2000). This activity translated well to the hamster cheek pouch model for oral mucositis in which the peptide significantly reduced oral bioburden; Intrabiotics went on to demonstrate safety and efficacy in Phase I and Phase II clinical trials. However, a clinical benefit could not be achieved in Phase III trials in patients having either radiotherapy or chemotherapy. The application of innate immunity peptides to oral mucositis would seem to be a very good fit based on the advantages of the class: microbicidality, low resistance threat, low systemic absorption, and activity in saliva. In addition, oral mucositis is an unmet medical need with no currently approved therapy. However, the extent and precise role of bacteria and fungi in the pathogenesis of this condition and the effectiveness of antimicrobial agents are still unknown (Simon et al., 2001; El-Sayed et al., 2002), and notably IB-367 did not have strong antifungal activity.

Amgen (Emoryville, California: www.amgen.com) has recently demonstrated significant benefit in Phase III trials in oral mucositis with recombinant keratinocyte growth factor (rHU-KGF) by protecting epithelial cells from damage. It may therefore be concluded that the effect of reducing microbial colonization of the oral cavity is less relevant than cellular damage (by immune as well as bacterial mechanisms) in the etiology of oral mucositis.

Iseganan is still under development by IntraBiotics for ventilator-associated pneumonia (VAP) and the CF lung. The status of these programs is uncertain as the company has significantly reduced its workforce and has stated that its focus for development candidates is broadening. The overall strategy of IntraBiotics is of significance for two reasons. First, the company primarily developed IB-367 for oral mucositis and then used the same peptide for further indications (line extensions). For a small company, the consolidation of resources around one peptide saves time, resources, and valuable money. However, this is at the expense of optimizing a peptide's properties for clinical indications as diverse as oral bioburden reduction and the CF lung. Second, the company moved away from peptides in order to build a product-development pipeline, which encompassed natural-product drug discovery and development. This was done to permit the company to compete in the systemic antibiotic arena, configuring the company as a pharmaceutical-development company rather than an innate immunity peptide company.

4.5.2. Micrologix Biotech Inc.

Micrologix (Vancouver, British Columbia: www.mbiotech.com) has developed clinical programs based around intellectual property initially licensed from the Hancock Laboratory at the University of British Columbia. These peptides are primarily analogs of indolicidin, first isolated from bovine neutrophils by Michael Selsted's group (Selsted et al., 1992). As with IntraBiotics, Micrologix has concentrated efforts on topical applications, including prevention of sepsis through reduction in central-line catheter contamination (MBI 226) and acne (MBI 594AN). Both of these peptides are believed to be analogs of indolicidin:

$$\text{Indolicidin,}\quad \text{ILPWKWPWWPWRR-NH}_2.$$

Micrologix announced in January 2003 the initiation of enrollment of patients in a Phase IIb clinical trial of MBI 594AN for the treatment of acne. This was a 12-week study in which a 2.5% or 1.25% gel formulation of the indolicidin analog was used. This program is advancing based on the encouraging results of the company's Phase IIa trial in which a significant reduction in both inflammatory and noninflammatory acne was observed. In February 2003, Micrologix announced, in collaboration with Fujisawa Healthcare Inc. (Deerfield, Illinois), the completion of patient enrollment in the Phase III clinical trial of MBI 226 for the prevention of CVC-related bloodstream infections. Total enrollment exceeded 1,400 subjects, the largest trial conducted to date for this indication. Results for the trial are expected to be available during the third quarter of calendar 2003. The objective of this trial is to demonstrate that MBI 226, administered at CVC insertion sites, reduces bacterial and fungal colonization of the catheters and prevents subsequent bloodstream infections. In the Phase III trial, MBI 226 is being compared with the standard of care in a randomized, multicenter study. The corporate partner for this trial, Fujisawa Healthcare Inc., is a subsidiary of Fujisawa Pharmaceutical Co., Ltd., based in Osaka, Japan. Fujisawa Pharmaceutical Co., Ltd., founded in 1894, markets a broad range of products in North America in the areas of anti-infectives, dermatology, transplantation, and cardiovascular.

As with IntraBiotics, Micrologix has made a corporate decision to diversify from antimicrobial peptides as the company builds its drug-development pipeline. This has included the in-licensing of two natural-product programs in 2002 for the systemic treatment of infectious diseases and the acquisition of antiviral technology based on antisense technology.

4.5.3. Entomed SA

Based in Strasbourg (France), Entomed (www.entomed.com) was founded in 1999 to capitalize on the work of Professor Jules Hoffman at the Centre National de la Recherche Scientifigue Molecular and Cellular Biology Institute in Strasbourg. Entomed has developed a number of validated technologies to rapidly identify and characterize novel molecules and genes and assay their specific biological activity. The company's discovery platform is designed to identify the most promising drug candidates from the battery of molecules synthesized by powerful defense systems and cellular control processes that insects have evolved (Dimarcq and Hunneyball, 2003).

The lead compounds that Entomed is developing are based on natural peptides and small molecules derived from insects. For instance, the company has been successful in identifying biologically important classes of peptides and small molecules that will provide a stream of drug candidates for treating severe, life-threatening hospital-acquired infections and other pathologies of high medical need. To date, Entomed has purified and identified hundreds of novel molecules that have a broad spectrum of activity against fungi and bacteria, or potent antiproliferative effects.

Entomed's lead peptide-based therapeutic is ETD-151 (44 amino acids containing three disulphide bonds), a novel analog of heliomicin (G20A/N19R/D17N) which is a naturally occurring antifungal peptide from the hemolymph of the lepidopteran *Heliothis viriscens* (Lamberty et al., 2001).

The target indication for this peptide is as a systemic antifungal. In vitro spectrum of activity includes all major fungal pathogens, with the exception of *C. glabrata*. ETD-151 has demonstrated efficacy in a 5-day murine *C. albicans* survival model, with doses of 5 and 30 mg/kg at 6, 24, and 48 h postinfection. The higher dose gave almost complete protection, despite a half-life of 5 min in the mouse. There are three factors that distinguish Entomed:

1. The company is focusing on systemically delivered therapeutics.
2. The company is focusing primarily on insect-derived bioactive molecules.
3. The company has secured the production of its lead peptide by recombinant production in *Saccharomyces cerevisiae*.

Entomed is currently seeking partners to exploit its proprietary technologies and product candidates through alliances, partnerships, and licensing agreements with pharmaceutical and biotechnology companies in multiple therapeutic areas.

4.5.4. Genaera Corporation

Genaera (Plymouth, Pennsylvania: www.genaera.com) is now the owner of the peptide-development programs of Magainin Pharmaceuticals. Magainin Pharmaceuticals was originally developed to take forward the first commercial innate immunity antimicrobial peptide candidate based on the initial discoveries of Michael Zasloff (Zasloff, 1987). However, the future of Locilex, the 22-amino-acid magainin peptide developed for infections of diabetic foot ulcers, seems unclear. This development candidate did not obtain FDA approval in 1999 and, in conjunction with GlaxoSmithKline, a decision on its future has not been made. In 2002 Genaera entered a 3-year option agreement for its antimicrobial peptide intellectual property with DuPont (E.I. du Pont de Nemours):

Locilex, GIGKFLKKAKKFGKAFVKILKK.

4.5.5. Xoma

Xoma (US) LLC (Berkeley, California; www.xoma.com) has developed the product Neuprex, an injectable formulation of $rBPI_{21}$, a modified recombinant fragment of a bactericidal/permeability-increasing protein (BPI). The BPI is a human host defense protein made by polymorphonuclear (PMN) leukocytes. The native protein was discovered in 1978 by Elsbach and Weiss (Weiss et al., 1978) at the New York University School of Medicine. The company has also developed several smaller peptide-sized derivatives.

BPI is composed of two functionally distinct structural domains: a potently antibacterial and antiendotoxin ~20-kd amino-terminal half and an opsonic carboxy-terminal portion. In multiple animal models, a recombinant amino-terminal fragment of BPI [$rBPI_{21}$] is nontoxic and protects against Gram-negative bacteria and endotoxin. In humans, $rBPI_{21}$ is also nontoxic and nonimmunogenic and has undergone phase II/III clinical trials with apparent therapeutic benefit.

A key cellular component of the innate immune response is the neutrophil, whose cytoplasmic granules contain a variety of antimicrobial proteins and peptides. Among these is the BPI, a cationic 55-kd protein whose selective anti-infective action against Gram-negative bacteria is based on its high (nanomolar) affinity for LPS (or "endotoxin"). Binding of BPI to Gram-negative bacteria results in growth inhibition, serves as an opsonin that enhances phagocytosis of bacteria, and inhibits bacteria-induced inflammatory responses by blocking the interaction of LPS with host proinflammatory pathways. Expression of BPI appears to be developmentally regulated as human newborns apparently have lower neutrophil BPI levels than adults. BPI expression has also recently been demonstrated in human epithelial cells where it appears to be inducible by endogenous anti-inflammatory lipids (lipoxins). BPI's potent antiendotoxic activity against a broad range of Gram-negative bacterial pathogens is manifest in biological fluids and renders it an attractive template for pharmaceutical development. Indeed, $rBPI_{21}$, an active recombinant protein derived from human BPI, has proven safe in Phase I human trials, has shown promise in Phase II trials, and has recently completed a Phase III trial for severe meningococcaemia, but the peptide demonstrated only a trend toward an apparent benefit. Identification and evaluation of additional disease entities characterized by Gram-negative bacteremia and/or endotoxemia as possible targets for BPI therapy continue.

In 2000, Xoma licensed certain rights to the Neuprex formulation to Baxter Healthcare Corporation. Baxter will fund future development costs, and Xoma will be entitled to certain milestone payments and a royalty on future product sales. Baxter has committed to development in multiple indications. These include retinopathies, Crohn's disease, bacterial infections, and endotoxin-related diseases.

4.5.6. AM-Pharma

AM-Pharma (Bilthoven, Holland: www.am-pharma.com) is developing human antimicrobial peptides based on histatin-derived and lactoferrin-derived sequences. These peptides possess both anti-infective and immunostimulating activity at the same time. Their mechanism of action involves binding of the positively charged peptides to the negatively charged microbial membrane structures. Subsequently, in fungi they enter the cell and bind to mitochondria, thereby interfering with the cell's energy supply.

The company's research has demonstrated that its proprietary compounds specifically accumulate at the site of infection. Together with their broad therapeutic window, this makes them highly suitable for therapeutic use after intravenous administration. In addition, the lead peptides are small and can be synthesized by well-established organic chemistry methods. There are a number of manufacturers with licensed peptide products on the market, which are able to supply pharmaceutical-grade peptides. Two main areas have been identified for product development by AM-Pharma. The company is testing lead peptides for the treatment of Hepatitis C. Simultaneously, the compound is in development for the prevention and treatment of infections that may develop after orthopedic and trauma surgery, those caused by multiple resistant bacteria (such as methicillin-resistant *S. aureus*) in particular. Two additional preclinical programs have recently been advanced into in vivo infection studies, exploring the use of antimicrobial peptides as antifungals and as antibacterials in combination with standard treatment.

4.5.7. Demegen Inc.

Demegen (Pittsburgh, Pennsylvania: www.demegen.com) has attempted to develop therapeutics based on two distinct classes of peptide. Their development candidates included D2A21 (a 22-residue α-helical peptide) and P-113 (a 12-residue portion of histatins, compounds found naturally in human saliva):

$$\text{D2A21, FAKKFAKKFKKFAKKFAKFA-NH}_2.$$
$$\text{P-113 AKRHHGYKRKFH-NH}_2.$$

The applications pursued by Demegen include reduction in bacterial bioburden in the CF lung by P-113D, the D-amino acid form of P-113. P-113D demonstrated activity against CF patient clinical isolates of bacteria that are resistant to traditional antibiotics. The compound was stable and maintained activity in the sputum of patients with CF. In 2002 the company announced it had been granted orphan drug status for the peptide for the treatment of CF infections by the U.S. FDA. Orphan drug designation is granted to applicants when the prevalence of a disease occurs in less than 200,000 patients in the United States. Under the orphan drug regulations, the FDA establishes procedures intended to encourage sponsors to develop drugs for patients with rare diseases. The FDA also works with sponsors to facilitate the development process.

Demegen is also developing a P-113L in a rinse formulation for the treatment and prevention of oral candidiasis (Rothstein et al., 2001). This condition, also known as oral thrush, is an AIDS-defining opportunistic illness that is highly prevalent in HIV-infected individuals. Up to 40% of patients with AIDS experience symptoms of oral candidiasis. Other immunocompromised individuals, including patients with cancer who are receiving chemotherapy, are also susceptible to oral fungal infections. Oropharyngeal candidiasis also develops in up to 30% of people with asthma who use inhaled corticosteriods. In vitro testing demonstrated that P-113L has very potent activity against a variety of isolates of *C. albicans, C. glabrata, C. parapsilosis,* and *C. tropicalis,* all with minimum inhibitory concentration (MIC) values of 3.1 μg/ml or less. Resistance to fluconazole had no bearing on the sensitivity of these organisms to P-113L. The initial market for Demegen's product to prevent or treat oral candidiasis is estimated to be up to $400 million.

Toxicology studies and experience in the clinic with a P-113 mouth rinse for the treatment of gingivitis demonstrated the peptide to be safe and well tolerated. The gingivitis product progressed through Phase II clinical studies and has been tested in over 300 people.

Demegen is developing the peptide D2A21, in a gel formulation (Demegel), as a wound-healing product to treat infected burns and wounds (Ballweber et al., 2002; Chalekson et al., 2002). These wounds are typically colonized with highly infectious pathogens that are often resistant to antibiotics. Demegel is a better treatment for these wounds, or at least as good as the current standards of care, as indicated by in vitro and in vivo results obtained by Demegen and its collaborators. Demegel contains a very low concentration of the active ingredient, D2A21, a 23-amino-acid amphipathic peptide. D2A21 is active against many infectious agents, including multidrug-resistant strains of *P. aeruginosa* and *S. aureus.* Also, antifungal activity has been demonstrated against *C. albicans, A. niger, Mucor* sp., and *Trichophyton mentagrophytes.*

The initial safety profile of D2A21 has been established. The peptide does not inhibit wound healing and is not cytotoxic to cultured keratinocyte skin cells. It is not a dermal irritant, nor does it induce contact sensitization. The peptide is not a mammalian or bacterial mutagen. Acute and chronic systemic toxicity studies have established a safe dose, which is a multiple of the efficacious dose for Demegel.

The Demegel development program is at the Investigational New Drug stage, at which Phase I/II clinical trials in infected burns and wounds can be initiated. Demegen intends to obtain a pharmaceutical partner to complete its development plan, obtain regulatory approval, market, and distribute the product. D2A21 is protected with composition and use patents in the United States and abroad. Hospital-related infected burns and wounds affect over 2 million people per year in the United States alone, representing a market of $250 million. The entire topical anti-infectives market is estimated at $2 billion. As of September 2002, despite cost reductions, Demegen did not have sufficient cash to move any of the projects forward without outside help (Shareholder Meeting Remarks) and the company believed new investment was unlikely. As opposed to the other companies previously described, Demegen did not expand the company's pipeline beyond peptides but rather expanded the applications of its technology, for example, into cancer therapy and agriculture.

4.5.8. Helix BioMedix Inc.

Helix BioMedix (Bothell, Washington: www.helixbiomedix.com) has assembled intellectual property covering a wide range of peptides of the innate immune system. Unlike other companies in the field, Helix has a diversity of structural classes including α-helical, β-sheet, linear and looped peptides in-licensed from Louisiana State University, The Hancock Laboratory at the University of British Columbia, and intellectual property developed in-house. The company's library includes peptides based on crab polyphemusins, insect cecropins and melittins, cattle bactenecins, and fish pleurocidins, and consists of over 100,000 distinct peptide sequences covered by over 40 patents.

The underlying properties of the peptides being exploited are both antimicrobial activity and modulation of the local immune system. The company has created a diverse and extensive library of peptides in order to optimize the required attributes for each clinical application. In this way the company believes the most appropriate peptide as defined by sequence, structure and bioactivity can be developed for a specific application. This development strategy can be successfully carried out only with the resource of an extensive and diverse library of peptides. In addition, Helix has developed peptides specifically around sequences that are cost effective to synthesize, thus enabling those peptides to compete with current therapies.

HB107 has demonstrated efficacy in both rat burn wound and mouse acute wound models and is the lead in the Helix BioMedix wound-healing program. The peptide has completed its first gel (0.01% HB107) formulation trial and has proven safe and significantly better than current therapies in preclinical testing. The peptide exhibits no antimicrobial activity or cytotoxicity and has an intravenous $LD_{50} > 100$ mg/kg in the mouse.

HB50 is a broad-spectrum antimicrobial peptide that has the potential to be applied to a wide range of topical applications such as burns, wounds and skin infections. This market includes, but is not limited to, the Bactroban (Mupirocin, GlaxoSmithKline) market. Bactroban has no Gram-negative or fungal coverage, and resistance is an ever-increasing problem, for example, Mupirocin/Methicillin-resistant *S. aureus*. HB50 is now entering a number of preclinical trials, exhibiting activity against multiply resistant *Pseudomonas, Staphylococcus,* and *Candida* species.

Two Helix peptides have demonstrated efficacy in the rat chronic lung infection model (reduction of 2 log orders in viable *P. aeruginosa* in 3 days). Following a screening program of over 200 antimicrobial peptides from the Helix library, five lead candidates have now been identified and are currently in preclinical testing. All lead candidates are active against multiple *P. aeruginosa,* and specific peptides also demonstrate an anti-inflammatory effect. Once lead and backup peptides have been identified, more rigorous preclinical testing, including toxicology and inflammation end points, are planned.

Certain peptides in the Helix library have been identified that have broad-spectrum activity against bacteria and yeast associated with STDs and vaginitis. These pathogens include *Candida* sp. (including azole-resistant strains), *H. ducreyei, N.gonnorhoeae,* and *C. trachomatis.* The lead peptides maintain their activity and cidality in gel formulation.

4.5.9. Inimex Pharmaceuticals Inc.

Inimex Pharmaceuticals (Vancouver, British Columbia: www.inimexpharma.com) was founded in 2001 to further develop and commercialize new discoveries made by the laboratories of Bob Hancock and Brett Finlay at the University of British Columbia. This initiative involves the proprietary understanding of the functional genomics associated with the upregulation and control of the innate immune response. It is believed by the company that modulating the innate immune response can provide potential new therapeutic strategies for the treatment of a number of diseases, including bacterial infections, viral infections, and cancer, as well as novel anti-inflammatory treatments. Inimex is currently screening for pharmaceutical peptides and small molecules, based on peptides of the innate immune system, with the potential to selectively upregulate elements of innate immunity while avoiding or limiting the damage caused by inflammation.

Inimex has demonstrated that its prospective lead peptide compounds can protect against bacterial infections in animal models. Inimex has also studied the mechanism of action for these novel peptides, demonstrating that they induce an upregulation of genes that determine chemokines and chemokine receptors, without stimulating the production of proinflammatory cytokines. Historic attempts to utilize the immune response have resulted in a coboosting of the inflammatory response.

The main focus of Inimex's activities over the next 3 years will be on lead optimization and Investigational New Drug studies. The company's initial product-development efforts will be based on current lead peptides delivered in combination with, or in addition to, existing antibiotic therapies for the treatment of nosocomial infections such as pneumonia. Subsequent product-development efforts will focus on the development of treatments for community-acquired infections and preventative medicines.

4.6. THE FUTURE

Xoma and Entomed are leading the way in the systemic application of innate immunity peptides. However, it is clear that true innovation will be required for fully adapting a wide range of peptides to systemic use in order to overcome the issues of stability, toxicity, and delivery. To date, the standard approaches of formulation have not yielded an effective systemic therapeutic. Innovation in the areas of polymer matrix delivery systems (Kipper et al., 2002), peptide targeting (Chen et al., 2001), and even gene therapy (Pelisek, Armeanu, and Nikol, 2001) could well provide a future answer, but not without substantial investment of research resources.

The rationale for the development of topical applications for peptides of the innate immune system is rooted in the fact that they have evolved in part to protect exposed biological surfaces and exhibit attributes specifically for that purpose, for example, fast-acting, cidal, readily degraded, low level of resistance emergence and activity against a wide range of pathogens (dependent on sequence). The present-day clinical relevance of using peptides in such a way provides a distinct separation of topical and systemic anti-infectives. Because of the lack of cross resistance between antimicrobial peptides and traditional antibiotics, there is certainly a chance of reducing the requirement for last-line-of-defense anti-infectives until completely necessary. Using innate immunity peptides as the first line of treatment for topical applications is now an attainable goal.

The range of applications for innate immunity peptides is increasing, partly because of a reduction in the cost of peptide synthesis. This cost is coming down for a number of reasons. First, improved cost-effective methods of production, such as solution phase synthesis of peptide subunits, are being used. Second, peptide-development companies are advancing therapeutic candidates that are relatively short, do not require folding (e.g., with disulphide bonds) and are required only in low concentrations to be effective in vivo. Third, recombinant production companies are now competing with synthetic peptide manufacturers for business, driving a lowering of costs of production. These factors have enabled peptide applications such as acne, topical antiseptics, antibiotics, and microbicides to be cost effective. In addition, new applications such as wound healing and anti-inflammatories are now emerging where current therapies are expensive or have significant side effects. In addition, as the cost of production comes down, such peptides are also likely to find their way into nonclinical products, for example, in the protection of polymer surfaces used in medical devices (Haynie, Crum, and Doele, 1995).

There is a significant therapeutic role that peptides of the innate immune system can play. The issues that need to be overcome for that to happen are not trivial, that is, the cost of goods, stability, toxicity, and delivery. However, these issues are the same for any drug-development candidate, and, with advances in drug delivery, reductions in the cost of peptide synthesis and a greater understanding of the limitations as well as the attributes of innate immunity peptides, these are likely to be overcome. The driving force for such development will be the ever-decreasing number of novel bioactive molecules with the potential to be therapeutics.

REFERENCES

Aarbiou, J., Ertmann, M., van Wetering, S., van Noort, P., Rook, D., Rabe, K. F., Litvinov, S. V., van Krieken, J. H., de Boer, W. I., and Hiemstra, P. S. (2002). Human neutrophil defensins induce lung epithelial cell proliferation in vitro. *Journal of Leukocyte Biology*, 72, 167–74.

Agerberth, B., Charo, J., Werr, J., Olsson, B., Idali, F., Lindbom, L., Kiessling, R., Jornvall, H., Wigzell, H., and Gudmundsson, G. H. (2000). The human antimicrobial and chemotactic peptides LL-37 and alpha-defensins are expressed by specific lymphocyte and monocyte populations. *Blood*, 96, 3086–93.

Ali, R. S., Falconer, A., Ikram, M., Bissett, C. E., Cerio, R., and Quinn, A. G. (2001). Expression of the peptide antibiotics human beta defensin-1 and human beta defensin-2 in normal human skin. *Journal of Investigative Dermatology*, 117, 106–11.

Ayabe, T., Satchell, D. P., Pesendorfer, P., Tanabe, H., Wilson, C. L., Hagen, S. J., and Ouellette, A. J. (2002a). Activation of Paneth cell alpha-defensins in mouse small intestine. *Journal of Biological Chemistry*, 277, 5219–28.

Ayabe, T., Wulff, H., Darmoul, D., Cahalan, M. D., Chandy, K. G., and Ouellette, A. J. (2002b). Modulation of mouse Paneth cell alpha-defensin secretion by mIKCa1, a Ca^{2+}-activated, intermediate conductance potassium channel. *Journal of Biological Chemistry*, 277, 3793–3800.

Ballweber, L. M., Jaynes, J. E., Stamm, W. E., and Lampe, M. F. (2002). In vitro microbicidal activities of cecropin peptides D2A21 and D4E1 and gel formulations containing 0.1 to 2% D2A21 against *Chlamydia trachomatis*. *Antimicrobial Agents and Chemotherapy*, 46, 34–41.

Bals, R., Wang, X., Zasloff, M., and Wilson, J. M. (1998). The peptide antibiotic LL-37/hCAP-18 is expressed in epithelia of the human lung where it has broad antimicrobial activity at the airway surface. *Proceedings of the National Academy of Sciences USA*, 95, 9541–6.

Becker, M. N., Diamond, G., Verghese, M. W., and Randell, S. H. (2000). CD14-dependent lipopolysaccharide-induced beta-defensin-2 expression in human tracheobronchial epithelium. *Journal of Biological Chemistry*, 275, 29731–6.

Birchler, T., Seibl, R., Buchner, K., Loeliger, S., Seger, R., Hossle, J. P., Aguzzi, A., and Lauener, R. P. (2001). Human Toll-like receptor 2 mediates induction of the antimicrobial peptide human beta-defensin 2 in response to bacterial lipoprotein. *European Journal of Immunology*, 31, 3131–7.

Boman, H. G. (1996). Peptide antibiotics: Holy or heretic grails of innate immunity? *Scandinavian Journal of Immunology*, 43, 475–82.

Bulow, E., Bengtsson, N., Calafat, J., Gullberg, U., and Olsson, I. (2002). Sorting of neutrophil-specific granule protein human cathelicidin, hCAP-18, when constitutively expressed in myeloid cells. *Journal of Leukocyte Biology*, 72, 147–53.

Cameron, D. W., Simonsen, J. N., D'Costa, L. J., Ronald, A. R., Maitha, G. M., Gakinya, M. N., Cheang, M., Ndinya-Achola, J. O., Piot, P., Brunham, R. C., and Plummer, F. A. (1989). Female to male transmission of human immunodeficiency virus type 1: Risk factors for seroconversion in men. *Lancet*, 2, 403–7.

Chalekson, C. P., Neumeister, M. W., and Jaynes, J. (2002). Improvement in burn wound infection and survival with antimicrobial peptide D2A21 (Demegel). *Plastic and Reconstructive Surgery*, 109, 1338–43.

Chaly, Y. V., Paleolog, E. M., Kolesnikova, T. S., Tikhonov, I. I., Petratchenko, E. V., and Voitenok, N. N. (2000). Neutrophil alpha-defensin human neutrophil peptide modulates cytokine production in human monocytes and adhesion molecule expression in endothelial cells. *European Cytokine Network*, 11, 257–66.

Chan, Y. R. and Gallo, R. L. (1998). PR-39, a syndecan-inducing antimicrobial peptide, binds and affects p130(Cas). *Journal of Biological Chemistry*, 273, 28978–85.

Chen, J., Falla, T. J., Liu, H., Hurst, M. A., Fujii, C. A., Mosca, D. A., Embree, J. R., Loury, D. J., Radel, P. A., Cheng Chang, C., Gu, L., and Fiddes, J. C. (2000). Development of protegrins for the treatment and prevention of oral mucositis: Structure–activity relationships of synthetic protegrin analogues. *Biopolymers*, 55, 88–98.

Chen, J., Cheng, Z., Owen, N. K., Hoffman, T. J., Miao, Y., Jurisson, S. S., and Quinn, T. P. (2001). Evaluation of an (111)In-DOTA-rhenium cyclized alpha-MSH analog: A novel cyclic-peptide analog with improved tumor-targeting properties. *Journal of Nuclear Medicine*, 42, 1847–55.

Chertov, O., Yang, D., Howard, O. M., and Oppenheim, J. J. (2000). Leukocyte granule proteins mobilize innate host defenses and adaptive immune responses. *Immunological Reviews*, 177, 68–78.

Cole, A. M., Hong, T., Boo, L. M., Nguyen, T., Zhao, C., Bristol, G., Zack, J. A., Waring, A. J., Yang, O. O., and Lehrer, R. I. (2002). Retrocyclin: A primate peptide that protects cells from infection by T- and M-tropic strains of HIV-1. *Proceedings of the National Academy of Sciences USA*, 99, 1813–18.

Com, E., Bourgeon, F., Evrard, B., Ganz, T., Colleu, D., Jegou, B., and Pineau, C. (2003). Expression of antimicrobial defensins in the male reproductive tract of rats, mice, and humans. *Biology of Reproduction*, 68, 95–104.

Cunliffe, R. N., Rose, F. R., Keyte, J., Abberley, L., Chan, W. C., and Mahida, Y. R. (2001). Human defensin 5 is stored in precursor form in normal Paneth cells and is expressed by some villous epithelial cells and by metaplastic Paneth cells in the colon in inflammatory bowel disease. *Gut*, 48, 176–85.

Dauletbaev, N., Gropp, R., Frye, M., Loitsch, S., Wagner, T. O., and Bargon, J. (2002). Expression of human beta defensin (HBD-1 and HBD-2) mRNA in nasal epithelia of adult cystic fibrosis patients, healthy individuals, and individuals with acute cold. *Respiration*, 69, 46–51.

De, Y., Chen, Q., Schmidt, A. P., Anderson, G. M., Wang, J. M., Wooters, J., Oppenheim, J. J., and Chertov, O. (2000). LL-37, the neutrophil granule- and epithelial cell-derived cathelicidin, utilizes formyl peptide receptor-like 1 (FPRL1) as a receptor to chemoattract human peripheral blood neutrophils, monocytes, and T cells. *Journal of Experimental Medicine*, 192, 1069–74.

Del Sal, G., Storici, P., Schneider, C., Romeo, D., and Zanetti, M. (1992). cDNA cloning of the neutrophil bactericidal peptide indolicidin. *Biochemical and Biophysical Research Communications*, 187, 467–72.

Desai, T. R., Hancock, R. E. W., and Finlay, W. H. (2003). Delivery of liposomes in dry powder form: Aerodynamic dispersion properties. *European Journal of Pharmaceutical Sciences*. In press.

Diamond, G., Zasloff, M., Eck, H., Brasseur, M., Maloy, W. L., and Bevins, C. L. (1991). Tracheal antimicrobial peptide, a cysteine-rich peptide from mammalian tracheal mucosa: Peptide isolation and cloning of a cDNA. *Proceedings of the National Academy of Sciences USA*, 88, 3952–6.

Dimarcq, J. L. and Hunneyball, I. (2003). Pharma-entomology: When bugs become drugs. *Drug Discovery Today*, 8, 107–10.

Dorschner, R. A., Pestonjamasp, V. K., Tamakuwala, S., Ohtake, T., Rudisill, J., Nizet, V., Agerberth, B., Gudmundsson, G. H., and Gallo, R. L. (2001). Cutaneous injury induces the release of cathelicidin anti-microbial peptides active against group A Streptococcus. *Journal of Investigative Dermatology*, 117, 91–7.

Duits, L. A., Ravensbergen, B., Rademaker, M., Hiemstra, P. S., and Nibbering, P. H. (2002). Expression of beta-defensin 1 and 2 mRNA by human monocytes, macrophages and dendritic cells. *Immunology*, 106, 517–25.

Durrant, C. (2001). The responsibility of the pharmaceutical industry. *Clinical Microbiology and Infection*, 7, Suppl. 6, 2–4.

Eastman, M. E., Khorsand, M., Maki, D. G., Williams, E. C., Kim, K., Sondel, P. M., Schiller, J. H., and Albertini, M. R. (2001). Central venous device-related infection and thrombosis in patients treated with moderate dose continuous-infusion interleukin-2. *Cancer*, 91, 806–14.

El-Sayed, S., Nabid, A., Shelley, W., Hay, J., Balogh, J., Gelinas, M., MacKenzie, R., Read, N., Berthelet, E., Lau, H., Epstein, J., Delvecchio, P., Ganguly, P. K., Wong, F., Burns, P., Tu, D., and Pater, J. (2002). Prophylaxis of radiation-associated mucositis in conventionally treated patients with head and neck cancer: a double-blind, phase III, randomized, controlled trial evaluating the clinical efficacy of an antimicrobial lozenge using a validated mucositis scoring system. *Journal of Clinical Oncology*, 20, 3956–63.

Falla, T. J., Karunaratne, D. N., and Hancock, R. E. (1996). Mode of action of the antimicrobial peptide indolicidin. *Journal of Biological Chemistry*, 271, 19298–303.

Fidel Jr., P. L. and Sobel, J. D. (1996). Immunopathogenesis of recurrent vulvovaginal candidiasis. *Clinical Microbiology Reviews*, 9, 335–48.

Flad, T., Bogumil, R., Tolson, J., Schittek, B., Garbe, C., Deeg, M., Mueller, C. A., and Kalbacher, H. (2002). Detection of dermcidin-derived peptides in sweat by ProteinChip technology. *Journal of Immunological Methods*, 270, 53–62.

Ford, C. W., Zurenko, G. E., and Barbachyn, M. R. (2001). The discovery of linezolid, the first oxazolidinone antibacterial agent. *Current Drug Targets Infectious Disorders*, 1, 181–99.

Fortney, K., Totten, P. A., Lehrer, R. I., and Spinola, S. M. (1998). Haemophilus ducreyi is susceptible to protegrin. *Antimicrobial Agents and Chemotherapy*, 42, 2690–3.

Frank, R. W., Gennaro, R., Schneider, K., Przybylski, M., and Romeo, D. (1990). Amino acid sequences of two proline-rich bactenecins. Antimicrobial peptides of bovine neutrophils. *Journal of Biological Chemistry*, 265, 18871–4.

Friedrich, C. L., Rozek, A., Patrzykat, A., and Hancock, R. E. (2001). Structure and mechanism of action of an indolicidin peptide derivative with improved activity against Gram-positive bacteria. *Journal of Biological Chemistry*, 276, 24015–22.

Frohm Nilsson, M., Sandstedt, B., Sorensen, O., Weber, G., Borregaard, N., and Stahle-Backdahl, M. (1999). The human cationic antimicrobial protein (hCAP18), a peptide antibiotic, is widely expressed in human squamous epithelia and colocalizes with interleukin-6. *Infection and Immunity*, 67, 2561–6.

Frye, M., Bargon, J., Dauletbaev, N., Weber, A., Wagner, T. O., and Gropp, R. (2000). Expression of human alpha-defensin 5 (HD5) mRNA in nasal and bronchial epithelial cells. *Journal of Clinical Pathology*, 53, 770–3.

Ganz, T. and Lehrer, R. I. (1997). Antimicrobial peptides of leukocytes. *Current Opinion in Hematology*, 4, 53–8.

Ganz, T. and Lehrer, R. I. (1998). Antimicrobial peptides of vertebrates. *Current Opinion in Immunology*, 10, 41–4.

Ganz, T., Oren, A., and Lehrer, R. I. (1992). Defensins: Microbicidal and cytotoxic peptides of mammalian host defense cells. *Medical Microbiology and Immunology*, 181, 99–105.

Ganz, T., Selsted, M. E., and Lehrer, R. I. (1990). Defensins. *European Journal of Haematology*, 44, 1–8.

Ganz, T. and Weiss, J. (1997). Antimicrobial peptides of phagocytes and epithelia. *Seminars in Hematology*, 34, 343–54.

Gao, Y., Lecker, S., Post, M. J., Hietaranta, A. J., Li, J., Volk, R., Li, M., Sato, K., Saluja, A. K., Steer, M. L., Goldberg, A. L., and Simons, M. (2000). Inhibition of ubiquitin-proteasome pathway-mediated I kappa B alpha degradation by a naturally occurring antibacterial peptide. *Journal of Clinical Investigation*, 106, 439–48.

Gennaro, R. and Zanetti, M. (2000). Structural features and biological activities of the cathelicidin-derived antimicrobial peptides. *Biopolymers*, 55, 31–49.

Gudmundsson, G. H., Magnusson, K. P., Chowdhary, B. P., Johansson, M., Andersson, L., and Boman, H. G. (1995). Structure of the gene for porcine peptide antibiotic PR-39, a cathelin gene family member: Comparative mapping of the locus for the human peptide antibiotic FALL-39. *Proceedings of the National Academy of Sciences USA*, 92, 7085–9.

Hancock, R. E. (2001). Cationic peptides: Effectors in innate immunity and novel antimicrobials. *Lancet Infectious Diseases*, 1, 156–64.

Hancock, R. E. and Lehrer, R. (1998). Cationic peptides: A new source of antibiotics. *Trends in Biotechnology*, 16, 82–8.

Hancock, R. E. and Scott, M. G. (2000). The role of antimicrobial peptides in animal defenses. *Proceedings of the National Academy of Sciences USA*, 97, 8856–61.

Harder, J., Bartels, J., Christophers, E., and Schroder, J. M. (1997). A peptide antibiotic from human skin. *Nature (London)*, 387, 861.

Harder, J., Bartels, J., Christophers, E., and Schroder, J. M. (2001). Isolation and characterization of human beta -defensin-3, a novel human inducible peptide antibiotic. *Journal of Biological Chemistry*, 276, 5707–13.

Hase, K., Eckmann, L., Leopard, J. D., Varki, N., and Kagnoff, M. F. (2002). Cell differentiation is a key determinant of cathelicidin LL-37/human cationic antimicrobial protein 18 expression by human colon epithelium. *Infection and Immunity*, 70, 953–63.

Haynie, S. L., Crum, G. A., and Doele, B. A. (1995). Antimicrobial activities of amphiphilic peptides covalently bonded to a water-insoluble resin. *Antimicrobial Agents and Chemotherapy*, 39, 301–7.

Helmerhorst, E. J., Breeuwer, P., van't Hof, W., Walgreen-Weterings, E., Oomen, L. C., Veerman, E. C., Amerongen, A. V., and Abee, T. (1999). The cellular target of histatin 5 on Candida albicans is the energized mitochondrion. *Journal of Biological Chemistry*, 274, 7286–91.

Helmerhorst, E. J., van't Hof, W., Breeuwer, P., Veerman, E. C., Abee, T., Troxler, R. F., Amerongen, A. V., and Oppenheim, F. G. (2001). Characterization of histatin 5 with respect to amphipathicity, hydrophobicity, and effects on cell and mitochondrial membrane integrity excludes a candidacidal mechanism of pore formation. *Journal of Biological Chemistry*, 276, 5643–9.

Hiemstra, P. S., van Wetering, S., and Stolk, J. (1998). Neutrophil serine proteinases and defensins in chronic obstructive pulmonary disease: Effects on pulmonary epithelium. *European Respiratory Journal*, 12, 1200–8.

Hoffmann, J. A. and Reichhart, J. M. (2002). Drosophila innate immunity: An evolutionary perspective. *Nature Immunology*, 3, 121–6.

Hoffmeyer, M. R., Scalia, R., Ross, C. R., Jones, S. P., and Lefer, D. J. (2000). PR-39, a potent neutrophil inhibitor, attenuates myocardial ischemia-reperfusion injury in mice. *American Journal of Physiology: Heart and Circulatory Physiology*, 279, H2824–8.

Hong, R. W., Shchepetov, M., Weiser, J. N., and Axelsen, P. H. (2003). Transcriptional profile of the *Escherichia coli* response to the antimicrobial insect peptide cecropin A. *Antimicrobial Agents and Chemotherapy*, 47, 1–6.

Jia, H. P., Mills, J. N., Barahmand-Pour, F., Nishimura, D., Mallampali, R. K., Wang, G., Wiles, K., Tack, B. F., Bevins, C. L., and McCray Jr., P. B. (1999). Molecular cloning and characterization of rat genes encoding homologues of human beta-defensins. *Infection and Immunity*, 67, 4827–33.

Jia, H. P., Wowk, S. A., Schutte, B. C., Lee, S. K., Vivado, A., Tack, B. F., Bevins, C. L., and McCray, P. B., Jr. (2000). A novel murine beta-defensin expressed in tongue, esophagus, and trachea. *Journal of Biological Chemistry*, 275, 33314–20.

Jia, H. P., Schutte, B. C., Schudy, A., Linzmeier, R., Guthmiller, J. M., Johnson, G. K., Tack, B. F., Mitros, J. P., Rosenthal, A., Ganz, T., and McCray Jr., P. B. (2001). Discovery of new human beta-defensins using a genomics-based approach. *Gene*, 263, 211–18.

Johansson, J., Gudmundsson, G. H., Rottenberg, M. E., Berndt, K. D., and Agerberth, B. (1998). Conformation-dependent antibacterial activity of the naturally occurring human peptide LL-37. *Journal of Biological Chemistry*, 273, 3718–24.

Jones, D. E. and Bevins, C. L. (1992). Paneth cells of the human small intestine express an antimicrobial peptide gene. *Journal of Biological Chemistry*, 267, 23216–25.

Kipper, M. J., Shen, E., Determan, A., and Narasimhan, B. (2002). Design of an injectable system based on bioerodible polyanhydride microspheres for sustained drug delivery. *Biomaterials*, 23, 4405–12.

Koczulla, A. R. and Bals, R. (2003). Antimicrobial peptides: Current status and therapeutic potential. *Drugs*, 63, 389–406.

Kokryakov, V. N., Harwig, S. S., Panyutich, E. A., Shevchenko, A. A., Aleshina, G. M., Shamova, O. V., Korneva, H. A., and Lehrer, R. I. (1993). Protegrins: Leukocyte antimicrobial peptides that combine features of corticostatic defensins and tachyplesins. *FEBS Letters*, 327, 231–6.

Korthuis, R. J., Gute, D. C., Blecha, F., and Ross, C. R. (1999). PR-39, a proline/arginine-rich antimicrobial peptide, prevents postischemic microvascular dysfunction. *American Journal of Physiology*, 277, H1007–13.

Lamberty, M., Caille, A., Landon, C., Tassin-Moindrot, S., Hetru, C., Bulet, P., and Vovelle, F. (2001). Solution structures of the antifungal heliomicin and a selected variant with both antibacterial and antifungal activities. *Biochemistry*, 40, 11995–12003.

Lehrer, R. I. and Ganz, T. (1992). Defensins: endogenous antibiotic peptides from human leukocytes. *Ciba Foundation Symposium*, 171, 276–290; discussion, 290–293.

Lehrer, R. I. and Ganz, T. (1996). Endogenous vertebrate antibiotics. Defensins, protegrins, and other cysteine-rich antimicrobial peptides. *Annals of the New York Academy of Sciences*, 797, 228–39.

Lehrer, R. I. and Ganz, T. (2002a). Cathelicidins: A family of endogenous antimicrobial peptides. *Current Opinion in Hematology*, 9, 18–22.

Lehrer, R. I. and Ganz, T. (2002b). Defensins of vertebrate animals. *Current Opinion in Immunology*, 14, 96–102.

Li, P., Chan, H. C., He, B., So, S. C., Chung, Y. W., Shang, Q., Zhang, Y. D., and Zhang, Y. L. (2001). An antimicrobial peptide gene found in the male reproductive system of rats. *Science*, 291, 1783–5.

Liu, L., Wang, L., Jia, H. P., Zhao, C., Heng, H. H., Schutte, B. C., McCray Jr., P. B., and Ganz, T. (1998). Structure and mapping of the human beta-defensin HBD-2 gene and its expression at sites of inflammation. *Gene*, 222, 237–44.

Ludtke, S. J., He, K., Heller, W. T., Harroun, T. A., Yang, L., and Huang, H. W. (1996). Membrane pores induced by magainin. *Biochemistry*, 35, 13723–8.

Michel, M. and Gutmann, L. (1997). Methicillin-resistant *Staphylococcus aureus* and vancomycin-resistant enterococci: Therapeutic realities and possibilities. *Lancet*, 349, 1901–6.

Morse, S. A. (1989). Chancroid and *Haemophilus ducreyi*. *Clinical Microbiology Reviews*, 2, 137–57.

Mosca, D. A., Hurst, M. A., So, W., Viajar, B. S., Fujii, C. A., and Falla, T. J. (2000). IB-367, a protegrin peptide with in vitro and in vivo activities against the microflora associated with oral mucositis. *Antimicrobial Agents and Chemotherapy*, 44, 1803–8.

Moser, C., Weiner, D. J., Lysenko, E., Bals, R., Weiser, J. N., and Wilson, J. M. (2002). beta-Defensin 1 contributes to pulmonary innate immunity in mice. *Infection and Immunity*, 70, 3068–72.

Nadel, J. A. and Stockley, R. A. (1998). Proteolytic enzymes and airway diseases. *European Respiratory Journal*, 12, 1250–1.

Nagaoka, I., Hirota, S., Niyonsaba, F., Hirata, M., Adachi, Y., Tamura, H., Tanaka, S., and Heumann, D. (2002). Augmentation of the lipopolysaccharide-neutralizing activities of human cathelicidin CAP18/LL-37-derived antimicrobial peptides by replacement with hydrophobic and cationic amino acid residues. *Clinical and Diagnostic Laboratory Immunology*, 9, 972–82.

Nagpal, S., Gupta, V., Kaur, K. J., and Salunke, D. M. (1999). Structure-function analysis of tritrypticin, an antibacterial peptide of innate immune origin. *Journal of Biological Chemistry*, 274, 23296–304.

Nicolas, G., Bennoun, M., Porteu, A., Mativet, S., Beaumont, C., Grandchamp, B., Sirito, M., Sawadogo, M., Kahn, A., and Vaulont, S. (2002a). Severe iron deficiency anemia in transgenic mice expressing liver hepcidin. *Proceedings of the National Academy of Sciences USA*, 99, 4596–601.

Nicolas, G., Chauvet, C., Viatte, L., Danan, J. L., Bigard, X., Devaux, I., Beaumont, C., Kahn, A., and Vaulont, S. (2002b). The gene encoding the iron regulatory peptide hepcidin is regulated by anemia, hypoxia, and inflammation. *Journal of Clinical Investigation*, 110, 1037–44.

Nicolas, G., Viatte, L., Bennoun, M., Beaumont, C., Kahn, A., and Vaulont, S. (2002c). Hepcidin, a new iron regulatory peptide. *Blood Cells, Molecules and Diseases*, 29, 327–35.

Nicolas, P. and Mor, A. (1995). Peptides as weapons against microorganisms in the chemical defense system of vertebrates. *Annual Review of Microbiology*, 49, 277–304.

Niyonsaba, F., Iwabuchi, K., Matsuda, H., Ogawa, H., and Nagaoka, I. (2002a). Epithelial cell-derived human beta-defensin-2 acts as a chemotaxin for mast cells through a pertussis toxin-sensitive and phospholipase C-dependent pathway. *International Immunology*, 14, 421–6.

Niyonsaba, F., Iwabuchi, K., Someya, A., Hirata, M., Matsuda, H., Ogawa, H., and Nagaoka, I. (2002b). A cathelicidin family of human antibacterial peptide LL-37 induces mast cell chemotaxis. *Immunology*, 106, 20–6.

Okrent, D. G., Lichtenstein, A. K., and Ganz, T. (1990). Direct cytotoxicity of polymorphonuclear leukocyte granule proteins to human lung-derived cells and endothelial cells. *American Review of Respiratory Disease*, 141, 179–85.

O'Neil, D. A., Cole, S. P., Martin-Porter, E., Housley, M. P., Liu, L., Ganz, T., and Kagnoff, M. F. (2000). Regulation of human beta-defensins by gastric epithelial cells in response to infection with *Helicobacter pylori* or stimulation with interleukin-1. *Infection and Immunity*, 68, 5412–15.

Ong, P. Y., Ohtake, T., Brandt, C., Strickland, I., Boguniewicz, M., Ganz, T., Gallo, R. L., and Leung, D. Y. (2002). Endogenous antimicrobial peptides and skin infections in atopic dermatitis. *New England Journal of Medicine*, 347, 1151–60.

Oppenheim, F. G., Xu, T., McMillian, F. M., Levitz, S. M., Diamond, R. D., Offner, G. D., and Troxler, R. F. (1988). Histatins, a novel family of histidine-rich proteins in human parotid secretion. Isolation, characterization, primary structure, and fungistatic effects on Candida albicans. *Journal of Biological Chemistry*, 263, 7472–7.

Oren, Z. and Shai, Y. (1998). Mode of action of linear amphipathic alpha-helical antimicrobial peptides. *Biopolymers*, 47, 451–63.

Ortega, M. R., Ganz, T., and Milner, S. M. (2000). Human beta defensin is absent in burn blister fluid. *Burns*, 26, 724–6.

Ouellette, A. J. (1997). Paneth cells and innate immunity in the crypt microenvironment. *Gastroenterology*, 113, 1779–84.

Ouellette, A. J. (1999). IV. Paneth cell antimicrobial peptides and the biology of the mucosal barrier. *American Journal of Physiology*, 277, G257–61.

Panes, J. and Granger, D. N. (1998). Leukocyte-endothelial cell interactions: Molecular mechanisms and implications in gastrointestinal disease. *Gastroenterology*, 114, 1066–90.

Paquette, D. W., Waters, G. S., Stefanidou, V. L., Lawrence, H. P., Friden, P. M., O'Connor, S. M., Sperati, J. D., Oppenheim, F. G., Hutchens, L. H., and Williams, R. C. (1997). Inhibition of experimental gingivitis in beagle dogs with topical salivary histatins. *Journal of Clinical Periodontology*, 24, 216–22.

Paquette, D. W., Simpson, D. M., Friden, P., Braman, V., and Williams, R. C. (2002). Safety and clinical effects of topical histatin gels in humans with experimental gingivitis. *Journal of Clinical Periodontology*, 29, 1051–8.

Park, C. B., Kim, H. S., and Kim, S. C. (1998). Mechanism of action of the antimicrobial peptide Buforin II: Buforin II kills microorganisms by penetrating the cell membrane and inhibiting cellular functions. *Biochemical and Biophysical Research Communications*, 244, 253–7.

Park, C. H., Valore, E. V., Waring, A. J., and Ganz, T. (2001). Hepcidin, a urinary antimicrobial peptide synthesized in the liver. *Journal of Biological Chemistry*, 276, 7806–10.

Pelisek, J., Armeanu, S., and Nikol, S. (2001). Quiescence, cell viability, apoptosis and necrosis of smooth muscle cells using different growth inhibitors. *Cell Proliferation*, 34, 305–20.

Plummer, F. A., Simonsen, J. N., Cameron, D. W., Ndinya-Achola, J. O., Kreiss, J. K., Gakinya, M. N., Waiyaki, P., Cheang, M., Piot, P., Ronald, A. R. (1991). Cofactors in male-female sexual transmission of human immunodeficiency virus type 1. *Journal of Infectious Diseases*, 163, 233–9.

Proost, P., Wuyts, A., and van Damme, J. (1996). The role of chemokines in inflammation. *International Journal of Clinical and Laboratory Research*, 26, 211–23.

Putsep, K., Carlsson, G., Boman, H. G., and Andersson, M. (2002). Deficiency of antibacterial peptides in patients with morbus Kostmann: An observation study. *Lancet*, 360, 1144–9.

Qu, X. D., Harwig, S. S., Oren, A. M., Shafer, W. M., and Lehrer, R. I. (1996). Susceptibility of *Neisseria gonorrhoeae* to protegrins. *Infection and Immunity*, 64, 1240–5.

Ramanathan, B., Davis, E. G., Ross, C. R., and Blecha, F. (2002). Cathelicidins: Microbicidal activity, mechanisms of action, and roles in innate immunity. *Microbes and Infection*, 4, 361–72.

Robinson Jr., W. E., McDougall, B., Tran, D., and Selsted, M. E. (1998). Anti-HIV-1 activity of indolicidin, an antimicrobial peptide from neutrophils. *Journal of Leukocyte Biology*, 63, 94–100.

Romeo, D., Skerlavaj, B., Bolognesi, M., and Gennaro, R. (1988). Structure and bactericidal activity of an antibiotic dodecapeptide purified from bovine neutrophils. *Journal of Biological Chemistry*, 263, 9573–5.

Rothstein, D. M., Spacciapoli, P., Tran, L. T., Xu, T., Roberts, F. D., Dalla Serra, M., Buxton, D. K., Oppenheim, F. G., and Friden, P. (2001). Anticandida activity is retained in P-113, a 12-amino-acid fragment of histatin 5. *Antimicrobial Agents and Chemotherapy*, 45, 1367–73.

Schittek, B., Hipfel, R., Sauer, B., Bauer, J., Kalbacher, H., Stevanovic, S., Schirle, M., Schroeder, K., Blin, N., Meier, F., Rassner, G., and Garbe, C. (2001). Dermcidin: A novel human antibiotic peptide secreted by sweat glands. *Nature Immunology*, 2, 1133–7.

Schroder, N. W., Opitz, B., Lamping, N., Michelsen, K. S., Zahringer, U., Gobel, U. B., and Schumann, R. R. (2000). Involvement of lipopolysaccharide binding protein, CD14, and Toll-like receptors in the initiation of innate immune responses by Treponema glycolipids. *Journal of Immunology*, 165, 2683–93.

Schutte, B. C., Mitros, J. P., Bartlett, J. A., Walters, J. D., Jia, H. P., Welsh, M. J., Casavant, T. L., and McCray Jr., P. B. (2002). Discovery of five conserved beta – defensin gene clusters using a computational search strategy. *Proceedings of the National Academy of Sciences USA*, 99, 2129–33.

Scott, M. G., Davidson, D. J., Gold, M. R., Bowdish, D., and Hancock, R .E. (2002). The human antimicrobial peptide LL-37 is a multifunctional modulator of innate immune responses. *Journal of Immunology*, 169, 3883–91.

Selsted, M. E., Novotny, M. J., Morris, W. L., Tang, Y. Q., Smith, W., and Cullor, J. S. (1992). Indolicidin, a novel bactericidal tridecapeptide amide from neutrophils. *Journal of Biological Chemistry*, 267, 4292–5.

Shai, Y. (1999). Mechanism of the binding, insertion and destabilization of phospholipid bilayer membranes by alpha-helical antimicrobial and cell non-selective membrane-lytic peptides. *Biochimica et Biophysica Acta*, 1462, 55–70.

Shi, J., Ross, C. R., Leto, T. L., and Blecha, F. (1996). PR-39, a proline-rich antibacterial peptide that inhibits phagocyte NADPH oxidase activity by binding to Src homology 3 domains of p47 phox. *Proceedings of the National Academy of Sciences USA*, 93, 6014–18.

Shike, H., Lauth, X., Westerman, M. E., Ostland, V. E., Carlberg, J. M., Van Olst, J. C., Shimizu, C., Bulet, P., and Burns, J. C. (2002). Bass hepcidin is a novel antimicrobial peptide induced by bacterial challenge. *European Journal of Biochemistry*, 269, 2232–7.

Simon, A., Fleischhack, G., Marklein, G., and Ritter, J. (2001). Antimicrobial prophylaxis of bacterial infections in pediatric oncology patients [in German]. *Klinische Padiatrie*, 213, Suppl. 1, A22–37.

Singh, P. K., Jia, H. P., Wiles, K., Hesselberth, J., Liu, L., Conway, B. A., Greenberg, E. P., Valore, E. V., Welsh, M. J., Ganz, T., Tack, B. F., and McCray Jr., P. B. (1998). Production of beta-defensins by human airway epithelia. *Proceedings of the National Academy of Sciences USA*, 95, 14961–6.

Skerlavaj, B., Gennaro, R., Bagella, L., Merluzzi, L., Risso, A., and Zanetti, M. (1996). Biological characterization of two novel cathelicidin-derived peptides and identification of structural requirements for their antimicrobial and cell lytic activities. *Journal of Biological Chemistry*, 271, 28375–81.

Sobel, J. D. (1988a). Candida infections in the intensive care unit. *Critical Care Clinics*, 4, 325–44.

Sobel, J. D. (1988b). Pathogenesis and epidemiology of vulvovaginal candidiasis. *Annals of the New York Academy of Sciences*, 544, 547–57.

Sorensen, O., Arnljots, K., Cowland, J. B., Bainton, D. F., and Borregaard, N. (1997). The human antibacterial cathelicidin, hCAP-18, is synthesized in myelocytes and metamyelocytes and localized to specific granules in neutrophils. *Blood*, 90, 2796–803.

Sorensen, O. E., Follin, P., Johnsen, A. H., Calafat, J., Tjabringa, G. S., Hiemstra, P. S., and Borregaard, N. (2001). Human cathelicidin, hCAP-18, is processed to the antimicrobial peptide LL-37 by extracellular cleavage with proteinase 3. *Blood*, 97, 3951–9.

Steinstraesser, L., Klein, R. D., Aminlari, A., Fan, M. H., Khilanani, V., Remick, D. G., Su, G. L., and Wang, S. C. (2001). Protegrin-1 enhances bacterial killing in thermally injured skin. *Critical Care Medicine*, 29, 1431–7.

Svinarich, D. M., Wolf, N. A., Gomez, R., Gonik, B., and Romero, R. (1997). Detection of human defensin 5 in reproductive tissues. *American Journal of Obstetrics and Gynecology*, 176, 470–5.

Tack, B. F., Sawai, M. V., Kearney, W. R., Robertson, A. D., Sherman, M. A., Wang, W., Hong, T., Boo, L. M., Wu, H., Waring, A. J., and Lehrer, R. I. (2002). SMAP-29 has two LPS-binding sites and a central hinge. *European Journal of Biochemistry*, 269, 1181–9.

Tenover, F. C. (1999). Implications of vancomycin-resistant *Staphylococcus aureus*. *Journal of Hospital Infection*, 43, Suppl., S3–7.

Tossi, A., Scocchi, M., Zanetti, M., Storici, P., and Gennaro, R. (1995). PMAP-37, a novel antibacterial peptide from pig myeloid cells. cDNA cloning, chemical synthesis and activity. *European Journal of Biochemistry*, 228, 941–6.

Trabi, M., Schirra, H. J., and Craik, D. J. (2001). Three-dimensional structure of RTD-1, a cyclic antimicrobial defensin from Rhesus macaque leukocytes. *Biochemistry*, 40, 4211–21.

Travis, S. M., Singh, P. K., and Welsh, M. J. (2001). Antimicrobial peptides and proteins in the innate defense of the airway surface. *Current Opinion in Immunology*, 13, 89–85.

Triantafilou, M. and Triantafilou, K. (2002). Lipopolysaccharide recognition: CD14, TLRs and the LPS-activation cluster. *Trends in Immunology*, 23, 301–4.

Uehara, N., Yagihashi, A., Kondoh, K., Tsuji, N., Fujita, T., Hamada, H., and Watanabe, N. (2003). Human beta-defensin-2 induction in *Helicobacter pylori*-infected gastric mucosal tissues: antimicrobial effect of overexpression. *Journal of Medical Microbiology*, 52, 41–5.

Verghese, S. L., Padmaja, P., and Koshi, G. (1998). Central venous catheter related infections in a tertiary care hospital. *Journal of the Association of Physicians of India*, 46, 445–7.

Wasserheit, J. N. (1992). Epidemiological synergy. Interrelationships between human immunodeficiency virus infection and other sexually transmitted diseases. *Sexually Transmitted Diseases*, 19, 61–77.

Weiss, J., Elsbach, P., Olsson, I., and Odeberg, H. (1978). Purification and characterization of a potent bactericidal and membrane active protein from the granules of human polymorphonuclear leukocytes. *Journal of Biological Chemistry*, 253, 2664–72.

Wiesenfeld, H. C., Heine, R. P., Krohn, M. A., Hillier, S. L., Amortegui, A. A., Nicolazzo, M., and Sweet, R. L. (2002). Association between elevated neutrophil defensin levels and endometritis. *Journal of Infectious Diseases*, 186, 792–7.

Wimley, W. C., Selsted, M. E., and White, S. H. (1994). Interactions between human defensins and lipid bilayers: Evidence for formation of multimeric pores. *Protein Science*, 3, 1362–73.

Workowski, K. A., Levine, W. C., and Wasserheit, J. N. (2002). U.S. Centers for Disease Control and Prevention guidelines for the treatment of sexually transmitted diseases: an opportunity to unify clinical and public health practice. *Annals of Internal Medicine*, 137, 255–62.

Wu, E. R., Daniel, R., and Bateman, A. (1998). RK-2: A novel rabbit kidney defensin and its implications for renal host defense. *Peptides*, 19, 793–9.

Wu, M. and Hancock, R. E. (1999). Interaction of the cyclic antimicrobial cationic peptide bactenecin with the outer and cytoplasmic membrane. *Journal of Biological Chemistry*, 274, 29–35.

Yamaguchi, Y., Fukuhara, S., Nagase, T., Tomita, T., Hitomi, S., Kimura, S., Kurihara, H., and Ouchi, Y. (2001). A novel mouse beta-defensin, mBD-6, predominantly expressed in skeletal muscle. *Journal of Biological Chemistry*, 276, 31510–14.

Yang, D., Chen, Q., Chertov, O., and Oppenheim, J. J. (2000). Human neutrophil defensins selectively chemoattract naive T and immature dendritic cells. *Journal of Leukocyte Biology*, 68, 9–14.

Yang, D., Chertov, O., and Oppenheim, J. J. (2001a). The role of mammalian antimicrobial peptides and proteins in awakening of innate host defenses and adaptive immunity. *Cellular and Molecular Life Sciences*, 58, 978–89.

Yang, D., Chertov, O., and Oppenheim, J. J. (2001b). Participation of mammalian defensins and cathelicidins in anti-microbial immunity: Receptors and activities of human defensins and cathelicidin (LL-37). *Journal of Leukocyte Biology*, 69, 691–7.

Yasin, B., Harwig, S. S., Lehrer, R. I., and Wagar, E. A. (1996). Susceptibility of *Chlamydia trachomatis* to protegrins and defensins. *Infection and Immunity*, 64, 709–13.

Zaiou, M. and Gallo, R. L. (2002). Cathelicidins, essential gene-encoded mammalian antibiotics. *Journal of Molecular Medicine*, 80, 549–61.

Zanetti, M., Gennaro, R., and Romeo, D. (1995). Cathelicidins: A novel protein family with a common proregion and a variable C-terminal antimicrobial domain. *FEBS Letters*, 374, 1–5.

Zarember, K. A., Katz, S. S., Tack, B. F., Doukhan, L., Weiss, J., and Elsbach, P. (2002). Host defense functions of proteolytically processed and parent (unprocessed) cathelicidins of rabbit granulocytes. *Infection and Immunity*, 70, 569–76.

Zasloff, M. (1987). Magainins, a class of antimicrobial peptides from Xenopus skin: Isolation, characterization of two active forms, and partial cDNA sequence of a precursor. *Proceedings of the National Academy of Sciences USA*, 84, 5449–53.

Zhang, L., Rozek, A., and Hancock, R. E. (2001). Interaction of cationic antimicrobial peptides with model membranes. *Journal of Biological Chemistry*, 276, 35714–22.

Zhao, C., Nguyen, T., Boo, L. M., Hong, T., Espiritu, C., Orlov, D., Wang, W., Waring, A., and Lehrer, R. I. (2001). RL-37, an alpha-helical antimicrobial peptide of the rhesus monkey. *Antimicrobial Agents and Chemotherapy*, 45, 2695–702.

CHAPTER 5

Mammalian β-defensins in mucosal defense

Gill Diamond, Danielle Laube, and Marcia Klein-Patel

5.1. INTRODUCTION

Mucosal surfaces of the body represent the initial site of interaction with environmental microbes. Thus a well-developed innate defense mechanism must be in place to prevent colonization by pathogenic organisms, which could lead to life-threatening infections. Indeed, in addition to a wide range of other defense-related molecules, all epithelial mucosa express β-defensins. These broad-spectrum antimicrobial peptides, which also exhibit potent chemotactic activity, are found in both constitutively expressed and inducible forms. Upregulation of inducible β-defensin genes occurs in response to both inflammatory mediators and infectious agents, as we subsequently describe in detail. This suggests that the mucosal epithelia are in a state of readiness to recognize potential pathogenic microorganisms and mount a defense response. However, each mucosal surface interacts with a range of bacteria amidst radically different environments. Some, like the airway, are relatively sterile, whereas others, like the oral cavity or the intestines, are host to many commensal microorganisms.

How does each surface modulate the complement of β-defensins to maintain either sterility or bacterial homeostasis? To address this, we examine the expression of β-defensins in several mammalian mucosal epithelia. It is our hypothesis that each mucosal surface has developed different mechanisms to regulate β-defensin gene expression based on its particular environment. These mechanisms include the use of different receptors and signal transduction pathways, leading to β-defensin gene expression.

The first β-defensins were discovered as part of a study to identify antibacterial components of the mammalian airway. Tracheal antimicrobial peptide (TAP) was isolated from the tracheal mucosa of the cow and

TAP	NPVSCVRNKGICVPIRCPGSMKQIGTCVGRAVKCCRKK
β-DEFENSIN CONSENSUS	-C----G-C----C-------G-C------CC-
α-DEFENSIN CONSENSUS	--C-CR---C---ER--G-C---G-----CC---

Figure 5.1. Primary amino acid sequence of TAP aligned with the consensus sequence of α- and β-defensins.

characterized as a defensinlike, broad-spectrum antimicrobial peptide (Diamond et al., 1991). Although TAP shares sequence similarity with the human neutrophil defensins (HNPs) (Lehrer and Ganz, 1990), it has a novel six-cysteine motif, unlike the one characteristic of the HNPs. Subsequently, thirteen peptides were discovered in the bovine neutrophil, exhibiting this alternative six-cysteine motif with TAP (Selsted et al., 1993) (Fig. 5.1). Further identification of peptides homologous to TAP in humans, monkeys, mice, and other species allowed the definition of this large gene family, now known as β-defensins.

The first human β-defensin (HBD1) was identified by purification from hemofiltrate (Bensch et al., 1995) and subsequently found to be expressed in epithelial cells throughout the body (Zhao, Wang, and Lehrer, 1996). In contrast, the second human β-defensin (HBD2) was found in psoriatic (but not normal) skin (Harder et al., 1997), whereas HBD1 is expressed even in normal skin (Ali et al., 2001). This characteristic expression pattern, whereby HBD1 is constitutively expressed and HBD2 expression is variable, is a paradigm almost uniform throughout the body (see Table 5.1). Other HBDs have been identified more recently, with expression studies on the six genes shown in Fig. 5.2. As we discuss in the next section, the mechanisms that control defensin expression are directly related to their potential role in host defense at mucosal surfaces. Gene mapping studies have determined that β-defensins are localized to clusters on syntenic chromosomes in human, mouse, and cow. Sequence analysis suggests that the β-defensin family derived from an ancestral gene that underwent evolutionary divergence and duplication (Bevins et al., 1996; Liu et al., 1997). The result is a diverse family of defensin peptides in each mammalian species (Fig. 5.3), the evolution of which may have occurred in response to the particular pathogens encountered. Using a computer-based search method, Schutte et al. (2002) identified 28 β-defensin genes in the human genome. We believe that this gene family, together with

Table 5.1. *Tissue-specific expression of human mucosal β-defensins*

Tissue	Defensin	Cell type	Induction[1]
Airway	HBD1	Ciliated epithelium	Constitutive
	HBD2	Ciliated epithelium	Live bacteria, LPS, IL-1β, TNF-α, PMA
	HBD3	Ciliated epithelium	IFN-γ
	HBD4	Ciliated epithelium	*Pseudomonas aeruginosa Strep. pneumonieae*
Oral Cavity	HBD1	Gingival epithelium, tongue, salivary glands, buccal mucosa	Constitutive
	HBD2	Gingival epithelium, tongue, salivary glands, buccal mucosa	IL-1β, TNF-α, *Fusobacterium nucleatum* cell-wall extracts
	HBD3	Gingival keratinocytes, tongue, salivary glands	Undetermined
Gastrointe-stinal	HBD1	Gastric epithelium	Constitutive, and by IL-1β, live *Helicobactor pylori*
	HBD2	Gastric epithelium	IL-1α, live *H. pylori*, other live enteroinvasive bacteria
	HBD1	Intestinal epithelium	Constitutive
	HBD2	Intestinal epithelium	IL-1α, live enteroinvasive bacteria
	EBD	Intestinal epithelium	*Cryptosporidium parvum*
Female repro-ductive tract	HD5	Endocervix	Varies with hormonal levels
	HBD1	Fallopian tubes, endometrium, endocervix, ectocervix	Constitutive
	HBD2	Endometrium	IFN-γ, IL-1β

(*Cont.*)

Table 5.1. (*Cont.*)

Tissue	Defensin	Cell type	Induction
Male reproductive tract	HBD4, 5, 6	Epididymis	Undetermined
Kidney	HBD1	Loops of Henle, distal tubules, collecting ducts	Constitutive
	HBD2	Loops of Henle, distal tubules, collecting ducts	Bacteria, IL-1α, TNF-α, LPS
Eye	HBD1	Conjunctiva	Constitutive
	HBD2	Conjunctiva	Constitutive, and infection, inflammation
Ear	HBD1	Tympanic membrane	Constitutive
	HBD2	Middle ear mucosa	IL-1α, TNF-α, LPS
Skin	HBD1	Epidermis, follicle, pilosebaceous duct	Constitutive
	HBD2	Epidermis, follicle, pilosebaceous duct	Psoriasis, infection, IL-1

[1] LPS, lipopolysaccharide; PMA, phorbol myristate acetate; TNF-α, tumor necrosis factor-α; IFN, Interferon.

the functional studies subsequently described, is potentially one of the most important in host defense.

5.2. TISSUE-SPECIFIC EXPRESSION OF β-DEFENSINS

5.2.1. The Respiratory Tract

Respiratory epithelial cells undergo a constant barrage from inspired microbial pathogens, and yet the uncompromised airway remains sterile. The huge surface area and thin physical barrier of gas-exchange epithelium necessitates a potent innate immune system. The proximal airway, composed primarily of ciliated columnar epithelial cells, provides for this defense in three broad ways: (1) as a physical barrier, (2) through the recruitment of immune cells, and (3) through the expression of a variety of pathogen receptors and antimicrobial products. In the respiratory epithelium, ciliary movement, coughing, and sneezing all help remove particles trapped in the mucous.

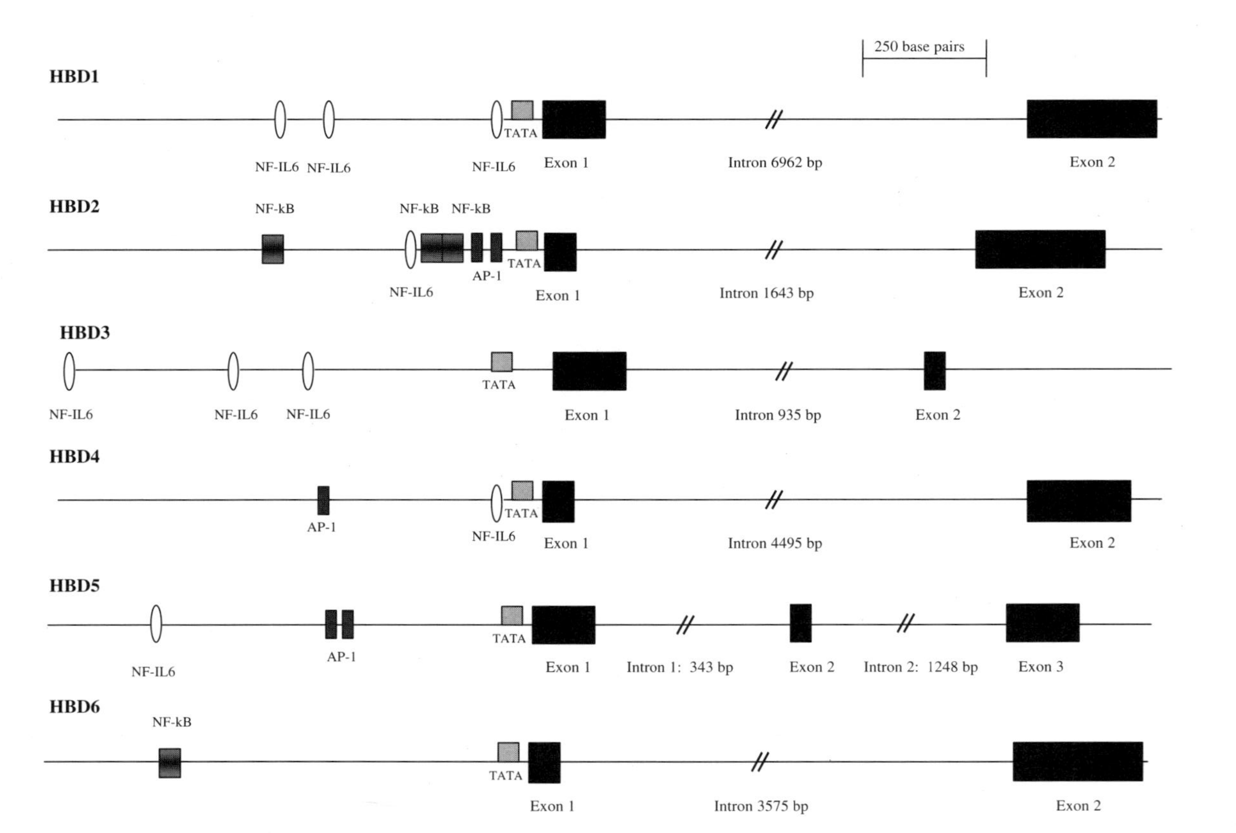

Figure 5.2. Gene and promoter structure of hBDs. Exon structure and putative transcription factor binding sites are shown for HBD1–HBD6.

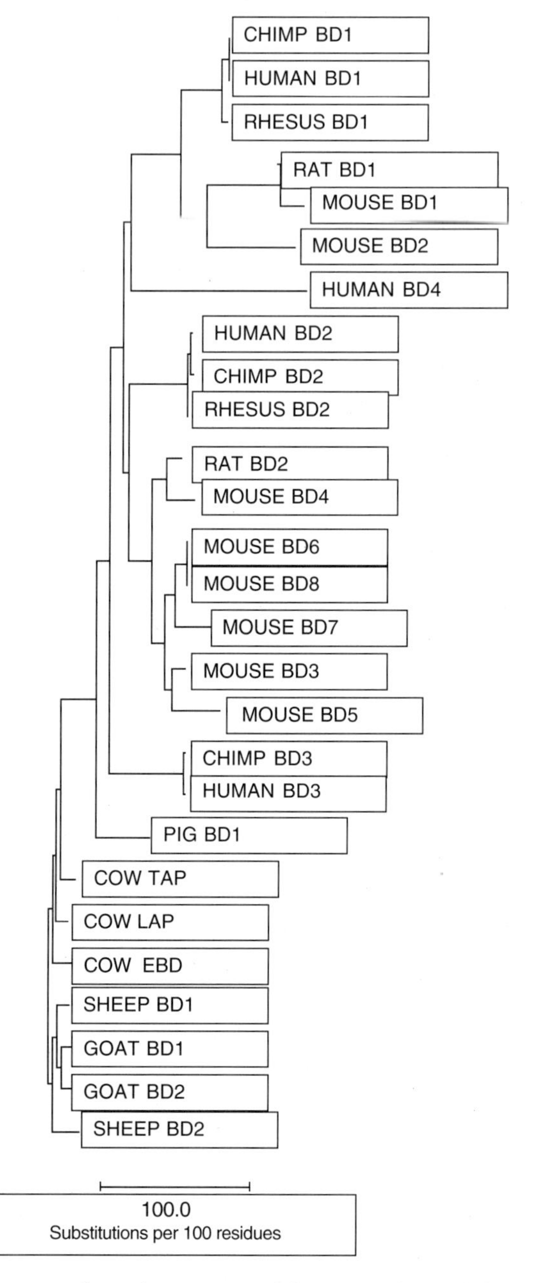

Figure 5.3. Evolutionary relationships among β-defensin peptides. A phylogenetic tree of β-defensin peptide sequences was created from a distance matrix by the neighbor-joining method (Growtree program, GCG software). Bootstrap analysis was used to test the significance of the phylogenetic relationships of the β-defensins.

Immune cells such as phagocytes, natural-killer (NK) cells, γ/δ T cells, and B-1 cells recognize infected host cells and bacterial antigens without previous or repeated exposure. The mucosa of the respiratory tract expresses pattern-recognition receptors, which provide for the early detection of invading organisms, and secretes a variety of antimicrobial products into airway surface fluid (ASF), including lactoferrin, lysozyme, and β-defensins.

HBD1–HBD4 are all expressed, at least minimally, in the trachea and lung, with distribution greatest in the large airways and decreasing distally. The two best-studied human defensins, HBD1 and HBD2, differ in both expression pattern and ability to respond to inflammatory stimuli. HBD1 and its mouse (mBD), rat (rBD), rhesus monkey (rhBD), and goat (gBD) orthologs [mBD-1 (Bals et al., 1998a), rBD-1 (Jia et al., 1999), rhBD-1 (Bals et al., 2001), and gBD-2 (Zhao et al., 1999)] are primarily expressed in urogenital tissue (Bensch et al., 1995; Valore et al., 1998; Zhao et al., 1999). They are observed in all other mucosal epithelia as well, including the respiratory tract (Zhao et al., 1996; McCray and Bentley, 1997), and can be found in bronchial alveolar lavage fluid (BALF) (Singh et al., 1998; Agerberth et al., 1999). HBD1 is not inducible by inflammatory mediators or bacterial products, including lipopolysaccharide (LPS), IL-6, phorbol myristate acetate (PMA), Interferon-γ (IFN-γ), tumor necrosis factor-α (TNF-α) (Zhao et al., 1996), or IL-1α (Singh et al., 1998). This pattern of constitutive expression is expected as the promoter region does not contain the necessary transcription factor binding sites to respond to these stimuli (see Fig. 5.2 and subsequent discussion) and suggests that HBD1 transcription is coordinately regulated with other epithelium-specific genes. Therefore the primary role of HBD1 is most likely to act as a baseline defense against microbial infection.

However, HBD2 and its orthologs [bovine TAP (Diamond et al., 1991), mBD2 and mBD3 (Bals et al., 1998b; Morrison, Davidson, and Dorin, 1999; Burd et al., 2002), rHBD2 (Bals et al., 2001; Boniotto, Tossi, and Crovella, 2002), rBD2 (Jia et al., 1999), and gBD1 (Zhao et al., 1999)] are expressed in greatest abundance in the airway. This group of β-defensins is also remarkable for the fact that they are inducible in response to microbial and host factors. The first mammalian studies on β-defensin gene regulation were conducted in bovine tracheal epithelial cells (TECs). When stimulated with LPS, a 13-fold induction of TAP mRNA expression was observed (Diamond, Russell, and Bevins, 1996). Interestingly, the amount of stimulus necessary for maximal hBD-2 upregulation increases distally. Harder et al. (2000) showed that nasal epithelial cells produced maximum amounts of HBD2 mRNA at 10^4 colonies/ml of *Pseudomonas aeruginosa*, whereas tracheal and bronchial epithelial cells required 10^5 colonies/ml to reach this level of expression.

The inducibility of HBD2 and its orthologs is due to the presence of one or multiple NF-κB consensus binding sites in the 5′ flanking region (see Fig. 5.2). In responsive cell types such as the airway epithelium, recognition of bacteria, bacterial products, and proinflammatory cytokines (Becker et al., 2000; Diamond et al., 2000) results in NF-κB activation and HBD2 transcription. HBD2 and TAP also contain consensus sites for activator protein-1 (AP-1) and NF-IL-6, and are responsive to TNF-α and IL-1β (Becker et al., 2000; Diamond et al., 2000). In bovine TEC, both NF-κB and NF-IL6 sequences appear necessary for TAP expression. The observation that HBD2 is upregulated by a wide variety of pathogenic and inflammatory stimuli suggests that HBD2 is an important first response to infection by the innate immune system of the respiratory epithelium.

The recently described HBD3 and HBD4 appear to have different patterns of expression and regulation compared with those of HBD1 and HBD2. HBD-3 is expressed primarily in the placenta and testes and has low expression in other epithelia, including the trachea (Garcia et al., 2001a). The regulation of HBD3 and its orthologs, pig BD-1 (Zhang et al., 1998) and perhaps mBD6, (Jia et al., 2000) is distinctly different from the other β-defensins; they have AP-1 and NF-IL-6, but not NF-κB, consensus sequences, and GM colony-stimulating factor and IFN-γ response element sites (Jia et al., 2001). Interestingly, HBD3 has been shown to be upregulated only by IFN-γ, a product of activated macrophages, and not by TNF-α, IL-1α, IL-6, PMA or heat-inactivated *P. aeruginosa* (Garcia et al., 2001a). HBD3 expression is inhibited by corticosteroids, whereas HBD1 and HBD2 are not (Duits et al., 2001). These results implicate HBD3 in the response of the airway to systemic infection or a previously activated immune system. In opposition to HBD3, HBD4 is primarily expressed in the lung and is inducible by heat inactivated *P. aeruginosa* and *Strep. pneumoniae,* but not inflammatory mediators IL-1β, INF-γ, or TNF-α (Garcia et al., 2001b), illustrating, again, the importance of the lung as a primary site for the recognition and elimination of pathogenic organisms.

Soon after the discovery of TAP and lingual antimicrobial peptide (LAP) in the bovine airway (Diamond et al., 1991; Schonwetter, Stolzenberg, and Zasloff, 1995; Diamond et al., 1996; Russell et al., 1996), Smith et al. (Smith et al., 1996) demonstrated that ASF from primary normal human epithelial cells is microbicidal to *P. aeruginosa,* whereas ASF from primary cultures from patients with cystic fibrosis (CF) is not. This led them to discover that increasing the salt concentration of normal ASF decreases its bactericidal activity, whereas decreasing the salt concentration of CF ASF increases its activity. This suggests that the antimicrobial component of ASF is salt sensitive and perhaps its inactivation could account for the cycle of infection and inflammation seen in patients with CF. Subsequently it was shown that HBD1, HBD2, and HBD4 and their murine orthologs are salt sensitive (Goldman et al., 1997; Bals et al., 1998a, 1998b; Garcia et al., 2001b; Harder et al., 2001), suggesting that β-defensins are the antimicrobial components whose activity is inhibited in CF sputum.

Additional support for this theory came from antisense DNA experiments that demonstrated that inhibition of HBD1 mRNA diminished the antipseudomonal activity of normal bronchial xenograft ASF (Goldman et al., 1997). Although HBD1 and HBD2 are thought to be inactivated in CF ASF, both are present in BALF, whereas only HBD1 is found in normal BALF (Singh et al., 1998). Similar clinical examinations have also suggested the importance of β-defensins in maintaining sterility of the airway and preventing and responding to respiratory infection. Schaller-Balls, Schulze, and Bals (2002) have demonstrated increased levels of HBD1 and HBD2 in the BALF of newborn infants with pulmonary and/or systemic infection. HBD2 levels in both plasma and BALF of patients with *Mycobacterium avium* intracellulare (MAI) infection are increased. In addition, disease severity, as determined by the presence of cavity formation and bronchiectasis, correlated with higher levels of HBD2. Once these patients were treated for MAI infection, HBD2 levels decreased to control levels (Ashitani et al., 2001).

Further evidence for the importance of β-defensins in airway innate immunity comes from studies indicating that different antimicrobial factors at the airway surface mucosa interact, increasing their activity. Bals et al. (1998b) showed that the minimal inhibitory concentrations of HBD2 against *Escherichia coli* and *Staphylococcus aureus* decreased two to four times in the presence of nonbacteriostatic concentrations of lactoferrin and twofold in the presence of nonbacteriostatic concentrations of lysozyme against *E. coli, P. aeruginosa*, and *Enterococcus faecalis*. In contrast, it appears that HBD1 and HBD2 are not synergistic with lysozyme and lactoferrin at bacteriocidal concentrations (Singh et al., 2000). The newly described HBD4 (but not HBD3) has also demonstrated synergy with lysozyme against *Staphylococcus carnosus* (Garcia et al., 2001b). Regardless, the multiple antimicrobial components found in ASF may combine to provide an efficient defense mechanism.

Although the HBD2 family is the primary group of β-defensins in the respiratory tract and the most abundantly identified, functional studies have recently been conducted on the importance of the constitutively expressed HBD1 family. Moser et al. (2002) recently created an mBD1 knockout infection model and showed that these mice, when infected intranasally with *Haemophilus influenzae*, exhibited decreased clearance of bacteria. Interestingly, mBD-1$^{-/-}$ mice showed a normal initial inflammatory response as measured by cytokine levels and neutrophils in the BALF, and no difference in the clearance of *S. pneumoniae*. This suggests that β-defensins have both redundant and nonoverlapping functions, which is further supported by mBD-1 knockout studies (Morrison et al., 2002). Mice inoculated intranasally with *S. aureus* showed no increase in respiratory infection or decrease in inflammatory mediators. However, these animals had higher levels of *S. aureus* and nonstaphylococcal urinary tract infections (see as discussed later). Thus we can see from the constitutive expression of HBD1 and the differential upregulation of hBD2, HBD3, and HBD4 that the airway has developed a sensitive innate immune response to maintain the sterility of the tissue.

5.2.2. The Oral Cavity

The oral cavity is colonized by a community of microorganisms that are nonpathogenic in individuals with healthy periodontium. This unique environment poses special challenges for the gingival epithelium. The bacteria that make up the oral flora are considered pathogenic when found in other parts of the body (Gendron, Grenier, and Maheu-Robert, 2000) and are implicated in conditions such as infective endocarditis (Bayliss et al., 1986), brain abscesses (Pallasch and Slots, 1996), and chronic conjunctivitis (van Winkelhoff et al., 1991). *Fusobacterium nucleatum*, a Gram-negative commensal organism considered to be nonpathogenic in the oral cavity, is seen with high frequency in amniotic fluid cultures from women with preterm labor (Hill, 1998). The mouth may also be an important reservoir of pathogens responsible for chronic infection of the respiratory tract in patients with CF (Komiyama et al., 1985).

Recent studies have described the expression of β-defensins in the human oral cavity, suggesting they contribute to the innate host defenses of this tissue (reviewed in Dale and Krisanaprakornkit, 2001). It has been proposed that defensins are secreted by the epithelial cells of the mouth and, by associating with salivary mucins, are distributed throughout the oral cavity (Sahasrabudhe et al., 2000). The concentrations of defensin in the saliva are believed to be sufficient for microbicidal activity against some organisms (Mathews et al., 1999).

HBD1 and HBD2 are similar in their cell-type-specific expression patterns in the oral cavity. Both are expressed in gingival tissue, salivary glands, tongue, and buccal mucosa (Zhao et al., 1996; Krisanaprakornkit et al.,

1998; Bonass et al., 1999; Mathews et al., 1999; Krisanaprakornkit et al., 2000; Dunsche et al., 2001). In the gingival epithelium, HBD1 is constitutive (Krisanaprakornkit et al., 1998), whereas HBD2 expression is increased by IL-1β, TNF-α, and certain bacterial products, although also present in uninflamed gingival tissue (Mathews et al., 1999; Krisanaprakornkit et al., 2000). HBD3 is seen in cultured gingival keratinocytes, tongue, and salivary glands (Jia et al., 2000; Garcia et al., 2001a; Harder et al., 2001; Dunsche et al., 2002), suggesting that its expression pattern may also resemble that of HBD1 and HBD2 in the oral cavity.

By use of immunohistochemistry, localization of β-defensins to specific strata within the oral epithelium has been observed. β-defensins are expressed at highest concentrations in the gingival margin, a surface adjacent to the region of plaque formation on the tooth surface, and in inflamed sulcular epithelium adjacent to the junctional epithelium, but not in the less-differentiated junctional epithelium itself (Dale and Krisanaprakornkit, 2001). Specifically, HBD1 and HBD2 mRNAs are detected in the spinous layer, whereas defensin peptides are observed in the upper spinous granular and cornified layers. It therefore appears that expression of defensins in the oral cavity is limited to well-differentiated stratified squamous epithelium.

As opposed to the airway, however, the oral cavity is host to a range of bacterial species. The concentrations of these bacteria, or their components such as LPS, would be sufficient to induce the expression of HBD2 in airway epithelium (Becker et al., 2000), yet the response in the gingival epithelium differs. Specifically, cell wall extracts of *Porphyromonas gingivalis,* a periodontal pathogen, fail to induce an HBD2 response, and adding concentrations of up to 10 μg/mL of *F. nucleatum* or *E. coli* LPS to human gingival epithelial cultures resulted in only a slight increase in HBD2 expression (Krisanaprakornkit et al., 2000). Upregulation is seen in this model only by crude cell wall extracts from *F. nucleatum.*

In a study to elucidate this mechanism, it was observed that after stimulation with *F. nucleatum* cell-wall extracts, the p65 subunit of NF-κB is translocated to the nucleus, and levels of this and a second subunit, p50, increase in nuclear extracts of stimulated cells within a few hours of treatment, as expected if this pathway was responsible for HBD2 induction. However, pretreatment with two different inhibitors of NF-κB, pyrrolidine dithiocarbamate (PDTC) and MG132, failed to reduce the levels of HBD2 expression (Krisanaprakornkit, Kimball, and Dale, 2002). Similar work has showed MG132 inhibition of HBD2 induction in an intestinal epithelial cell line (O'Neil et al., 1999), indicating that different mechanisms are involved in this signaling pathway. NF-κB translocation to the nucleus is seen concomitant with HBD2 expression, although it is not necessarily sufficient or required for HBD2 induction in all cell types. This implicates the involvement of another transcription factor in the upregulation of HBD2 in the mouth. (Krisanaprakornkit et al., 2002).

In addition to three NF-κB consensus sequences (Liu et al., 1998), the promoter region of HBD2 has several binding sites for AP-1 (Harder et al., 2000). The AP-1 transcription factor family is involved in many cell processes, including proliferation, maturation, and differentiation, and is mediated by three mitogen-activated protein (MAP) kinase-signaling cascades. The extracellular signal-regulated kinase (ERK) pathway is activated by growth factors, whereas the c-Jun NH_2-terminal kinase (JNK), and the p38 MAP kinase pathways are activated by environmental stresses and proinflammatory cytokines (Cobb and Goldsmith, 1995; Seger and Krebs, 1995; Kyriakis and Avruch, 1996; Su and Karin, 1996).

To establish the involvement of the MAP kinase pathways, inhibitors of the ERK, p38, and JNK pathways were used (Krisanaprakornkit et al., 2002). The ERK inhibitors PD98059 and U0126 did not alter HBD2 expression. However, pretreatment with SB203580, a specific inhibitor of p38, and tyrphostin AG126, a JNK inhibitor, showed a different result. Both SB203580 and tyrphostin AG126 partially blocked induction of HBD2, and, when used in concert, these inhibitors prevented HBD2 upregulation entirely. Together these results indicate that induction of β-defensin gene expression in the oral cavity is complex, involving numerous pathways. The factors that induce these pathways and how they play a role in maintaining the overall homeostasis of nonpathogenic bacteria remain to be elucidated.

5.2.3. The Gastrointestinal Tract

The normal flora of the gastrointestinal (GI) tract varies greatly along its length. The gastric mucosa provides a uniquely unfriendly environment to microorganisms, with only one bacterial species, *Helicobacter pylori,* adapted to survive in its acidic environment (Dunn, Cohen, and Blaser, 1997). In distinct contrast to what is observed in other tissues, a recent study has reported inducible expression of HBD1 (Bajaj-Elliott et al., 2002). An increase in HBD1 was observed in histological sections of *H. pylori*-infected tissues over control tissues. In the gastric cell line MKN7, treatment with *H. pylori* and with IL-1β also resulted in an increase in HBD1 expression. Similar to other epithelia, HBD2, which is bacteriostatic against *H. pylori* (Hamanaka et al., 2001), was induced in infected gastric tissue and in cultured gastric epithelial cells challenged with bacteria, including *H. pylori, Salmonella enterica* (serotype Dublin) (Wada et al., 1999; O'Neil et al., 2000), *S. typhimurium, S. enteridis,* and *S. typhi* (Wada et al., 1999), or cytokines IL-1α (O'Neil et al., 2000) and IL-1β (Bajaj-Elliott et al., 2002). However, no effect was seen when either heat-killed bacteria or purified LPS was used as a stimulus (O'Neil et al., 2000), suggesting that, in contrast to what is observed in the respiratory epithelium, either an active component of the bacterium or the infection itself is responsible for this induction. Similar to the airway, stimulation of HBD2 in gastric cells occurs by means of the NF-κB pathway, as deleting or mutating the NF-κB site abolishes HBD2 upregulation by *H. pylori* (Wada et al., 2001). Overexpression of HBD2 in a cell line results in the inhibition of *H. pylori* growth, although the activity of this peptide in the unusual environment of the stomach has yet to be characterized (Uehara et al., 2003).

In contrast to the stomach, the intestines play host to a wide range of bacterial species. Initially, it was discovered that α-defensins were expressed in the Paneth cells of the small intestine (Ouellette et al., 1989), suggesting that defensins serve an important antibacterial role in the maintenance of the sterility of the crypts. Subsequently, constitutive HBD1 expression was observed in the surface and crypt epithelium of the colon and the villous and crypt epithelium of the small intestine (O'Neil et al., 1999).

HBD2 mRNA expression is induced in intestinal epithelial cell lines by IL-1α and by enteroinvasive bacteria such as *S. enteritidis* (Ogushi et al., 2001) and *E. coli* O29:NM (O'Neil et al., 1999), primarily through activation of NF-κB (O'Neil et al., 1999). The bacterial agent responsible for this induction, similar to what is observed in the oral cavity, is not LPS. This is consistent with the hypothesis that cells in constant contact with bacteria have a different response mechanism for achieving activation on interaction only with potentially pathogenic organisms. Enteroinvasive bacteria are known to stimulate the NF-κB pathway in colonic epithelia (Elewaut et al., 1999), which is one of the pathways responsible for HBD2 upregulation in this system.

Expression of bovine enteric β-defensin (EBD) increases distally along the small intestine, with maximal expression in the proximal colon (Tarver et al., 1998). EBD has a high degree of sequence similarity with both TAP and LAP (see Fig. 5.1). Although its gene does not contain an upstream NF-κB site, calves infected with the intestinal parasite *Cryptosporidium parvum* show a fivefold–tenfold increase in EBD in the distal small intestine and the colon. In the small intestine, this expression was more intense in the crypt cells than in cells covering the villi, where *C. parvum* colonizes.

In an attempt to characterize the specific bacterial components that upregulate defensins in the GI tract, the major flagella filament protein of *S. enteritidis*, FliC, has been proposed to induce HBD2 (Ogushi et al., 2001). FliC induces NF-κB translocation to the nucleus (Takahashi et al., 2001). However, O'Neil et al. (1999) contended that NF-κB is not adequate for higher levels of HBD2 induction in this system, because, although TNF-α is known to activate NF-κB, it is a poor inducer of HBD2 in the intestines. It is possible that the AP-1 pathway, which appears to be involved in HBD2 signaling in the oral cavity, might also be involved in host response to pathogenic organisms in the GI tract.

In addition to transcriptional regulation, control at other levels may regulate β-defensin expression. A second isoform of HBD2, larger than the 41-amino-acid peptide previously observed in other tissues, has been found in the colon epithelial cell line Caco-2 (O'Neil et al., 1999) and was subsequently also described in the gastric epithelial cell line AGS (Hamanaka et al., 2001). These observations, as well as those discussed in the genitourinary tract, suggest that defensins exist as multiple isoforms. It has been speculated that the 41-amino-acid peptide may arise as a result of posttranslational NH_2-terminal proteolytic processing of the larger peptide, as occurs with HBD1 in the urinary tract.

Furthermore, an unusually high level of sequence identity in the 5′-untranslated region (UTR) and the signal sequence coding region between defensins, despite wide species boundaries, suggests that there may be yet another level of regulation. Jurevic et al. (2003) have identified a single-nucleotide polymorphism (SNP) in the 5′ UTR of hBD-1 that is associated with differential levels of *Candida albicans* colonization. This suggests that HBD1 may also be regulated at the posttranscriptional level. It seems possible that individuals carrying this SNP have reduced levels of HBD1 that are due to a decrease in translation, resulting in an increased susceptibility to infection.

5.2.4. The Genitourinary Tract

Female Reproductive Tract

As the primary site of entry for sexually transmitted pathogens that can occur in semen, the female reproductive tract must exhibit strong innate mucosal host defenses (for review, see Quayle, 2002). Yet it must simultaneously be receptive to sperm, and therefore must demonstrate unique mechanisms of defensin gene regulation. Similar to the GI tract, the female reproductive tract exhibits varying levels of sterility along its length. The vagina and the vaginal portion of the cervix (the ectocervix) maintain a high concentration of commensal microorganisms, which themselves contribute to the innate defenses. β-defensins are expressed along the entire length of the reproductive tract (Valore et al., 1998). However, the endocervix has the highest and most consistent level of expression among the tissues sampled, whereas the endometrium is the most variable. Indeed, primary cultures of endometrial cells express low levels of HBD1 and increased levels of HBD2 mRNA in response to IFN-γ and IL-1β (King et al., 2002).

In contrast to this lower portion of the tract, the endocervix acts as both physical and chemical barriers to pathogenic infections of the upper portion (the uterus and fallopian tubes), which could have deleterious effects on reproduction. This upper portion, however, is open to colonization from the lower portion only during the preovulatory phase of the menstrual cycle, when the cervical plug degrades and endocervical secretions alter to allow penetration of sperm. Expression of defensins parallels this change (Quayle et al., 1998). Expression of the α-defensin HD5 appears to be regulated during the menstrual cycle, whereby highest expression is seen during the mid–late proliferative and early secretory phases. Thus, rather than responding to the presence of pathogenic microorganisms as with the other epithelia previously described, the epithelial cells of the female reproductive tract exhibit a mechanism of defensin gene regulation appropriate to its microbial exposure.

Translational modification of defensin peptides has also been observed in the genitourinary tract. Isolation of HBD1 from human tissues, urine, and vaginal secretions has resulted in the identification of numerous β-defensin isotypes, which appear to be variants differing in length (Valore et al., 1998; O'Neil et al., 1999; Hiratsuka et al., 2000). Vaginal lavage fluid contains two isoforms of HBD1, one of 39/40 amino acids in length and the other 44 amino acids (Valore et al., 1998). Minor changes to the sequence of the HBD1 amino-terminus had a noticeable effect on bacterial killing, although all of the isoforms identified had some level of antimicrobial activity, suggesting a mechanism for creating a broad spectrum of activity from only a few peptides (Valore et al., 1998).

Male Reproductive Tract

In distinct contrast to other epithelial surfaces, the male reproductive tract has been a source of unique β-defensins. Although low levels of HBD1 expression have been observed in testes, mRNA encoding HBD4 (Garcia et al., 2001b) were first described in the testes and then later found to be colocalized with mRNAs for two new human β-defensins, HBD5 and HBD6, in the epididymis. Similarly, mBD4 was found to be colocalized with two novel mouse homologues, mBD11 and mBD12 (Yamaguchi et al., 2002). A defensinlike gene that is predicted to encode a peptide the same size as previously described peptides, and containing the same six-cysteine motif, has been described in the epididymis of rats (Li et al., 2001). It has also recently become evident that the male reproductive tract expresses many more defensins and defensinlike peptides, the characterization of which will undoubtedly lead to a greater understanding of the role of defensins.

Although experimental analysis of gene regulation of these new β-defensins has not yet been reported, the promoter region of HBD6 (but not HBD5) includes consensus binding sites for NF-κB, suggesting differential regulation similar to the HBD1/HBD2 scheme seen in other sterile epithelia. Interestingly, research on the male rat reproductive tract has indicated that rBD1 is not upregulated by LPS, but is upregulated by androgens (Palladino, Mallonga, and Mishra, 2003). These results, taken together with the cyclic expression of defensins in the female menstrual cycle indicate the evolution of a unique regulatory mechanism for defensins in order to provide protection against introduced pathogens in the reproductive tract.

Urinary Tract

The highest levels of β-defensin mRNA in any human tissue are that of HBD1 in the urinary tract. Specifically, HBD1 is expressed in the loops of Henle, the distal tubules, and the collecting ducts of the kidney (Schnapp, Reid, and Harris, 1998; Valore et al., 1998). HBD2 is expressed in the same tissues. However, as with other epithelia, this peptide is only detectable in infected or inflamed tissue (Lehmann et al., 2002; Nitschke et al., 2002). In vitro induction of HBD2 can be observed with IL-1α or bacteria, and to a lesser extent with TNF-α or *E. coli* LPS (Nitschke et al., 2002).

Concentrations of HBD1 in urine range from 10 to 100 μg/ml (Valore et al., 1998), which, although below the minimal inhibitory concentration necessary for antibiotic activity, may reflect higher localized levels at mucosal surfaces. Six isoforms of HBD1 with variable NH_2-terminal truncations have been isolated from urine (Valore et al., 1998). Male urine predominantly contains one 44-amino-acid HBD1 peptide, whereas two isoforms, including the 44-residue peptide, are the predominant isoforms found in both female urine and vaginal secretions.

Furthermore, HBD1 levels in urine are increased during pregnancy. Serum and urine levels of HBD1 peptide increase threefold in patients with pyelonephritis, whereas levels of the rat homolog (rBD1) were lower in a rat model of diabetes (Hiratsuka et al., 2001). This expression follows the general pattern for a tissue that is sterile and provides a crucial barrier to environmental pathogens. Support for this role in urinary tract defense is seen in the mBD1 knockout mouse model, in which deficiency of the murine homolog of HBD1, mBD1, resulted in high levels of *Staphylococcus* spp. in the urine (Morrison et al., 2002).

5.2.5. Other epithelia

Eye

The eyes are under constant barrage by environmental pathogens, but are resistant to infection because of multiple defense mechanisms, both on the ocular surface and within the aqueous humor of the eye. The ocular surface comprises the conjunctiva and the corneal epithelium. At this site, HBD1 is constitutively expressed, whereas HBD2 is expressed both in uncompromised tissue and inflamed or bacterially challenged samples (Gottsch et al., 1998; Hattenbach, Gumbel, and Kippenberger, 1998; Haynes, Tighe, and Dua, 1998; McNamara et al., 1999; Lehmann, Hussain, and Watt, 2000). A similar profile of expression is seen in the conjunctiva of rhesus monkeys and cows (Schonwetter et al., 1995; Bals et al., 2001). This shows that the eye has a particularly high basal level of defensin expression, as seen in the oral cavity.

Furthermore, both partially and fully regenerated corneal epithelia show a large increase in HBD2 peptide when compared with presurgical levels (McDermott, Redfern, and Zhang, 2001), indicating a mechanism for responding to mechanical injury. Bacterial access to the intraocular region is limited; only HBD1 has been found, specifically in the vitreous and aqueous humors (Haynes et al., 2000) and the tissues of the iris and the lens (Lehmann et al., 2000). In the germ-free nasolacrimal ducts, the expression of HBD1 and HBD2 (Paulsen et al., 2001) is the same as seen in other sterile mucosal surfaces, such as the respiratory tract.

Ear

There are three bacterial requirements for successful colonization of the middle ear, leading to otitis media. The pathogen must first adhere to the nasopharyngeal epithelium, then enter and traverse the eustachian tube to the middle ear, and overcome host defense in the middle ear. Moon et al. (2002) have shown that β-defensin 2 plays a role in the protection of the normally sterile middle ear. HBD2 mRNA and peptide are found in the middle-ear mucosa only in patients with otitis media, not in healthy subjects. In a human middle-ear cell line, HBD2 is upregulated to the greatest degree by IL-1α and TNF-α, but LPS also causes upregulation of this response. These results are mirrored in experiments with rats, in which middle-ear inoculation of IL-1α and LPS are used to stimulate an rBD2 response.

The inhibitors used in gingival epithelial cells to ascertain the signaling pathway leading to defensin upregulation in the oral cavity (Krisanaprakornkit et al., 2002) were also employed in the human middle-ear cell line (Moon et al., 2002). Here, the ERK inhibitor decreased the IL-1α-induced hBD-2 response in a dose-dependent manner, causing up to 70% inhibition. Results were comparable when the JNK inhibitor was used; however, the p38 inhibitor had no effect. Defensin expression has also been seen in the outer ear. Keratinocytes of the tympanic membrane and the external auditory canal express HBD1 in normal subjects (Bøe et al., 1999).

5.3. SUMMARY

The expression of β-defensins follows a paradigm whereby HBD1 is generally expressed constitutively, possibly to provide an initial line of defense against bacterial growth in sterile tissues and homeostasis in colonized tissues. In cases in which the tissue is relatively sterile, such as the airway or the middle ear, simple contact with bacteria or bacterial products such as LPS is sufficient to stimulate high-level expression of HBD2 to kill those microbes that survive the constitutive defense. Other tissues, such as the oral mucosa or the intestines, which are in constant contact with nonpathogenic microorganisms, have developed a more complex method for antimicrobial peptide-based defense. In these cases, posttranslational regulation may be the mechanism to maintain homeostatic levels of bacteria, whereas the upregulation of HBD2 appears to be a result of an early inflammatory response to pathogenic colonization.

Furthermore, the use of multiple signal transduction pathways may signify a method to selectively target a response to microbes with pathogenic potential. The result of the varied regulatory mechanisms may be the selective production of a peptide with activity against only the pathogen. The evolution of the diverse β-defensin family members, combined with the different, tightly controlled regulatory pathways, has produced a unique host defense system, which belies the "primitive" description of innate immunity and peptide-based host defense systems.

REFERENCES

Agerberth, B., Grunewald, J., Castanos, V. E., Olsson, B., H, J., Wigzell, H., Eklund, A., and Gudmundsson, G. H. (1999). Antibacterial components in bronchoalveolar lavage fluid from healthy individuals and sarcoidosis patients. *American Journal of Respiratory and Critical Care Medicine*, 160, 283–90.

Ali, R. S., Falconer, A., Ikram, M., Bissett, C. E., Cerio, R., and Quinn, A. G. (2001). Expression of the peptide antibiotics human β defensin-1 and human β defensin-2 in normal human skin. *Journal of Investigative Dermatology*, 117, 106–11.

Ashitani, J., Mukae, H., Hiratsuka, T., Nakazato, M., Kumamoto, K., and Matsukura, S. (2001). Plasma and BAL fluid concentrations of antimicrobial peptides in patients with *Mycobacterium avium*-intracellulare infection. *Chest*, 119, 1131–7.

Bajaj-Elliott, M., Fedeli, P., Smith, G. V., Domizio, P., Maher, L., Ali, R. S., Quinn, A. G., and Farthing, M. J. (2002). Modulation of host antimicrobial peptide (β-defensins 1 and 2) expression during gastritis. *Gut*, 51, 356–61.

Bals, R., Goldman, M. J., and Wilson, J. M. (1998a). Mouse β-defensin 1 is a salt-sensitive antimicrobial peptide present in epithelia of the lung and urogenital tract. *Infection and Immunity*, 66, 1225–32.

Bals, R., Lang, C., Weiner, D. J., Vogelmeier, C., Welsch, U., and Wilson, J. M. (2001). Rhesus monkey (*Macaca mulatta*) mucosal antimicrobial peptides are close homologues of human molecules. *Clinical and Diagnostic Laboratory Immunology*, 8, 370–5.

Bals, R., Wang, X., Wu, Z., Freeman, T., Bafna, V., Zasloff, M., and Wilson, J. M. (1998b). Human β-defensin 2 is a salt-sensitive peptide antibiotic expressed in human lung. *Journal of Clinical Investigation*, 102, 874–80.

Bayliss, R., Clarke, C., Oakley, C. M., Somerville, W., Whitfield, A. G., and Young, S. E. (1986). Incidence, mortality and prevention of infective endocarditis. *Journal of the Royal College of Physicians London*, 20, 15–20.

Becker, M., Diamond, G., Verghese, M., and Randell, S. H. (2000). CD14-dependent LPS-induced β-defensin expression in human tracheobronchial epithelium. *Journal of Biological Chemistry*, 275, 29731–6.

Bensch, K. W., Raida, M., Magert, H.-J., Schulz-Knappe, P., and Forssmann, W.-G. (1995). hBD-1: a novel beta-defensin from human plasma. *FEBS Letters*, 368, 331–5.

Bevins, C. L., Jones, D. E., Dutra, A., Schaffzin, J., and Muenke, M. (1996). Human enteric defensin genes: Chromosomal map position and a model for possible evolutionary relationships. *Genomics*, 31, 95–106.

Bøe, R., Silvola, J., Yang, J., Moens, U., McCray, Jr., P. B., Stenfors, L. E., and Seljfelid, R. (1999). Human β-defensin-1 mRNA is transcribed in tympanic membrane and adjacent auditory canal epithelium. *Infection and Immunity*, 67, 4843–6.

Bonass, W. A., High, A. S., Owen, P. J., and Devine, D. A. (1999). Expression of β-defensin genes by human salivary glands. *Oral Microbiology and Immunology*, 14, 371–4.

Boniotto, M., Tossi, A., and Crovella, S. (2002). β-defensin 2 in the rhesus monkey (*Macaca mulatta*) and the long-tailed macaque (*M. fascicularis*). *Clinical and Diagnostic Laboratory Immunology*, 9, 503–4.

Burd, R. S., Furrer, J. L., Sullivan, J., and Smith, A. L. (2002). Murine β-defensin-3 is an inducible peptide with limited tissue expression and broad-spectrum antimicrobial activity. *Shock*, 18, 461–4.

Cobb, M. H. and Goldsmith, E. J. (1995). How MAP kinases are regulated. *Journal of Biological Chemistry*, 270, 14843–6.

Dale, B. A. and Krisanaprakornkit, S. (2001). Defensin antimicrobial peptides in the oral cavity. *Journal of Oral Pathology and Medicine*, 30, 321–7.

Diamond, G., Kaiser, V., Rhodes, J., Russell, J. P., and Bevins, C. L. (2000). Transcriptional regulation of β-defensin gene expression in tracheal epithelial cells. *Infection and Immunity*, 68, 113–9.

Diamond, G., Russell, J. P., and Bevins, C. L. (1996). Inducible expression of an antibiotic peptide gene in lipopolysaccharide-challenged tracheal epithelial cells. *Proceedings of the National Academy of Sciences USA*, 93, 5156–60.

Diamond, G., Zasloff, M., Eck, H., Brasseur, M., Maloy, W. L., and Bevins, C. L. (1991). Tracheal antimicrobial peptide, a novel cysteine-rich peptide from mammalian tracheal mucosa: Peptide isolation and cloning of a cDNA. *Proceedings of the National Academy of Sciences USA*, 88, 3952–6.

Duits, L. A., Rademaker, M., Ravensbergen, B., van Sterkenburg, M. A., van Strijen, E., Hiemstra, P. S., and Nibbering, P. H. (2001). Inhibition of hBD-3, but not hBD-1 and hBD-2, mRNA expression by corticosteroids. *Biochemistry Biophysics Research Communications*, 280, 522–5.

Dunn, B. E., Cohen, H., and Blaser, M. J. (1997). *Helicobacter pylori*. *Clinical Microbiology Reviews*, 10, 720–41.

Dunsche, A., Acil, Y., Dommisch, H., Siebert, R., Schroder, J. M., and Jepsen, S. (2002). The novel human β-defensin-3 is widely expressed in oral tissues. *European Journal of Oral Science*, 110, 121–4.

Dunsche, A., Acil, Y., Siebert, R., Harder, J., Schroder, J. M., and Jepsen, S. (2001). Expression profile of human defensins and antimicrobial proteins in oral tissues. *Journal of Oral Pathology and Medicine*, 30, 154–8.

Elewaut, D., DiDonato, J. A., Kim, J. M., Truong, F., Eckmann, L., and Kagnoff, M. F. (1999). NF-κ B is a central regulator of the intestinal epithelial cell innate immune response induced by infection with enteroinvasive bacteria. *Journal of Immunology*, 163, 1457–66.

Garcia, J. R., Jaumann, F., Schulz, S., Krause, A., Rodriguez-Jimenez, J., Forssmann, U., Adermann, K., Kluver, E., Vogelmeier, C., Becker, D., Hedrich, R., Forssmann, W. G., and Bals, R. (2001a). Identification of a novel, multifunctional β-defensin (human β- defensin 3) with specific antimicrobial activity. Its interaction with plasma membranes of *Xenopus* oocytes and the induction of macrophage chemoattraction. *Cell and Tissue Research*, 306, 257– 64.

Garcia, J. R., Krause, A., Schulz, S., Rodrigues-Jimenez, F. J., Kluver, E., Adermann, K., Forssmann, U., Frimpong-Boateng, A., Bals, R., and Forssmann, W. G. (2001b). Human β-defensin 4: A novel inducible peptide with a salt-sensitive spectrum of antimicrobial activity. *FASEB Journal*, 15, 1819–21.

Gendron, R., Grenier, D., and Maheu-Robert, L. (2000). The oral cavity as a reservoir of bacterial pathogens for focal infections. *Microbes and Infection*, 2, 897–906.

Goldman, M. J., Anderson, G. M., Stolzenberg, E. D., Kari, U. P., Zasloff, M., and Wilson, J. M. (1997). Human β-defensin-1 is a salt-sensitive antibiotic in lung that is inactivated in cystic fibrosis. *Cell*, 88, 553–60.

Gottsch, J. D., Li, Q., Ashraf, M. F., O'Brien, T. P., Stark, W. J., and Liu, S. H. (1998). Defensin gene expression in the cornea. *Current Eye Research*, 17, 1082–6.

Hamanaka, Y., Nakashima, M., Wada, A., Ito, M., Kurazono, H., Hojo, H., Nakahara, Y., Kohno, S., Hirayama, T., and Sekine, I. (2001). Expression of human β-defensin 2 (hBD-2) in *Helicobacter pylori* induced gastritis: Antibacterial effect of hBD-2 against *Helicobacter pylori*. *Gut*, 49, 481–7.

Harder, J., Bartels, J., Christophers, E., and Schroder, J.-M. (1997). A peptide antibiotic from human skin. *Nature* (London), 387, 861.

Harder, J., Bartels, J., Christophers, E., and Schroder, J. M. (2001). Isolation and characterization of human β-defensin 3, a novel human inducible peptide antibiotic. *Journal of Biological Chemistry*, 276, 5707–13.

Harder, J., Meyer-Hoffert, U., Teran, L. M., Schwichtenberg, L., Bartels, J., Maune, S., and Schroder, J. M. (2000). Mucoid *Pseudomonas aeruginosa*, TNF-α, and IL-1β, but not IL-6, induce human β-defensin-2 in respiratory epithelia. *American Journal of Respiratory Cell and Molecular Biology*, 22, 714–21.

Hattenbach, L. O., Gumbel, H., and Kippenberger, S. (1998). Identification of β-defensins in human conjunctiva. *Antimicrobial Agents and Chemotherapy*, 42, 3332.

Haynes, R. J., McElveen, J. E., Dua, H. S., Tighe, P. J., and Liversidge, J. (2000). Expression of human β-defensins in intraocular tissues. *Investigative Ophthalmology and Visual Science*, 41, 3026–31.

Haynes, R. J., Tighe, P. J., and Dua, H. S. (1998). Innate defense of the eye by antimicrobial defensin peptides [letter]. *The Lancet*, 352(9126), 451–2.

Hill, G. B. (1998). Preterm birth: Associations with genital and possibly oral microflora. *Annals of Periodontology*, 3, 222–32.

Hiratsuka, T., Nakazato, M., Date, Y., Mukae, H., and Matsukura, S. (2001). Nucleotide sequence and expression of rat β-defensin-1: Its significance in diabetic rodent models. *Nephron*, 88, 65–70.

Hiratsuka, T., Nakazato, M., Ihi, T., Minematsu, T., Chino, N., Nakanishi, T., Shimizu, A., Kangawa, K., and Matsukura, S. (2000). Structural analysis of human β-defensin-1 and its significance in urinary tract infection. *Nephron*, 85, 34–40.

Jia, H. P., Mills, J. N., Barahmand-Pour, F., Nishimura, D., Mallampali, R. K., Wang, G., Wiles, K., Tack, B. F., Bevins, C. L., and McCray Jr., P. B. (1999). Molecular cloning and characterization of rat genes encoding homologues of human β-defensins. *Infection and Immunity*, 67, 4827–33.

Jia, H. P., Schutte, B. C., Schundy, A., Linzmeier, R., Guthmiller, J. M., Johnson, G. K., Tack, B. F., Mitros, J. P., Rosenthal, A., Ganz, T., and McCray Jr., P. B. (2001). Discovery of new human β-defensins using a genomics-based approach. *Gene*, 263, 211–8.

Jia, H. P., Wowk, S. A., Schutte, B. C., Lee, S. K., Vivado, A., Tack, B. F., Bevins, C. L., and McCray Jr., P. B. (2000). A novel murine β-defensin expressed in tongue, esophagus, and trachea. *Journal of Biological Chemistry*, 275, 33314–20.

Jurevic, R. J., Bai, M., Chadwick, R. B., White, T. C., and Dale, B. A. (2003). Single-nucleotide polymorphisms (SNPs) in human β-defensin 1: High-throughput SNP assays and association with *Candida* carriage in type I diabetics and nondiabetic controls. *Journal of Clinical Microbiology*, 41, 90–6.

King, A. E., Fleming, D. C., Critchley, H. O., and Kelly, R. W. (2002). Regulation of natural antibiotic expression by inflammatory mediators and mimics of infection in human endometrial epithelial cells. *Molecular Human Reproduction*, 8, 341–9.

Komiyama, K., Tynan, J. J., Habbick, B. F., Duncan, D. E., and Liepert, D. J. (1985). *Pseudomonas aeruginosa* in the oral cavity and sputum of patients with cystic fibrosis. *Oral Surgery, Oral Medicine, Oral Pathology, Oral Radiology and Endodontics*, 59, 590–4.

Krisanaprakornkit, S., Kimball, J. R., and Dale, B. A. (2002). Regulation of human β-defensin-2 in gingival epithelial cells: The involvement of mitogen-activated protein kinase pathways, but not the NF-κB transcription factor family. *Journal of Immunology*, 168, 316–24.

Krisanaprakornkit, S., Kimball, J. R., Weinberg, A., Darveau, R. P., Bainbridge, B. W., and Dale, B. A. (2000). Inducible expression of human β-defensin 2 by *Fusobacterium nucleatum* in oral epithelial cells: Multiple signaling pathways and role of commensal bacteria in innate immunity and the epithelial barrier. *Infection and Immunity*, 68, 2907–15.

Krisanaprakornkit, S., Weinberg, A., Perez, C. N., and Dale, B. A. (1998). Expression of the peptide antibiotic human β-defensin 1 in cultured gingival epithelial cells and gingival tissue. *Infection and Immunity*, 66, 4222–8.

Kyriakis, J. M. and Avruch, J. (1996). Sounding the alarm: Protein kinase cascades activated by stress and inflammation. *Journal of Biological Chemistry*, 271, 24313–6.

Lehmann, J., Retz, M., Harder, J., Krams, M., Kellner, U., Hartmann, J., Hohgrawe, K., Raffenberg, U., Gerber, M., Loch, T., Weichert-Jacobsen, K., and Stockle, M. (2002). Expression of human β-defensins 1 and 2 in kidneys with chronic bacterial infection. *BioMed Central (BMC) Infectious Diseases*, 2, 20.

Lehmann, O. J., Hussain, I. R., and Watt, P. J. (2000). Investigation of β-defensin gene expression in the ocular anterior segment by semiquantitative RT-PCR. *British Journal of Ophthalmology*, 84, 523–6.

Lehrer, R. I. and Ganz, T. (1990). Antimicrobial polypeptides of human neutrophils. *Blood*, 76, 2169–81.

Li, P., Chan, H. C., He, B., So, S. C., Chung, Y. W., Shang, Q., Zhang, Y. D., and Zhang, Y. L. (2001). An antimicrobial peptide gene found in the male reproductive system of rats. *Science*, 291, 1783–5.

Liu, L., Wang, L., Jia, H. P., Zhao, C., Heng, H. H. Q., Schutte, B. C., McCray, P. B., and Ganz, T. (1998). Structure and mapping of the human β-defensin HBD-2 gene and its expression at sites of inflammation. *Gene*, 222, 237–44.

Liu, L., Zhao, C., Heng, H. H. Q., and Ganz, T. (1997). The human β-defensin-1 and α-defensins are encoded by adjacent genes: Two peptide families with differing disulfide topology share a common ancestry. *Genomics*, 43, 316–20.

Mathews, M., Jia, H. P., Guthmiller, J. M., Losh, G., Graham, S., Johnson, G. K., Tack, B. F., and McCray Jr., P. B. (1999). Production of β-defensin antimicrobial peptides by the oral mucosa and salivary glands. *Infection and Immunity*, 67, 2740–5.

McCray Jr., P. B. and Bentley, L. (1997). Human airway epithelia express a β-defensin. *American Journal of Respiratory Cell and Molecular Biology*, 16, 343–9.

McDermott, A. M., Redfern, R. L., and Zhang, B. (2001). Human β-defensin 2 is up-regulated during re-epithelialization of the cornea. *Current Eye Research*, 22, 64–7.

McNamara, N., Van, R., Tuchin, O. S., and Fleiszig, S. M. (1999). Ocular surface epithelia express mRNA for human β defensin-2. *Experimental Eye Research*, 69, 483–90.

Moon, S. K., Lee, H. Y., Li, J. D., Nagura, M., Kang, S. H., Chun, Y. M., Linthicum, F. H., Ganz, T., Andalibi, A., and Lim, D. J. (2002). Activation of a Src-dependent Raf-MEK1/2-ERK signaling pathway is required for IL-1α-induced upregulation of β-defensin 2 in human middle ear epithelial cells. *Biochimica et Biophysica Acta*, 1590, 41–51.

Morrison, G., Kilanowski, F., Davidson, D., and Dorin, J. (2002). Characterization of the mouse β defensin 1, Defb1, mutant mouse model. *Infection and Immunity*, 70, 3053–60.

Morrison, G. M., Davidson, D. J., and Dorin, J. R. (1999). A novel mouse β-defensin, Defb2, which is upregulated in the airways by lipopolysaccharide. *FEBS Letters.*, 442, 112–6.

Moser, C., Weiner, D. J., Lysenko, E., Bals, R., Weiser, J. N., and Wilson, J. M. (2002). β-Defensin 1 contributes to pulmonary innate immunity in mice. *Infection and Immunity*, 70, 3068–72.

Nitschke, M., Wiehl, S., Baer, P. C., and Kreft, B. (2002). Bactericidal activity of renal tubular cells: The putative role of human β-defensins. *Experimental Nephrology*, 10, 332–7.

O'Neil, D. A., Cole, S. P., Martin-Porter, E., Housley, M. P., Liu, L., Ganz, T., and Kagnoff, M. F. (2000). Regulation of human β-defensins by gastric epithelial cells in response to infection with *Helicobacter pylori* or stimulation with interleukin-1. *Infection and Immunity*, 68, 5412–5.

O'Neil, D. A., Porter, E. M., Elewaut, D., Anderson, G. M., Eckmann, L., Ganz, T., and Kagnoff, M. F. (1999). Expression and regulation of the human β-defensins hBD-1 and hBD-2 in intestinal epithelium. *Journal of Immunology*, 163, 6718–24.

Ogushi, K., Wada, A., Niidome, T., Mori, N., Oishi, K., Nagatake, T., Takahashi, A., Asakura, H., Makino, S., Hojo, H., Nakahara, Y., Ohsaki, M., Hatakeyama, T., Aoyagi, H., Kurazono, H., Moss, J., and Hirayama, T. (2001). *Salmonella enteritidis* FliC (flagella filament protein) induces human β-defensin-2 mRNA production by Caco-2 cells. *Journal of Biological Chemistry*, 276, 30521–6.

Ouellette, A. J., Greco, R. M., James, M., Frederick, D., Naftilan, J., and Fallon, J. T. (1989). Developmental regulation of cryptdin, a corticostatin/defensin precursor mRNA in mouse small intestinal crypt epithelium. *Journal of Cell Biology*, 108, 1687–95.

Palladino, M. A., Mallonga, T. A., and Mishra, M. S. (2003). Messenger RNA (mRNA) expression for the antimicrobial peptides β-defensin-1 and β-defensin-2 in the male rat reproductive tract: β-defensin-1 mRNA in initial segment and caput epididymidis is regulated by androgens and not bacterial lipopolysaccharides. *Biology of Reproduction*, 68, 509–15.

Pallasch, T. J. and Slots, J. (1996). Antibiotic prophylaxis and the medically compromised patient. *Periodontology 2000*, 10, 107–38.

Paulsen, F. P., Pufe, T., Schaudig, U., Held-Feindt, J., Lehmann, J., Schroder, J. M., and Tillmann, B. N. (2001). Detection of natural peptide antibiotics in human nasolacrimal ducts. *Investigative Ophthalmology and Visual Science*, 42, 2157–63.

Quayle, A. J. (2002). The innate and early immune response to pathogen challenge in the female genital tract and the pivotal role of epithelial cells. *Journal of Reproductive Immunology*, 57, 61–79.

Quayle, A. J., Martin Porter, E., Nussbaum, A. A., Wang, Y.-M., Brabec, C., and Mok, S. C. (1998). Gene expression, immunolocalization and secretion of human defensin-5 in human female genital tract. *American Journal of Pathology*, 152, 1247–58.

Russell, J. P., Diamond, G., Tarver, A. P., Scanlin, T. F., and Bevins, C. L. (1996). Coordinate induction of two antibiotic genes in tracheal epithelial cells exposed to the inflammatory mediators lipopolysaccharide and tumor necrosis factor-α. *Infection and Immunity*, 64, 1565–68.

Sahasrabudhe, K. S., Kimball, J. R., Morton, T. H., Weinberg, A., and Dale, B. A. (2000). Expression of the antimicrobial peptide, human β-defensin 1, in duct cells of minor salivary glands and detection in saliva. *Journal of Dental Research*, 79, 1669–74.

Schaller-Bals, S., Schulze, A., and Bals, R. (2002). Increased levels of antimicrobial peptides in tracheal aspirates of newborn infants during infection. *American Journal of Respiratory and Critical Care Medicine*, 165, 992–5.

Schnapp, D., Reid, C. J., and Harris, A. (1998). Localization of expression of human β defensin-1 in the pancreas and kidney. *Journal of Pathology*, 186, 99–103.

Schonwetter, B. S., Stolzenberg, E. D., and Zasloff, M. A. (1995). Epithelial antibiotics induced at sites of inflammation. *Science*, 267, 1645–8.

Schutte, B. C., Mitros, J. P., Bartlett, J. A., Walters, J. D., Jia, H. P., Welsh, M. J., Casavant, T. L., and McCray Jr., P. B. (2002). Discovery of five conserved β-defensin gene clusters using a computational search strategy. *Proceedings of the National Academy of Sciences USA*, 99, 2129–33.

Seger, R. and Krebs, E. G. (1995). The MAPK signaling cascade. *FASEB Journal*, 9, 726–35.

Selsted, M. E., Tang, Y.-Q., Morris, W. L., McGuire, P. A., Novotny, M. J., Smith, W., Henschen, A. H., and Cullor, J. S. (1993). Purification, primary structures, and antibacterial activities of β-defensins, a new family of antimicrobial peptides from bovine neutrophils. *Journal of Biological Chemistry*, 268, 6641–8.

Singh, P. K., Jia, H. P., Wiles, K., Hesselberth, J., Liu, L., Conway, B. A. D., Greenberg, E. P., Valore, E. V., Welsh, M. J., Ganz, T., Tack, B. F., and McCray Jr., P. B. (1998). Production of β-defensins by human airway epithelia. *Proceedings of the National Academy of Sciences USA*, 95, 14961–6.

Singh, P. K., Tack, B. F., McCray Jr., P. B., and Welsh, M. J. (2000). Synergistic and additive killing by antimicrobial factors found in human airway surface liquid. *American Journal of Physiology, Lung Cell and Molecular Physiology*, 279, L799–805.

Smith, J. J., Travis, S. M., Greenberg, E. P., and Welsh, M. J. (1996). Cystic fibrosis airway epithelia fail to kill bacteria because of abnormal airway surface fluid. *Cell*, 85, 229–36.

Su, B. and Karin, M. (1996). Mitogen-activated protein kinase cascades and regulation of gene expression. *Current Opinion in Immunology*, 8, 402–11.

Takahashi, A., Wada, A., Ogushi, K., Maeda, K., Kawahara, T., Mawatari, K., Kurazono, H., Moss, J., Hirayama, T., and Nakaya, Y. (2001). Production of β-defensin-2 by human colonic epithelial cells induced by *Salmonella enteritidis* flagella filament structural protein. *FEBS Letters*, 508, 484–8.

Tarver, A. P., Clark, D. P., Diamond, G., Russell, J. P., Erdjument-Bromage, H., Tempst, P., Cohen, K. S., Jones, D. E., Sweeney, R. W., Wines, M., Hwang, S., and Bevins, C. L. (1998). Enteric β-defensin: Molecular cloning and characterization of a gene with inducible intestinal epithelial cell expression associated with *Cryptosporidium parvum* infection. *Infection and Immunity*, 66, 1045–56.

Uehara, N., Yagihashi, A., Kondoh, K., Tsuji, N., Fujita, T., Hamada, H., and Watanabe, N. (2003). Human β-defensin-2 induction in *Helicobacter pylori*-infected gastric mucosal tissues: Antimicrobial effect of overexpression. *Journal of Medical Microbiology*, 52, 41–5.

Valore, E. V., Park, C. H., Quayle, A. J., Wiles, K. R., McCray Jr., P. B., and Ganz, T. (1998). Human β-defensin-1: An antimicrobial peptide of urogenital tissues. *Journal of Clinical Investigation*, 101, 1633–42.

van Winkelhoff, A. J., Abbas, F., Pavicic, M. J., and de Graaff, J. (1991). Chronic conjunctivitis caused by oral anaerobes and effectively treated with systemic metronidazole plus amoxicillin. *Journal of Clinical Microbiology*, 29, 723–5.

Wada, A., Mori, N., Oishi, K., Hojo, H., Nakahara, Y., Hamanaka, Y., Nagashima, M., Sekine, I., Ogushi, K., Niidome, T., Nagatake, T., Moss, J., and Hirayama, T. (1999). Induction of human β-defensin-2 mRNA expression by

Helicobacter pylori in human gastric cell line MKN45 cells on cag pathogenicity island. *Biochemistry Biophysics Research Communications*, 263, 770–4.

Wada, A., Ogushi, K., Kimura, T., Hojo, H., Mori, N., Suzuki, S., Kumatori, A., Se, M., Nakahara, Y., Nakamura, M., Moss, J., and Hirayama, T. (2001). *Helicobacter pylori*-mediated transcriptional regulation of the human β-defensin 2 gene requires NF-κB. *Cellular Microbiology*, 3, 115–23.

Yamaguchi, Y., Nagase, T., Makita, R., Fukuhara, S., Tomita, T., Tominaga, T., Kurihara, H., and Ouchi, Y. (2002). Identification of multiple novel epididymis-specific β-defensin isoforms in humans and mice. *Journal of Immunology*, 169, 2516–23.

Zhang, G., Wu, H., Shi, J., Ganz, T., Ross, C. R., and Blecha, F. (1998). Molecular cloning and tissue expression of porcine β-defensin-1. *FEBS Letters*, 424, 37–40.

Zhao, C., Nguyen, T., Liu, L., Shamova, O., Brogden, K., and Lehrer, R. I. (1999). Differential expression of caprine β-defensins in digestive and respiratory tissues. *Infection and Immunity*, 67, 6221–4.

Zhao, C., Wang, I., and Lehrer, R. I. (1996). Widespread expression of β-defensin hBD-1 in human secretory glands and epithelial cells. *FEBS Letters*, 396, 319–22.

CHAPTER 6

Biology and expression of the human cathelicidin LL-37

Gudmundur H. Gudmundsson and Birgitta Agerberth

6.1. INTRODUCTION

LL-37 is the single cathelicidin peptide in humans with an amphipathic α-helical structure. The initial activities for the peptide included bactericidal activity against both Gram-positive and Gram-negative bacteria and lipopolysaccharide- (LPS-) binding properties. Additional functions such as chemotaxis and immunomodulation have recently been detected for this peptide. The expression of LL-37 is widely distributed in the human body, with the most pronounced expression in polymorphonuclear blood cells, epithelial cells, and epididymis. B, natural-killer (NK), and γδT cells, in addition to monocytes/macrophages and mast cells, are also known to express LL-37. The cathelicidin antimicrobial peptide (CAMP) gene that encodes LL-37 is mapped to chromosome 3 (3p21.3). Detailed understanding of the regulation for the CAMP gene has not yet been established. During inflammatory skin disorders, LL-37 is induced in keratinocytes, most obviously in psoriatic lesions, whereas in atopic dermatitis the induction is less pronounced. Transcriptional upregulation has also been demonstrated in gut epithelia by short-chain fatty acid, and butyrate is the strongest inducer. This induction involves the MEK-MAP (mitogen activated protein) kinase/ERK (extracellular signal-regulated kinase) signal pathway. In contrast, LL-37 is downregulated in gut epithelia in severe infections such as *Shigella*, indicating that pathogens are able to rupture the constitutive barrier of antimicrobial peptides at mucosal surfaces.

In summary, LL-37 appears as a multifunctional peptide that is important in killing bacteria and neutralizing endotoxins, but also as a signaling molecule that attracts immune cells to infected sites and functions as a modulator for the total immune response.

6.2. CATHELICIDINS

Mammalian antimicrobial peptides are gene-encoded effector molecules of innate immunity that function as endogenous antibiotics. They are included in the first line of defense, induced or constitutively expressed by surface epithelia exposed to microbes. Furthermore, antimicrobial peptides or peptide antibiotics are an active part of the neutrophil armament, and neutrophils are the first cells recruited to a site of infection.

In mammals, two main families of antimicrobial peptides have been identified, the defensins and the cathelicidins. The defensins are divided into α-, β-, and the recently discovered circular θ-defensins (Tang et al., 1999; Lehrer and Ganz, 2002). The cathelicidins contain a highly conserved proregion called cathelin with a C-terminal antimicrobial domain (Zanetti, Gennaro, and Romeo, 1995), which is cleaved off by processing enzyme(s), liberating the antimicrobial peptide. Initially the cathelin protein was isolated from porcine leukocytes as a protease inhibitor, specifically as a cathepsin L inhibitor, hence the name cathelin (Ritonja et al., 1989). Through sequence information from the cDNA of bovine bactenecins (Zanetti et al., 1990) and the gene for the porcine antimicrobial peptide PR-39 (Gudmundsson et al., 1995), cathelin was clearly shown as a common proregion for structurally variable antimicrobial peptides.

The variation in the primary and secondary structures of the microbicidal peptides connected to cathelin is pronounced. They can fold into an amphipathic α-helical structure, as demonstrated by the human LL-37 (Agerberth et al., 1995), or form a loop with a β-sheet structure as the protegrins (Fahrner et al., 1996). An additional group of cathelicidins contain a high proportion of one or two special residues, Pro-Arg-rich peptides as in the porcine PR-39 (Agerberth et al., 1991) and the bovine bactenecins (Frank et al., 1990). The cow peptide indolicidin is another example with high tryptophan content, in which 5 residues out of 13 are tryptophan (Selsted et al., 1992). The mammalian cathelicidin gene structure is well conserved; the first three exons encode the signal sequence and the cathelin part, and the fourth exon encodes the processing site and the antimicrobial variable domain (Zanetti, Gennaro, and Romeo, 1995). Additional variation regarding this gene family concerns the number of cathelicidin genes. In pig and cow, there are several cathelicidin encoding genes, at least ten genes in the porcine genome, giving different peptides, and at least seven in the bovine genome, whereas in human, mouse and rat, only one cathelicidin gene is present.

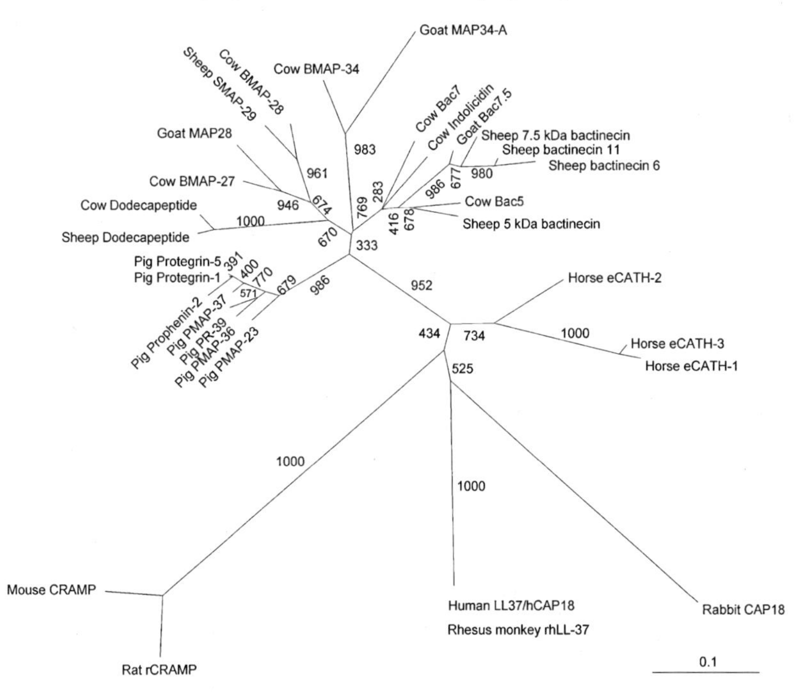

Figure 6.1. Unrooted neighbor-joining phylogenetic tree made from the alignment, bootstrapped 1,000 times. Many different trees can be constructed from the same alignment, and the bootstrap value at each branch represents in how many out of 1,000 constructed trees the branch places itself there. The higher the value, the more probable that true phylogeny has been found. Neighbor joining is a distance-based method, and, with the given scale that indicates 0.1 amino acid substitutions per site, the close relationship of the cathelicidins can be seen. The tree was made with the Phylip package version 3.5c and presented by TreeView 1.6.5 (Termén et al., 2003). (Reproduced with permission.)

A phylogenetic tree has been recently constructed according to the preproregion of the characterized cathelicidins (Termén et al., 2003) (Fig. 6.1). The horse cathelicidins group together in one branch, as do the pig cathelicidins, whereas cathelicidins of cow, sheep, and goat cluster together on a separate branch (Termén et al., 2003). This would suggest that gene duplication events followed by exon shuffling of exon four, encoding the antimicrobial domain, are species or lineage specific through evolution. In accordance with the lineage relation, primate sequences group together, with the human cathelin (linked to LL-37) (Gudmundsson et al., 1996) and the rhesus monkey cathelin (linked to RL-37) (Zhao et al., 2001) demonstrating 93.3% identity. Their closest characterized relative is the rabbit form [cationic antimicrobial protein of 18 kDa (CAP18)] and then the mouse cathelin-related antimicrobial peptide (CRAMP) and the rat rCRAMP (Termén et al., 2003). When only one cathelicidin is present, as in human, mouse, and rat, its peptide domain is of the amphipathic α-helical structure found in LL-37. It is conceivable that this conservation is linked to another function, such as receptor binding.

Considering evolutionary events, it is of interest to look at the cathelicidin variation relative to the radiation of the defensin family. In some mammalian species there has been a spatial replacement in expression of the two families. Human neutrophils are loaded with several α-defensins but have only one cathelicidin. In contrast, porcine neutrophils contain several cathelicidins, but no defensins have been detected. Detailed relationships between the two families in terms of numbers of genes and cell-specific expression is not possible at present, but in the future will certainly be one of several interesting investigations in this growing research field.

In summary, gene duplicatory events and exon shuffling have been selected in the mammalian lineages, resulting in a highly variable antimicrobial armament. The reason for this species variation may reflect an adaptive value for the host to be able to cope with rapid changes of microbial targets in the environment.

6.2.1. Cathelin

The cathelin protein was first isolated and characterized as a cathepsin L inhibitor from porcine leukocytes (Ritonja et al., 1989). A further support for its antiprotease assignment was derived from the fact that structural alignment showed it to be related to the cystatin superfamily of thiol protease inhibitors (Zaiou and Gallo, 2002). Therefore it was unexpected when the claimed protease-inhibiting activity of cathelin was retracted and found to be due to a contamination (Lenarcic et al., 1993). However, a subsequent study on the cathelin function again indicated protease inhibitory effects (Verbanac, Zanetti, and Romeo, 1993). Interestingly, a recent study showed that the cathelin domain can act as a cysteine protease inhibitor that also exhibits antibacterial activity distinct from the activity of the peptide LL-37. Thus it is conceivable that the cathelin proregion and the active peptide LL-37 should be viewed as complementary entities in innate defenses (Zaiou, Nizet, and Gallo, 2003). The three-dimensional structure of the porcine protegrin-3 cathelin has recently been resolved (Sanchez et al., 2002), and further studies can now be performed to elucidate additional functions of this well-conserved protein.

6.3. THE HUMAN CATHELICIDIN LL-37

6.3.1. Discovery

Our approach in the initial cloning of a cDNA encoding a human cathelicidin was based on a polymerase chain reaction (PCR) strategy, which used primers that covered conserved cathelin regions. The amplified fragment was sequenced and utilized as a probe to screen a human bone marrow library. Several clones were obtained and sequenced, all encoding cathelin linked to a putative microbicidal peptide. A potential dibasic processing site was identified, and a putative peptide was synthesized, which exhibited salt-dependent antibacterial activity (Agerberth et al., 1995). Antibodies were raised against this synthetic version, which were important tools in the characterization of the active mature peptide LL-37 from granulocytes (Gudmundsson et al., 1996). A second approach was to find a human homolog of the LPS-binding protein CAP18, originally identified in the rabbit (Larrick et al., 1991). A low stringency screen of a human bone marrow library, with the CAP18 clone as a probe, resulted in the cDNA characterization of the human cathelicidin that was named hCAP18 (for human CAP18) (Larrick et al., 1995a). Subsequently, Larrick and co-workers identified the LPS-binding region of hCAP18 and demonstrated that it coincided with the microbicidal part, that is, peptide LL-37 (Larrick et al., 1995b). A third approach aimed at identifying unique proteins in the specific granules of neutrophils. Information from the amino acid sequencing of one such protein was used to design primers, which were utilized to clone the corresponding cDNA from a chronic myeloid leukemia library and led to the characterization of hCAP18, the precursor protein of LL-37 (Cowland, Johnsen, and Borregaard, 1995).

Unfortunately, the terminology of this human cathelicidin has been confusing. Originally, we named the putative peptide FALL-39, a 39-residue peptide with Phe-Ala-Leu-Leu at the N-terminus. After isolation and characterization of the mature active peptide from granulocytes, we found the peptide to be two residues shorter at the N-terminus, and we then renamed it LL-37. The name hCAP18 refers to the whole precursor protein and is commonly used by several groups. When it is not possible or necessary to distinguish between the precursor protein and the microbicidal peptide LL-37, as in immunohistochemistry or expression at the mRNA level, the name hCAP18/LL-37 is logical.

6.3.2. Structure

Almost all antimicrobial peptides are cationic with an amphipathic character. An Edmundsson wheel prediction indicates that the putative peptide FALL-39 is amphipathic, with the basic residues at one side of the helix. In line with this observation, circular dicroism (CD) measurements demonstrated that an α-helical structure was induced by salt (Agerberth et al., 1995). In an extended study, it was confirmed that this secondary structure holds for the active peptide LL-37. A detailed investigation also demonstrated that the α-helical structure was required for optimal microbicidal activity and that divalent anions were the strongest inducers of the α-helical structure, also dependent on pH (Johansson et al., 1998). Two main mechanisms of action have been suggested for antimicrobial peptides: the barrel-stave mechanism, which results in pore formation in bacterial membranes, and the carpetlike mechanism, in which the membrane destruction/solubilization occurs by means of parallel binding of the peptides to the bacterial membrane in a carpetlike manner (Shai, 1995). Biophysical analyses of several α-helical antimicrobial peptides support their carpetlike mechanism, which also includes the mode of action for LL-37, both for negatively charged and zwitterionic membranes, mimicking bacterial and mammalian membranes, respectively (Oren et al., 1999). By Fourier transform infrared spectroscopy, it was found that the peptide LL-37 was oriented almost parallel also in zwitterionic membranes, supporting the carpetlike mechanism for the cytotoxic effect on host cells (see Section 6.4.3) (Oren et al., 1999).

6.3.3. Activities

LL-37 is a microbicidal peptide that can kill both Gram-positive and Gram-negative bacteria, although antifungal activity has not been detected. The most detailed study of the antimicrobial spectra for LL-37 has been performed by Lehrer's group (Turner et al., 1998). Furthermore, limited antiviral activity has been described for LL-37 against *Herpes simplex* virus type 1 and 2 (Yasin et al., 2000). It seems that LL-37 is a key player in the cutaneous protection, as the skin pathogen *Strep.* Group A is very sensitive to LL-37, with a minimum inhibitory concentration (MIC) value down to 1 μM or 4.5 μg/ml (Dorschner et al., 2001). In addition, the knockout mouse for the CRAMP gene, the homolog to the human CAMP gene encoding LL-37, is very sensitive to skin infections by *Strep.* Group A (Nizet et al., 2001). The infected lesions in the gene-deficient mice were shown to grow more rapidly, become larger, and to contain live bacteria for a longer time than infected lesions in normal littermates.

One of the original functional studies on LL-37 showed that the peptide bound various forms of LPS and could even protect mice that were injected with lethal quantities of LPS (Larrick et al., 1995a). A further support for the LPS-binding capacity was provided by studies on a truncated version of LL-37 (including 27 residues of LL-37 from F_6 to V_{32}). This peptide was shown to protect infected mice from antibiotic-induced endotoxin shock that was due to LPS-neutralizing activity rather than to antimicrobial activity (Kirikae et al., 1998). Furthermore, systemic overexpression of LL-37/hCAP18 from an adenoviral vector resulted in increased survival after injection of LPS and significantly decreased tumor necrosis factor-α (TNF-α) induction as compared with controls (Bals et al., 1999b). These results indicate that LL-37 not only contributes to bacterial clearance, but also has the capacity to protect against deleterious effects from microbial invasion.

Both Oppenheim's group and we have found that LL-37 exhibits chemotactic properties (Agerberth et al., 2000, Yang et al., 2000). In microchamber wells, we found that both polymorphonuclear cells and lymphocytes are attracted by LL-37, and CD4 T cells were identified as the subset of attracted lymphocytes (Agerberth et al., 2000). In Oppenheim's study the same cells and monocytes, cells that also mobilize Ca^{2+} after LL-37 stimulation were attracted by LL-37. Interestingly, the same group was also able to characterize the formyl peptide receptor-like 1 (FPRL1) as the receptor mediating this chemotacic activity for LL-37 (Yang et al., 2000). This receptor is known to be present on monocytes, neutrophils, and T cells (Yang, Chertov, and Oppenheim, 2001). Furthermore, it has been demonstrated that LL-37 can induce histamine release from rat mast cells (Niyonsaba et al., 2001). The facts that LL-37 affects the mobilization of intracellular Ca^{2+} in rat mast cells and that these cells can be attracted by LL-37 (Niyonsaba et al., 2002) indicate that mast cells are a functional target for LL-37.

In addition to chemotactic activation of immune cells, LL-37 has been shown to affect gene expression in a murine macrophage cell line (RAW 264.7). By analysis of gene expression arrays on LL-37 treatment, upregulated expression of several cytokine and chemokine genes was found (Scott et al., 2002). Downregulation of several genes was also observed. The experimental drawback was the heterogeneity of the system, that is, murine cells stimulated with a human peptide. However, similar effects were observed, when a homologous system was utilized. Thus, with a human epithelial cell line and whole human blood release of the chemokines MCP-1 and IL-8 after stimulation with LL-37 was measured. Therefore LL-37 seems to function also as a modulator of immune respones (Scott et al., 2002).

6.4. CELLULAR EXPRESSION

6.4.1. Polymorphonuclear Cells

According to Northern blot analysis, bone marrow is the main place for transcription of the gene encoding LL-37 (Agerberth et al., 1995). During differentiation of the hematopoetic lineage for granulocytes, the precursor of LL-37 (hCAP18) is mainly stored in specific granules of these cells (Sorensen et al., 1997a). Neutrophils are the cells that are recruited first to a site of infection, and in their granules several antimicrobial proteins and peptides, including hCAP18/LL-37, are stored. Upon stimulation of neutrophils, all these effectors are released to the exterior or in the phagolysosome, where the content from different granules meet and cooperate in the killing of microbes. The precursor protein hCAP18 is present in phogolysosomes, whereas the processed peptide LL-37 is not detectable. In contrast, hCAP18 is cleaved to generate LL-37 in exocytosed material (Sorensen et al., 2001). Thus LL-37 is a part of the bactericidal armament of exocytosis from neutrophils, where it works in synergy with other microbicidal components. In vitro, it has been demonstrated that LL-37 acts in synergy with lactoferrin and lysozyme (Bals et al., 1998, Singh et al., 2000). Notably, when α-defensins are released extracellularly, they are not able to function as antimicrobial components by themselves because of the salt concentration in plasma (Nagaoka et al., 2000). However, α-defensins can work synergistically with LL-37 to exert an antimicrobial activity in the extracellular milieu (Nagaoka et al., 2000).

In a recent study with the aim of evaluating the importance of LL-37 in the cutaneous defence of newborn infants, we demonstrated localization of LL-37 in EG2 expressing cells, that is, eosinophils, in lesions of the inflammatory disorder *Erythema toxicum neonatorum* (Marchini et al., 2002). This is the first observation of LL-37 in eosinophilic granulocytes, but CD1a expressing dendritic cells also colocalize with LL-37 in inflammatory lesions, as shown by confocal microscopic imaging (Marchini et al., 2002).

6.4.2. Mononuclear Cells

In the supernatant of IL-2 stimulated peripherial blood mononuclear cells (PBMCs), enriched for T and NK cells, we have isolated and characterized the α-defensins HNP1–HNP3 together with lysozyme, LL-37, and histone H2B (Agerberth et al., 2000). These results were confirmed by immunohistochemistry on PBMC double stained for LL-37 and the surface markers of B cells, NK cells, γδT cells, monocytes/macrophages, and CD3 T cells. All these subsets of mononuclear cells, except CD3 T cells, were found to contain LL-37. We also investigated different cell lines and/or clones originating from B, NK, αβT, and γδT cells and monocytes/macrophages for active transcription of LL-37. In agreement with the results from PBMC, all cell lines and clones except αβT cells expressed LL-37 as analyzed by RT-PCR (Agerberth et al., 2000).

Although the main function of LL-37 in the mononuclear cells is presumably bactericidal, such an activity has not been fully established. It was reported already in 1989 that NK cells and later that T cells were able to directly kill bacteria (Garcia-Penarrubia et al., 1989, Levitz, Mathews, and Murphy, 1995). However, the effector molecules mediating this activity were never characterized. Most likely, the activity originating from LL-37 and α-defensins is a part of the killing capacity of these cell types. However, a detailed analysis of the relative contribution from LL-37 and α-defensins to the antibacterial activity has not yet been performed. We have shown that LL-37 is liberated on stimulation with γ-Interferon (γ-IFN) and IL-6, and that the released material exhibited antibacterial activity (Agerberth et al., 2000).

6.4.3. Cytotoxicity and Scavenging

LL-37 is toxic also to mammalian cells, and the most sensitive cells analyzed are T cells, in which the lowest cytotoxic effect was detected at 12.5 μM, which is equal to 56.2 μg/ml (Johansson et al., 1998). Because the microbicidal effect of LL-37 is blocked in plasma by binding to apolipoprotein-A1 (Wang et al., 1998; Sorensen et al., 1999), blood cells are protected from the cytotoxic effect of LL-37. Furthermore, Borregaard's group reported that the precursor hCAP18 was found in plasma at 1.12 μg/ml (Sorensen et al., 1997b), which is a higher concentration than that of other proteins derived from the specific granules of neutrophils but well below the cytotoxic concentration. Proteinase 3 is the processing enzyme for the activation of LL-37 extracellularly, as shown by the same group (Sorensen et al., 2001). The signal for this activation is not known, but such a rapid activation by only one proteolytic cleavage, liberating a bactericidal effector or a LPS scavanger, must be an advantageous response for the host during infection. The role of Apo-A1 in this activation is unknown, as is the contribution of peptide LL-37 or the precursor hCAP18 from other blood cells. Notably, the chemotactic activity was not affected by serum at concentrations that inhibited the microbicidal activity (Yang et al., 2001), indicating that the receptor binding can be mediated in the presence of lipoproteins.

6.5. EXPRESSION IN EPITHELIAL CELLS

The epithelial expression of LL-37 is widely distributed. A careful study has been performed by Frohm Nilsson et al. (1999), demonstrating constitutive expression in human squamous epithelia of the tongue, mouth, esophagus, cervix, and vagina. In addition, peptide LL-37 and its precursur have been detected in human saliva (Murakami et al., 2002a). The intestinal tract, skin, and the lung are organs with large surfaces that are highly exposed to microorganisms. In the gut, particularly in the colon, the commensal flora is overwhelming. In all these organs, LL-37 is expressed together with β-defensins, and the expression can be altered during pathological conditions (O'Neil et al., 1999; Schroder, 1999; Bals, 2000; Singh et al., 2000; Islam et al., 2001; Gallo et al., 2002; Hase et al., 2002; Schauber et al., 2003).

6.5.1. Skin

LL-37 has a low constitutive expression in normal skin (Marchini et al., 2002). However, in inflammatory skin disorders there is a pronounced upregulation in keratinocytes of the epidermis (Frohm et al., 1997). This was demonstrated in biopsies from psoriatic lesions and areas from challenged nickel allergy, at both the mRNA and the peptide level (Frohm et al., 1997). Recently, a more quantitative study was conducted both for LL-37 and human β-defensin-2 (HBD2), showing that the upregulation in psoriasis is much higher than in atopic dermatitis (Ong et al., 2002). Clinically these findings are relevant because psoriatic lesions are rarely infected, whereas infections of *Staphylococcus aureus* are common in atopic excema. *S. aureus* is sensitive to both LL-37 and HBD2, and in synergy, these peptides are effective in killing this bacterium (Ong et al., 2002). In the skin of the newborn infant with the inflammatory reaction *Erythema toxicum*, we have found an influx of several immune cells harboring LL-37 (Marchini et al., 2002). In addition, *Vernix caseosa*, the white lipid-rich substance that covers the skin of neonates, exhibits antimicrobial activity, originating from several polypeptides, including LL-37 (Yoshio et al., 2003). Thus *Vernix* contributes to the strengthening of the antimicrobial skin-barrier of neonates. Similarly sweat seems to constitute an additional defense mechanism against bacteria on the skin surface, as LL-37 is expressed in sweat glands and is released constitutively into the sweat (Murakami et al., 2002b).

6.5.2. Lung

The first indication that failure in antimicrobial peptide function can be connected to disease was obtained when Welsh's group reported that the antimicrobial peptide activity in the lung mucosa of patients with cystic fibrosis (CF) is inactivated (Smith et al., 1996). CF results from a genetic defect in the gene encoding CF transmembrane-conductance regulator (CFTR). The microenvironment in the lung is altered by the high salt concentration, which can inactivate several antimicrobial peptides (Smith et al., 1996). LL-37 is expressed in the epithelia of the lung (Bals et al., 1998; Agerberth et al., 1999), but also in alveolar macrophages (Agerberth et al., 1999). The activity of LL-37 in vitro is dependent on salt. However, it is not known whether LL-37 is also inactivated in CF mucosa. In a xenograft model, antimicrobial activity can be restored after exposure of the CF xenograft to an adenovirus expressing the gene encoding LL-37 (Bals et al., 1999a). Furthermore, we have isolated several antimicrobial polypeptides, including LL-37, from bronchoalveolar lavage fluid of sarcoidosis patients. The total antimicrobial activity in this material was significantly induced compared with that of healthy controls (Agerberth et al., 1999). Increased concentrations of LL-37 are also found in aspirates of newborns during lung infections (Schaller-Bals, Schulze, and Bals, 2002). Thus the lung is protected against invading microbes by antimicrobial peptides of epithelial cells and alveolar macrophages, and their concentration level is enhanced during inflammation and infection, increasing the strength of the antimicrobial shield.

6.5.3. Gut

The upper part of the small intestine is almost sterile, but the amount of resident bacteria increases gradually toward the colon, with its huge number of commensal bacteria. A physiological role for the colonic microflora is well established, and a beneficial role of the microflora is true also for the host (Gustafsson, 1982; see also Chapter 2 by Devine). However, the interaction between the host and the microflora and their regulated interplay are far from clarified. More than a decade ago, it was suggested that antimicrobial peptides might be key players in this setting (Boman, 1991). We have demonstrated colonic epithelial expression of LL-37 in healthy individuals (Islam et al., 2001). Furthermore, we detected transcriptional downregulation of LL-37 and HBD1 early in *Shigella* infection. At the cellular level we demonstrated, by immunohistochemistry, epithelial downregulation of LL-37. The downregulation seems to be a direct effect, as live *Shigella*, infecting the cell lines HT-29 and U937, abolished the expression of LL-37 in 9 h. In addition, we have data that suggest that *Shigella* plasmid DNA could be a mediator of the downregulation (Islam et al., 2001). In contrast, the expression of LL-37 in colonic epithelial cells is upregulated by short-chain fatty acids (SCFAs), and most effectively by butyrate (Hase et al., 2002, Schauber et al., 2003). Notably, SCFAs are derived from bacterial fermentation of undigested dietary fibers in the colon (Scheppach, 1994) and exert important effects on colonic physiology (Scheppach et al., 1992). Interestingly, in a rabbit model of *Shigella* infections, infected rabbits showed reduced clinical symptoms after treatment with SCFAs, including butyrate (Rabbani et al., 1999). It is not far-fetched to suggest a relationship between the expression of LL-37 in the colonic epithelia and reduced symptoms in *Shigella*-infected rabbits after butyrate treatment. An important virulence element in general may be the weakening of the colonic epithelial barrier against microbicidal peptides. Escaping peptide activity may be a common mechanism for pathogenic bacteria. Several human pathogens have indeed been shown to synthesize proteases as a part of their virulence mechanism, and these proteases can degrade the active peptide LL-37 (Schmidtchen et al., 2002).

6.5.4. Testis

In our first report about the cloning of the cDNA for LL-37, we detected high expression in the testis by Northern blot analysis (Agerberth et al., 1995). In a detailed study, it was found that the expression is in the epididymis and that the sperm harbor a high amount of the peptide (Malm et al., 2000). Furthermore, the same research group has shown that the precursor protein hCAP18 was bound to prostasomes, membrane-bound vesicles derived from prostate epithelial cells. The speculation is that this defense molecule protects the sperm from bacteria in the female cervix and contributes to a sterile environment during fertilization (Andersson et al., 2002).

6.6. THE CAMP GENE AND ITS REGULATION

All cathelicidin genes characterized have the same organization as the CAMP gene, encoding LL-37 (Gudmundsson et al., 1996). The genes consist of four exons; the three first exons code for the signal peptide and the conserved cathelin domain (Zhao, Ganz, and Lehrer, 1995a, 1995b; Gudmundsson et al., 1996; Scocchi, Wang, and Zanetti; 1997; Pestonjamasp, Huttner, and Gallo, 2001). The CAMP gene is mapped to chromosome 3 (3p21.3) (Gudmundsson et al., 1995). This region has no known relationship to infection susceptibility, inflammatory disorders, or other diseases.

In the promoter of the CAMP gene, potential binding sites for transcription factors have been identified as NF-IL6 (nuclear factor for IL-6 expression) and acute phase response factor (APRF), now known as STAT3 (signal transducer and activator of transcription 3) (Gudmundsson et al., 1996). However, no functional studies have demonstrated an importance of these sites in the expression of the CAMP gene. NF-IL6 is equivalent to C/EBPβ (CCAAT/enhancer-binding protein β), thus belonging to the C/EBP family of transcription factors, which are recognized by a basic-leucine zipper domain. This domain is engaged in DNA binding and dimerization (Ramji and Foka, 2002). The members of the C/EBP family are involved in the regulation of diverse cellular responses, including inflammation, indicating a possible link to the expression of LL-37. A binding site for another C/EBP family member, C/EBPε, was later identified in the cathelicidin promoters of both human and mouse (Verbeek et al., 1999). Notably, in C/EBPε knockout mice, the expression of the mouse cathelicidin CRAMP in the bone marrow was almost absent, indicating a connection to the susceptibility of these mice to Gram-negative bacteria (Verbeek et al., 1999). Colocalization of LL-37 and IL-6 at several epithelial surfaces has been reported (Frohm Nilsson et al., 1999). However, it has not been clarified if IL-6 regulates the expression of LL-37. The induction of LL-37 and IL-6 could be a parallel regulation. To date, it has not been possible to affect the expression of LL-37 by IL-6 in several cell lines of different origin (C. Svanholm personal communication). However, the situation in tissues might be different. Interestingly, IL-1α has been shown to stimulate expression of LL-37 in composite keratinocyte grafts, whereas gene expression is not affected in keratinocyte cultures (Erdag and Morgan, 2002).

SCFAs are able to modulate the expression of LL-37, and the strongest inducer is butyrate (Hase et al., 2002, Schauber et al., 2003). Butyrate also stimulates the differentiation of colonic epithelial cells, and therefore LL-37 expression has been claimed to be differentiation dependent (Hase et al., 2002). By utilizing specific signal pathway inhibitors, we could resolve the butyrate route of stimulating LL-37 expression (Schauber et al., 2003). In colonic cell lines, the butyrate-induced expression of LL-37 could be blocked by the MEK [mitogen-activated protein (MAP) kinase/ERK kinase] inhibitor U0126. By blocking the p38/MAP kinase pathway with the inhibitor SB203580, butyrate-induced differentiation was prevented, whereas the butyrate induction of

LL-37 expression was unaffected. This would indicate that at least two main pathways are affected by butyrate; the activation through MEK/ERK leads to enhanced expression of LL-37, and differentiation is pushed forward through the p38/MAP kinase pathway. Notably in vivo these events would be in parallel. Interestingly, MEK/ERK activation is known to affect the DNA binding affinity of STAT3 (O'Rourke and Shepherd, 2002) and several binding sites for this transcription factor are present in the promoter of the gene encoding LL-37 (Gudmundsson et al., 1996).

6.7. LL-37 IN CONNECTION TO DISEASE

The expression of antimicrobial peptides has been shown to be altered in connection with infection and/or inflammation. The first observation of changes in the expression of LL-37 was in inflammatory skin disorders such as psoriasis and allergy (Frohm et al., 1997). The regulatory pathway for this induction is not known. However, a correlation has been shown between quantities of peptide and sensitivity to infections (Ong et al., 2002). In conformity with other genes, mice have served as a functional model system, and the mouse that is deficient in the CRAMP gene, the homolog of CAMP, is indeed sensitive to certain skin infections (Nizet et al., 2001). In the future, CRAMP knockout mice will certainly provide further information on the importance of antimicrobial genes.

A correlation with heritable susceptibility to infections, in situations in which antimicrobial peptide levels are reduced or absent, would underline the importance of these peptides in vivo. Recently, such a correlation was shown in a severe congenital neutropenia, morbus Kostmann, that has been a fatal syndrome (Putsep et al., 2002). The neutrophil levels in Kostmann's disease can be restored by treatment with granulocyte-colony-stimulating factor (G-CSF). Still the patients get recurrent infections and especially periodontal diseases, leading to long-lasting treatment with classical antibiotics. The amounts of the main neutrophilic antimicrobial peptides α-defensins were shown to decrease to 30% of normal levels, whereas the cathelicidin peptide LL-37 was hardly detected (Putsep et al., 2002). Only after bone marrow transplants was the concentration level of LL-37 restored. Both LL-37 and α-defensins are synthesized in the bone marrow, and during maturation of neutrophils they are localized in the granula of this cell type. The underlying defect in Kostmann's disease could be a mutated protein in a common signal pathway or in transcriptional regulation needed for the expression of these antimicrobial peptides. The protein in question would normally be required for LL-37 expression and enhanced α-defensin expression. Although the regulation of the expression of these neutrophilic antimicrobial peptides has not been resolved in detail, as such information develops it will lead to better tools in tracing mutations that can lead to pathological conditions of this type.

Advances within the field of antimicrobial peptides have been rapid during the past few years. Clarity has been brought to their importance in immunity and as powerful eliminators of microbial invaders, but also as modulators of immune responses. The future will involve attempts to utilize this system to fight serious infections by induction of relevant genes or direct administration of peptides.

ACKNOWLEDGMENTS

We thank Hans Jörnvall for critical reading of the manuscript. Figure 6.1 is reproduced with the permission from Cellular and Molecular Life Sciences. The authors are supported by grants from The Swedish Research Council, The Swedish Foundation for International Cooperation in Research and Higher Education, (STINT), Magnus Bergvall's Foundation, Petrus and Augusta Hedlund's Foundation, Ruth and Richard Julin's Foundation, and Prof. Nanna Svartz' Foundation.

REFERENCES

Agerberth, B., Charo, J., Werr, J., Olsson, B., Idali, F., Lindbom, L., Kiessling, R., Jornvall, H., Wigzell, H., and Gudmundsson, G. H. (2000). The human antimicrobial and chemotactic peptides LL-37 and alpha-defensins are expressed by specific lymphocyte and monocyte populations. *Blood*, 96, 3086–93.

Agerberth, B., Grunewald, J., Castanos-Velez, E., Olsson, B., Jornvall, H., Wigzell, H., Eklund, A., and Gudmundsson, G. H. (1999). Antibacterial components in bronchoalveolar lavage fluid from healthy individuals and sarcoidosis patients. *American Journal of Respiratory and Critical Care Medicine*, 160, 283–90.

Agerberth, B., Gunne, H., Odeberg, J., Kogner, P., Boman, H. G., and Gudmundsson, G. H. (1995). FALL-39, a putative human peptide antibiotic, is cysteine-free and expressed in bone marrow and testis. *Proceedings of the National Academy of Sciences USA*, 92, 195–9.

Agerberth, B., Lee, J. Y., Bergman, T., Carlquist, M., Boman, H. G., Mutt, V., and Jornvall, H. (1991). Amino acid sequence of PR-39. Isolation from pig intestine of a new member of the family of proline-arginine-rich antibacterial peptides. *European Journal of Biochemistry*, 202, 849–54.

Andersson, E., Sorensen, O. E., Frohm, B., Borregaard, N., Egesten, A., and Malm, J. (2002). Isolation of human cationic antimicrobial protein-18 from seminal plasma and its association with prostasomes. *Human Reproduction*, 17, 2529–34.

Bals, R. (2000). Epithelial antimicrobial peptides in host defense against infection. *Respiratory Research*, 1, 141–50.

Bals, R., Wang, X., Zasloff, M., and Wilson, J. M. (1998). The peptide antibiotic LL-37/hCAP-18 is expressed in epithelia of the human lung where it has broad antimicrobial activity at the airway surface. *Proceedings of the National Academy of Sciences USA*, 95, 9541–6.

Bals, R., Weiner, D. J., Meegalla, R. L., and Wilson, J. M. (1999b). Transfer of a cathelicidin peptide antibiotic gene restores bacterial killing in a cystic fibrosis xenograft model. *Journal of Clinical Investigation*, 103, 1113–7.

Bals, R., Weiner, D. J., Moscioni, A. D., Meegalla, R. L., and Wilson, J. M. (1999b). Augmentation of innate host defense by expression of a cathelicidin antimicrobial peptide. *Infection and Immunity*, 67, 6084–9.

Boman, H. G. (1991). Antibacterial peptides: Key components needed in immunity. *Cell*, 65, 205–7.

Cowland, J. B., Johnsen, A. H., and Borregaard, N. (1995). hCAP-18, a cathelin/pro-bactenecin-like protein of human neutrophil specific granules. *FEBS Letters*, 368, 173–6.

Dorschner, R. A., Pestonjamasp, V. K., Tamakuwala, S., Ohtake, T., Rudisill, J., Nizet, V., Agerberth, B., Gudmundsson, G. H., and Gallo, R. L. (2001). Cutaneous injury induces the release of cathelicidin anti-microbial peptides active against group A Strep. *Journal of Investigative Dermatology*, 117, 91–7.

Erdag, G. and Morgan, J. R. (2002). Interleukin-1alpha and interleukin-6 enhance the antibacterial properties of cultured composite keratinocyte grafts. *Annals of Surgery*, 235, 113–24.

Fahrner, R. L., Dieckmann, T., Harwig, S. S., Lehrer, R. I., Eisenberg, D., and Feigon, J. (1996). Solution structure of protegrin-1, a broad-spectrum antimicrobial peptide from porcine leukocytes. *Chemistry and Biology*, 3, 543–50.

Frank, R. W., Gennaro, R., Schneider, K., Przybylski, M., and Romeo, D. (1990). Amino acid sequences of two proline-rich bactenecins. Antimicrobial peptides of bovine neutrophils. *Journal of Biological Chemistry*, 265, 18871–4.

Frohm, M., Agerberth, B., Ahangari, G., Stahle-Backdahl, M., Liden, S., Wigzell, H., and Gudmundsson, G. H. (1997). The expression of the gene coding for the antibacterial peptide LL-37 is induced in human keratinocytes during inflammatory disorders. *Journal of Biological Chemistry*, 272, 15258–63.

Frohm Nilsson, M., Sandstedt, B., Sorensen, O., Weber, G., Borregaard, N., and Stahle-Backdahl, M. (1999). The human cationic antimicrobial protein (hCAP18), a peptide antibiotic, is widely expressed in human squamous epithelia and colocalizes with interleukin-6. *Infection and Immunity*, 67, 2561–6.

Gallo, R. L., Murakami, M., Ohtake, T., and Zaiou, M. (2002). Biology and clinical relevance of naturally occurring antimicrobial peptides. *Journal of Allergy and Clinical Immunology*, 110, 823–31.

Garcia-Penarrubia, P., Koster, F. T., Kelley, R. O., McDowell, T. D., and Bankhurst, A. D. (1989). Antibacterial activity of human natural killer cells. *Journal of Experimental Medicine*, 169, 99–113.

Gudmundsson, G. H., Agerberth, B., Odeberg, J., Bergman, T., Olsson, B., and Salcedo, R. (1996). The human gene FALL39 and processing of the cathelin precursor to the antibacterial peptide LL-37 in granulocytes. *European Journal of Biochemistry*, 238, 325–32.

Gudmundsson, G. H., Magnusson, K. P., Chowdhary, B. P., Johansson, M., Andersson, L., and Boman, H. G. (1995). Structure of the gene for porcine peptide antibiotic PR-39, a cathelin gene family member: comparative mapping of the locus for the human peptide antibiotic FALL-39. *Proceedings of the National Academy of Sciences USA*, 92, 7085–9.

Gustafsson, B. E. (1982). The physiological importance of the colonic microflora. *Scandinavian Journal of Gastroenterology*. (Suppl.), 77. 117–31.

Hase, K., Eckmann, L., Leopard, J. D., Varki, N., and Kagnoff, M. F. (2002). Cell differentiation is a key determinant of cathelicidin LL-37/human cationic antimicrobial protein 18 expression by human colon epithelium. *Infection and Immunity*, 70, 953–63.

Islam, D., Bandholtz, L., Nilsson, J., Wigzell, H., Christensson, B., Agerberth, B., and Gudmundsson, G. (2001). Downregulation of bactericidal peptides in enteric infections: A novel immune escape mechanism with bacterial DNA as a potential regulator. *Nature Medicine*, 7, 180–5.

Johansson, J., Gudmundsson, G. H., Rottenberg, M. E., Berndt, K. D., and Agerberth, B. (1998). Conformation-dependent antibacterial activity of the naturally occurring human peptide LL-37. *Journal of Biological Chemistry*, 273, 3718–24.

Kirikae, T., Hirata, M., Yamasu, H., Kirikae, F., Tamura, H., Kayama, F., Nakatsuka, K., Yokochi, T., and Nakano, M. (1998). Protective effects of a human 18-kilodalton cationic antimicrobial protein (CAP18)-derived peptide against murine endotoxemia. *Infection and Immunity*, 66, 1861–8.

Larrick, J. W., Hirata, M., Balint, R. F., Lee, J., Zhong, J., and Wright, S. C. (1995a). Human CAP18: A novel antimicrobial lipopolysaccharide-binding protein. *Infection and Immunity*, 63, 1291–7.

Larrick, J. W., Hirata, M., Zhong, J., and Wright, S. C. (1995b). Anti-microbial activity of human CAP18 peptides. *Immunotechnology*, 1, 65–72.

Larrick, J. W., Morgan, J. G., Palings, I., Hirata, M., and Yen, M. H. (1991). Complementary DNA sequence of rabbit CAP18 – a unique lipopolysaccharide binding protein. *Biochemical and Biophysical Research Communications*, 179, 170–5.

Lehrer, R. I. and Ganz, T. (2002). Defensins of vertebrate animals. *Current Opinion in Immunology*, 14, 96–102.

Lenarcic, B., Ritonja, A., Dolenc, I., Stoka, V., Berbic, S., Pungercar, J., Strukelj, B., and Turk, V. (1993). Pig leukocyte cysteine proteinase inhibitor (PLCPI), a new member of the stefin family. *FEBS Letters*, 336, 289–92.

Levitz, S. M., Mathews, H. L., and Murphy, J. W. (1995). Direct antimicrobial activity of T cells. *Immunology Today*, 16, 387–91.

Malm, J., Sorensen, O., Persson, T., Frohm-Nilsson, M., Johansson, B., Bjartell, A., Lilja, H., Stahle-Backdahl, M., Borregaard, N., and Egesten, A. (2000). The human cationic antimicrobial protein (hCAP-18) is expressed in the epithelium of human epididymis, is present in seminal plasma at high concentrations, and is attached to spermatozoa. *Infection and Immunity*, 68, 4297–302.

Marchini, G., Lindow, S., Brismar, H., Stabi, B., Berggren, V., Ulfgren, A. K., Lonne-Rahm, S., Agerberth, B., and Gudmundsson, G. H. (2002). The newborn infant is protected by an innate antimicrobial barrier: Peptide antibiotics are present in the skin and vernix caseosa. *British Journal of Dermatology*, 147, 1127–34.

Murakami, M., Ohtake, T., Dorschner, R. A., and Gallo, R. L. (2002a). Cathelicidin antimicrobial peptides are expressed in salivary glands and saliva. *Journal of Dental Research*, 81, 845–50.

Murakami, M., Ohtake, T., Dorschner, R. A., Schittek, B., Garbe, C., and Gallo, R. L. (2002b). Cathelicidin anti-microbial peptide expression in sweat, an innate defense system for the skin. *Journal of Investigative Dermatology*, 119, 1090–5.

Nagaoka, I., Hirota, S., Yomogida, S., Ohwada, A., and Hirata, M. (2000). Synergistic actions of antibacterial neutrophil defensins and cathelicidins. *Inflammation Research*, 49, 73–9.

Niyonsaba, F., Iwabuchi, K., Someya, A., Hirata, M., Matsuda, H., Ogawa, H., and Nagaoka, I. (2002). A cathelicidin family of human antibacterial peptide LL-37 induces mast cell chemotaxis. *Immunology*, 106, 20–6.

Niyonsaba, F., Someya, A., Hirata, M., Ogawa, H., and Nagaoka, I. (2001). Evaluation of the effects of peptide antibiotics human beta-defensins-1/-2 and LL-37 on histamine release and prostaglandin D(2) production from mast cells. *European Journal of Immunology*, 31, 1066–75.

Nizet, V., Ohtake, T., Lauth, X., Trowbridge, J., Rudisill, J., Dorschner, R. A., Pestonjamasp, V., Piraino, J., Huttner, K., and Gallo, R. L. (2001). Innate antimicrobial peptide protects the skin from invasive bacterial infection. *Nature (London)*, 414, 454–7.

O'Neil, D. A., Porter, E. M., Elewaut, D., Anderson, G. M., Eckmann, L., Ganz, T., and Kagnoff, M. F. (1999). Expression and regulation of the human beta-defensins hBD-1 and hBD-2 in intestinal epithelium. *Journal of Immunology*, 163, 6718–24.

Ong, P. Y., Ohtake, T., Brandt, C., Strickland, I., Boguniewicz, M., Ganz, T., Gallo, R. L., and Leung, D. Y. (2002). Endogenous antimicrobial peptides and skin infections in atopic dermatitis. *New England Journal of Medicine*, 347, 1151–60.

Oren, Z., Lerman, J. C., Gudmundsson, G. H., Agerberth, B., and Shai, Y. (1999). Structure and organization of the human antimicrobial peptide LL-37 in phospholipid membranes: Relevance to the molecular basis for its non-cell-selective activity. *Biochemical Journal*, 341, 501–13.

O'Rourke, L. and Shepherd, P. R. (2002). Biphasic regulation of extracellular-signal-regulated protein kinase by leptin in macrophages: Role in regulating STAT3 Ser727 phosphorylation and DNA binding. *Biochemical Journal*, 364, 875–9.

Pestonjamasp, V. K., Huttner, K. H., and Gallo, R. L. (2001). Processing site and gene structure for the murine antimicrobial peptide CRAMP. *Peptides*, 22, 1643–50.

Putsep, K., Carlsson, G., Boman, H. G., and Andersson, M. (2002). Deficiency of antibacterial peptides in patients with morbus Kostmann: An observation study. *Lancet*, 360, 1144–9.

Rabbani, G. H., Albert, M. J., Hamidur Rahman, A. S., Moyenul Isalm, M., Nasirul Islam, K. M., and Alam, K. (1999). Short-chain fatty acids improve clinical, pathologic, and microbiologic features of experimental shigellosis. *Journal of Infectious Diseases*, 179, 390–7.

Ramji, D. P. and Foka, P. (2002). CCAAT/enhancer-binding proteins: Structure, function and regulation. *Biochemical Journal*, 365, 561–75.

Ritonja, A., Kopitar, M., Jerala, R., and Turk, V. (1989). Primary structure of a new cysteine proteinase inhibitor from pig leucocytes. *FEBS Letters*, 255, 211–4.

Sanchez, J. F., Wojcik, F., Yang, Y. S., Strub, M. P., Strub, J. M., Van Dorsselaer, A., Martin, M., Lehrer, R., Ganz, T., Chavanieu, A., Calas, B., and Aumelas, A. (2002). Overexpression and structural study of the cathelicidin motif of the protegrin-3 precursor. *Biochemistry*, 41, 21–30.

Schaller-Bals, S., Schulze, A., and Bals, R. (2002). Increased levels of antimicrobial peptides in tracheal aspirates of newborn infants during infection. *American Journal of Respiratory and Critical Care Medicine*, 165, 992–5.

Schauber, J., Svanholm, C., Termén, S., Iffland, K., Menzel, T., Scheppach, W., Melcher, R., Agerberth, B., Luhrs, H., and Gudmundsson, G. (2003). Expression of the cathelicidin LL-37 is modulated by short chain fatty acids in colonocytes: Relevance of signalling pathways. *Gut*, 52, 735–41.

Scheppach, W. (1994). Effects of short chain fatty acids on gut morphology and function. *Gut*, 35, S35–8.

Scheppach, W., Sommer, H., Kirchner, T., Paganelli, G. M., Bartram, P., Christl, S., Richter, F., Dusel, G., and Kasper, H. (1992). Effect of butyrate enemas on the colonic mucosa in distal ulcerative colitis. *Gastroenterology*, 103, 51–6.

Schmidtchen, A., Frick, I. M., Andersson, E., Tapper, H., and Bjorck, L. (2002). Proteinases of common pathogenic bacteria degrade and inactivate the antibacterial peptide LL-37. *Molecular Microbiology*, 46, 157–68.

Schroder, J. M. (1999). Epithelial antimicrobial peptides: Innate local host response elements. *Cellular and Molecular Life Sciences*, 56, 32–46.

Scocchi, M., Wang, S., and Zanetti, M. (1997). Structural organization of the bovine cathelicidin gene family and identification of a novel member. *FEBS Letters*, 417, 311–5.

Scott, M. G., Davidson, D. J., Gold, M. R., Bowdish, D., and Hancock, R. E. (2002). The human antimicrobial peptide LL-37 is a multifunctional modulator of innate immune responses. *Journal of Immunology*, 169, 3883–91.

Selsted, M. E., Novotny, M. J., Morris, W. L., Tang, Y. Q., Smith, W., and Cullor, J. S. (1992). Indolicidin, a novel bactericidal tridecapeptide amide from neutrophils. *Journal of Biological Chemistry*, 267, 4292–5.

Shai, Y. (1995). Molecular recognition between membrane-spanning polypeptides. *Trends in Biochemical Sciences*, 20, 460–4.

Singh, P. K., Tack, B. F., McCray Jr., P. B., and Welsh, M. J. (2000). Synergistic and additive killing by antimicrobial factors found in human airway surface liquid. *American Journal of Physiology Lung Cellular Molecular Physiology*, 279, L799–805.

Smith, J. J., Travis, S. M., Greenberg, E. P., and Welsh, M. J. (1996). Cystic fibrosis airway epithelia fail to kill bacteria because of abnormal airway surface fluid. *Cell*, 85, 229–36.

Sorensen, O., Arnljots, K., Cowland, J. B., Bainton, D. F., and Borregaard, N. (1997a). The human antibacterial cathelicidin, hCAP-18, is synthesized in myelocytes and metamyelocytes and localized to specific granules in neutrophils. *Blood*, 90, 2796–803.

Sorensen, O., Bratt, T., Johnsen, A. H., Madsen, M. T., and Borregaard, N. (1999). The human antibacterial cathelicidin, hCAP-18, is bound to lipoproteins in plasma. *Journal of Biological Chemistry*, 274, 22445–51.

Sorensen, O., Cowland, J. B., Askaa, J., and Borregaard, N. (1997b). An ELISA for hCAP-18, the cathelicidin present in human neutrophils and plasma. *Journal of Immunological Methods*, 206, 53–9.

Sorensen, O. E., Follin, P., Johnsen, A. H., Calafat, J., Tjabringa, G. S., Hiemstra, P. S., and Borregaard, N. (2001). Human cathelicidin, hCAP-18, is processed to the antimicrobial peptide LL-37 by extracellular cleavage with proteinase 3. *Blood*, 97, 3951–9.

Tang, Y. Q., Yuan, J., Osapay, G., Osapay, K., Tran, D., Miller, C. J., Ouellette, A. J., and Selsted, M. E. (1999). A cyclic antimicrobial peptide produced in primate leukocytes by the ligation of two truncated alpha-defensins. *Science*, 286, 498–502.

Termén, S., Tollin, M., Olsson, B., Svenberg, T., Agerberth, B., and Gudmundsson, G. (2003). Phylogeny, processing and expression of the rat cathelicidin rCRAMP: A model for innate antimicrobial peptides. *Cellular and Molecular Life Sciences*, 60, 536–49.

Turner, J., Cho, Y., Dinh, N. N., Waring, A. J., and Lehrer, R. I. (1998). Activities of LL-37, a cathelin-associated antimicrobial peptide of human neutrophils. *Antimicrobial Agents and Chemotherapy*, 42, 2206–14.

Wang, Y., Agerberth, B., Lothgren, A., Almstedt, A., and Johansson, J. (1998). Apolipoprotein A-I binds and inhibits the human antibacterial/cytotoxic peptide LL-37. *Journal of Biological Chemistry*, 273, 33115–8.

Verbanac, D., Zanetti, M., and Romeo, D. (1993). Chemotactic and protease-inhibiting activities of antibiotic peptide precursors. *FEBS Letters*, 317, 255–8.

Verbeek, W., Lekstrom-Himes, J., Park, D. J., Dang, P. M., Vuong, P. T., Kawano, S., Babior, B. M., Xanthopoulos, K., and Koeffler, H. P. (1999). Myeloid transcription factor C/EBP epsilon is involved in the positive regulation of lactoferrin gene expression in neutrophils. *Blood*, 94, 3141–50.

Yang, D., Chen, Q., Schmidt, A. P., Anderson, G. M., Wang, J. M., Wooters, J., Oppenheim, J. J., and Chertov, O. (2000). LL-37, the neutrophil granule- and epithelial cell-derived cathelicidin, utilizes formyl peptide receptor-like 1 (FPRL1) as a receptor to chemoattract human peripheral blood neutrophils, monocytes, and T cells. *Journal of Experimental Medicine*, 192, 1069–74.

Yang, D., Chertov, O., and Oppenheim, J. J. (2001). Participation of mammalian defensins and cathelicidins in anti-microbial immunity: Receptors and activities of human defensins and cathelicidin (LL-37). *Journal of Leukocyte Biology*, 69, 691–7.

Yasin, B., Pang, M., Turner, J. S., Cho, Y., Dinh, N. N., Waring, A. J., Lehrer, R. I., and Wagar, E. A. (2000). Evaluation of the inactivation of infectious Herpes simplex virus by host-defense peptides. *European Journal of Clinical Microbiology and Infectious Diseases*, 19, 187–94.

Yoshio, H., Tollin, M., Gudmundsson, G. H., Lagercrantz, H., Jornvall, H., Marchini, G., and Agerberth, B. (2003). Antimicrobial polypeptides of human vernix caseosa and amniotic fluid: Implications for newborn innate defense. *Pediatric Research*, 53, 211–16.

Zaiou, M. and Gallo, R. L. (2002). Cathelicidins, essential gene-encoded mammalian antibiotics. *Journal of Molecular Medicine*, 80, 549–61.

Zaiou, M., Nizet, V., and Gallo, R. L. (2003). Antimicrobial and protease inhibitory functions of the human cathelicidin (hCAP18/LL-37) prosequence. *Journal of Investigative Dermatology*, 120, 810–6.

Zanetti, M., Gennaro, R., and Romeo, D. (1995). Cathelicidins: A novel protein family with a common proregion and a variable C-terminal antimicrobial domain. *FEBS Letters*, 374, 1–5.

Zanetti, M., Litteri, L., Gennaro, R., Horstmann, H., and Romeo, D. (1990). Bactenecins, defense polypeptides of bovine neutrophils, are generated from precursor molecules stored in the large granules. *Journal of Cell Biology*, 111, 1363–71.

Zhao, C., Ganz, T., and Lehrer, R. I. (1995a). The structure of porcine protegrin genes. *FEBS Letters*, 368, 197–202.

Zhao, C., Ganz, T., and Lehrer, R. I. (1995b). Structures of genes for two cathelin-associated antimicrobial peptides: Prophenin-2 and PR-39. *FEBS Letters*, 376, 130–4.

Zhao, C., Nguyen, T., Boo, L. M., Hong, T., Espiritu, C., Orlov, D., Wang, W., Waring, A., and Lehrer, R. I. (2001). RL-37, an alpha-helical antimicrobial peptide of the rhesus monkey. *Antimicrobial Agents and Chemotherapy*, 45, 2695–702.

CHAPTER 7

Antimicrobial peptides of the alimentary tract of mammals

Charles L. Bevins and Tomas Ganz

7.1. INTRODUCTION

The alimentary tract of animals is the site of a wide variety of host–microbe interactions. The numbers of commensal bacteria colonizing these epithelia range from massive numbers in the colon to far fewer bacteria in the adjacent small intestine and more proximal segments (Fig. 7.1). The digestive tract is exposed to a great variety of microbes originating from food and water, some of which are potentially virulent and others are less prone to cause disease. Each anatomical part of the digestive tract is lined with specialized epithelium suited for the specific function of that region. For example, the tongue and esophagus have a squamous epithelium specialized for rapid transfer of ingested food, whereas the stomach and intestine have columnar epithelia for transepithelial movement of ions, nutrients, and water. After a meal, the small intestine is filled with a nutrient-rich medium, potentially supporting the rapid growth of microbes. Because there are relatively few microbes in the small intestine, an array of defense measures must exist to limit the numbers of colonizing microbes. In fact, multiple mechanisms are employed to bring about effective host defense of this organ system, including both innate and adaptive immunity.

A large commitment of metabolic resources is allocated to acquired immune responses in the mammalian alimentary tract. Indeed, based on total numbers of lymphocytes, the mammalian gut may be considered the largest of lymphoid organs. This arm of immunity, initiated by specific antigens, provides both antibody and cellular responses that are crucial for health. For example, the memory responses of acquired immunity provide effective defense against infection by pathogens that have primed the system by immunization or prior infection. However, on first encounter, the development

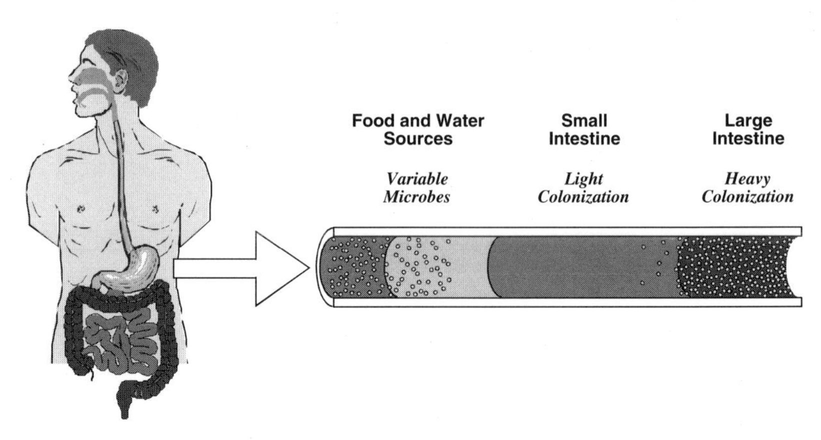

Figure 7.1. Exposure of the alimentary tract to microbial challenges. Schematic diagram of elongated alimentary tract highlighting the large numbers of colonizing microbes (indicated by dots) in the large intestine, fewer colonizing microbes in the small intestine, and variable numbers of microbes that enter the system by contaminated food and water. (Illustrated by D. Schumick, Department of Medical Illustration, Cleveland Clinic Foundation.)

of T-cell and B-cell responses may not be sufficiently rapid to control acute infections by pathogens. Therefore an exclusive reliance on adaptive immunity would allow microbes to grow and outpace the immune response. The ability to control microbial growth in the intestine predates the development of adaptive immunity because effective defense of the alimentary tract can be achieved in lower organisms in which lymphocyte-mediated immunity is lacking. Moreover, many lines of investigation support the idea that innate immune responses provide the first line of protection against deleterious microbes.

Innate defenses of the alimentary tract encompass both physical and chemical barriers. Many of the physiological processes associated with digestion likely contribute to mucosal defenses. For example, gastric acid, digestive enzymes, and bile salts may exert direct antimicrobial activity in addition to their key role in the breakdown of dietary nutrients. Other processes such as peristalsis and epithelial cell renewal likely impair the ability of microbes to colonize. Mucus and the tight junctions between epithelial cells constitute physical barriers. Finally, the resident commensal flora contributes effectively to the inhibition of exogenous pathogens.

In this chapter we focus on antimicrobial peptides, one component of the innate immune defense that evolutionarily predates both lymphocytes and immunoglobulins. This focus does imply, however, that the effector molecules of the innate immune system function separately from the acquired immune responses. To the contrary, as we gain further insights into the repertoire of biological activities of antimicrobial peptides, we are likely to learn that antimicrobial peptides provide molecular signals that influence the complex adaptive immune system in the mammalian gut.

7.2. MAMMALIAN ANTIMICROBIAL PEPTIDES

Similar to their counterparts in plants and invertebrates, mammalian antimicrobial peptides are gene-encoded molecules that exert microbicidal activity at micromolar concentrations against bacteria, fungi and some parasites (Boman, 1995; Hancock, Falla, and Brown, 1995; Lehrer, Bevins, and Ganz, 1998; Zasloff, 2002). On a molecular level, the mammalian antimicrobial peptides are generally amphipathic, cationic molecules that disrupt membrane structure and function of the target microbes. The impressive parallels in the structure and function of antimicrobial peptides from diverse species underscore that in a wide range of biological contexts antimicrobial peptides are effective defense molecules. The antimicrobial peptides of the alimentary tract are expressed in other organ systems, and in many cases information on their structure and activity is extrapolated from studies of peptides isolated from other sites. It is important to isolate the peptides from each site to confirm the peptide structure, as posttranslational modifications could alter activity.

Members of the two major families of mammalian antimicrobial peptides, defensins, and cathelicidins are expressed in the alimentary tract (Table 7.1). Defensin peptides contain three intramolecular disulfide bonds and have a largely β-sheet structure (Lehrer et al., 1998; Lehrer & Ganz, 2002a). There are three types of defensins: α-defensins that have 29–35 residues, β-defensins that have 34–47 residues, and θ-defensins, circular peptides that contain 18 residues and to date have been detected only in rhesus macaque neutrophils. In addition to size, the defensin types differ with respect to (1) spacing of their six cysteines and their disulfide-pairing pattern, (2) variation in the length of the prosegment of the precursor molecule, and (3) gene structure and chromosomal gene position. The disulfide bonds of α-defensins form between cysteines C1 and C6, C2 and C4, and C3 and C5, and for the larger β-defensins the pairing is between C1 and C5, C2 and C4, and C3 and C6 (Fig. 7.2). Although their cysteine linkages differ, the three-dimensional structure of these two groups of peptides is similar because the C5 and C6 cysteines, whose connectivity differs between α- and β-defensins, are adjacent. The structure of α- and β-defensins consists of three β-strands.

Table 7.1. *Examples of expression of antimicrobial peptides in the mammalian alimentary tract*

Anatomical site	Antimicrobial peptides[1]
Oral mucosa	β-defensins, cathelicidins
Periodontal sulcus	+high concentration of α-defensins from neutrophils
Tongue	β-defensins
Esophagus	β-defensins, α-defensins (Paneth cell metaplasia)
Stomach	β-defensins
Small intestine	α-defensins, lysozyme, sPLA
Appendix/cecum	lysozyme, sPLA
Colon and rectum	β-defensin, cathelicidins, α-defensins (PC metaplasia)

[1] sPLA, secretory phospholipase A_2; PC, *Paneth cell.*

In the alimentary tract of mammals, β-defensins are expressed at multiple sites, whereas α-defensins are largely confined to the small intestine.

The cathelicidin antimicrobial peptides are derived from precursor molecules with a bipartite structure, consisting of a conserved N-terminal propeptide domain and a variable C-terminal antimicrobial peptide domain (Fig. 7.3) (Zanetti, Gennaro, and Romeo, 1997; Lehrer and Ganz, 2002b; Zanetti et al., 2002). The approximately 100-amino-acid propeptide segment of these peptides was discovered independently as a cathepsin L inhibitor and was therefore given the shortened name cathelin. The antimicrobial peptides at the C-terminus of the precursor molecule are remarkably diverse in size (ranging from 12 to 100 amino acids) and chemical structure. The human cationic antimicrobial peptide LL-37, also named hCAP-18, is a cathelicidin peptide expressed predominantly in neutrophils but also in the alimentary tract in addition to many other sites (Fig. 7.3).

7.3. ACTIVITIES OF ALIMENTARY TRACT ANTIMICROBIAL PEPTIDES

Defensins and cathelicidins exert activity against Gram-negative and Gram-positive bacteria, and some have also been shown to have activity against fungi, viruses, and protozoa (Lehrer, Lichtenstein, and Gangz, 1993;

A. α – defensins

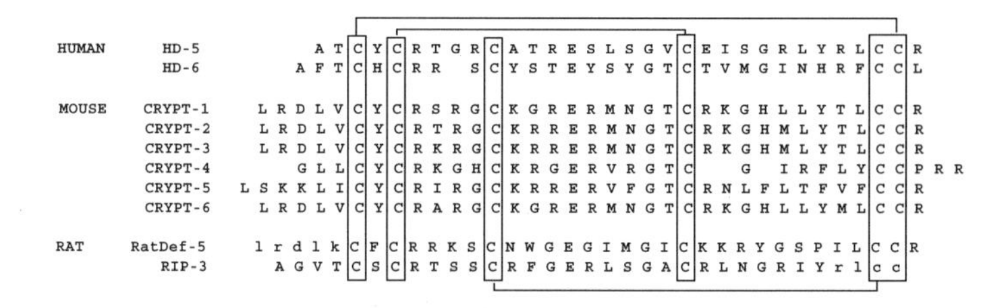

B. β – defensins

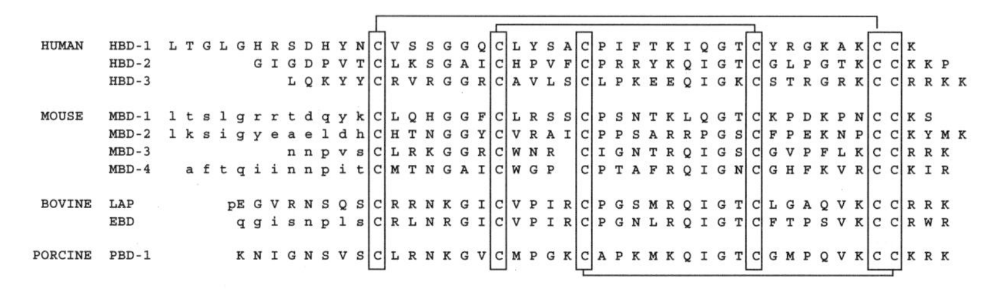

Figure 7.2. Primary structure and disulfide linkages of α- and β-defensins expressed in the alimentary tract. (a) Sequence alignment of selected α-defensins from human, mouse, and rat sources. Human defensin peptides [HD-5 and -6 (Porter et al., 1998)] and mouse crypt defensins (cryptdins) (Eisenhauer et al., 1992; Ouellette et al., 1992; Selsted et al., 1992) were isolated from small intestinal tissues and lumen. Rat sequences were deduced from cDNA cloned from small intestine [rat defensin-5 (Condon, 1999)] and peptide isolation [RIP-3 (Qu et al., 1996)]. (b) Sequence alignment of selected β-defensins from human, mouse, bovine, and porcine sources. Human β-defensin peptide sequences were determined from peptides isolated from tissues outside of the alimentary tract (Bensch et al., 1995; Harder et al., 1997; Valore et al., 1998; Zucht et al., 1998; Harder et al., 2001). Mouse β-defensins were from deduced cDNA sequences (Huttner, Kozak, and Bevins, 1997; Bals, Goldman, and Wilson et al., 1998; Morrison et al., 1998; Jia et al., 2000). Bovine sequences were from tongue [lingual antimicrobial peptide (LAP), (Schonwetter et al., 1995)] and colon [enteric β-defensin (EBD) (Tarver et al., 1998)] and porcine sequences were from tongue [pBD-1 (Shi et al., 1999)]. Lowercase letters indicate deduced sequences for which peptide data are unavailable or incomplete. Disulfide linkages are shown in (a) and (b) as brackets (Selsted and Harwig, 1989; Tang and Selsted, 1993).

Hancock, Falla, and Brown, 1995; Zanetti et al., 2002). These cationic antimicrobial peptides are thought to kill target microbes through disruption of membrane integrity by a generic mechanism coined the Shai–Matsuzaki–Huang model (Zasloff, 2002). According to this model (Huang, 1999; Matsuzaki, 1999; Shai, 1999; Zasloff, 2002), the cationic peptides interact through

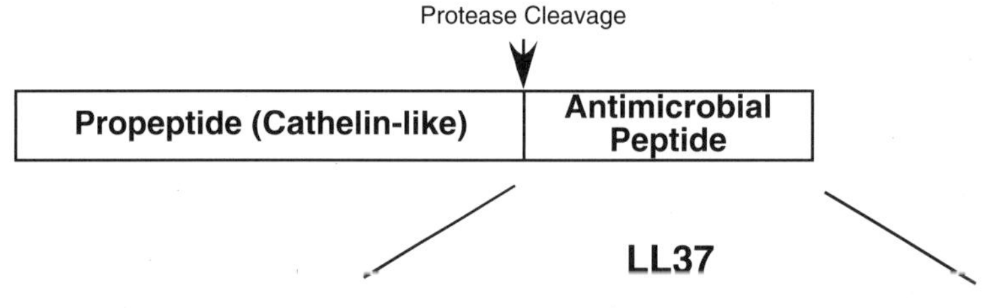

Figure 7.3. Domain structure of cathelicidin precursors. A diverse collection of antimicrobial peptides is synthesized at the carboxyl-terminal portion of precursors with a highly conserved amino-terminal domain, approximately 100-amino-acid residues in size, termed "cathelin." The active peptide is released following proteolytic cleavage (Zanetti et al., 1997; Lehrer and Ganz, 2002b; Zanetti et al., 2002). Within the alimentary tract, human LL-37 is expressed.

electrostatic attraction with negatively charged phospholipid head groups on the outer leaflet of the target cell's membrane. The peptides accumulate on the outer leaflet of the bilayer and interpose themselves between the head groups, displacing lipids by their presence. This accumulation weakens the membrane and causes strain in the bilayer as the surface area of the outer leaflet expands relative to the inner. This strain is relieved by a phase transition in which peptides reorient across the membrane, along with associated lipids, and form toroidal pores. As peptides continue to accumulate on at the outer leaflet of the bilayer, the process continues and the membrane begins to disintegrate into vesicles carpeted with peptides.

Antimicrobial peptides expressed in the alimentary tract have activity profiles and expression patterns that overlap with other mammalian peptides. Under optimized pH and ionic conditions, mouse Paneth cell α-defensins, also called cryptdins, were found to have microbicidal activity against *Escherichia coli, Staphylococcus aureus,* and *Salmonella typhimurium* (Ouellette et al., 1994). An earlier report indicated that cryptdin-1 was active against only an attenuated, not the parental, virulent strain of *S. typhimurium* (Selsted et al., 1992). Among the six mouse Paneth cell-defensins, cryptdins 4 and 5 are the more potent (Ouellette et al., 1994). Similarly, human Paneth cell HD5 is active against several bacterial species including *Listerid monocytogenes, E. coli,* and *S. typhimurium,* and also against the fungus *Candida albicans* (Porter, van Dam et al., 1997; Ghosh et al., 2002).

Primary structure of α-defensins affects their antimicrobial activities in vitro. For example, cryptdin 3, which differs from cryptdin 2 only by a single K/T substitution in position 10, was a more potent microbicide against *E. coli* (Ouellette et al., 1994). Against the intestinal protozoan *Giardia lamblia* trophozoites, cryptdins 2 and 3 were highly active in vitro, but cryptdins 1 and 6 had little effect on trophozoite survival. Although cryptdin 4 was the most cationic and potent of the cryptdins recovered from the intestinal lumen, a variant of cryptdin 4, which lacked the N-terminal Gly residue, was found to have greatly attenuated antibacterial activity against *E. coli* and *S. typhimurium* (Ouellette and Selsted, 1996). Thus it appears that the composition of N-terminus may be an important factor in the ability of this peptide class to kill certain microbial targets.

The β-defensins also have in vitro activity against bacteria. Human β-defensin 1 (HBD1) has activity against *Pseudomonas aeruginosa, L. monocytogenes,* and *E. coli* (Valore et al., 1998), but it is not a particularly potent defensin in vitro. Except for the kidney and a few other tissues in which high-level expression is observed, interpretation of its direct antimicrobial role in the wide distribution of tissues with low-level expression requires some caution. HBD2 is microbicidal against *P. aeruginosa, E. coli,* and, *C. albicans* (Harder, Bartels et al., 1997; Bals, Wang, Wu, Freeman, Bafna, Zasloff, and Wilson, 1998; Singh et al., 1998), but shows less activity against Gram-positive *S. aureus* (Harder, Bartels et al., 1997). In contrast, the more cationic HBD3 has activity against *S. aureus* (Harder et al., 2001) and is less sensitive to the ion composition of the assay medium.

Cathelicidins are structurally very diverse, and the expression patterns and activities of many peptides have been characterized (Zanetti et al., 1997; Lehrer and Ganz, 2002b; Zanetti et al., 2002). Within the alimentary tract, human LL-37 has been the focus of extensive study. This peptide has a broad spectrum of antibacterial activity, including *L. monocytogenes, S. aureus, P. aeruginosa, E. coli,* and *S. typhimurium* (Frohm et al., 1997; Bals et al., 1998; Turner et al., 1998). It does not have activity against *C. albicans* (Turner et al., 1998).

There is evidence that some antimicrobial peptides, in addition to their antimicrobial activities, may act as chemokines (Yang et al., 1999; Yang et al., 2000). For example, HBD2 has chemoattractant activity for dendritic cells through interaction with the chemokine receptor CCR-6 (Yang et al., 1999) and could serve as a bridge between the innate and adaptive immune systems (Yang, Chertov, and Oppenheim, 2001). Interestingly, recent studies on MIP-3α/CCL20, a chemokine with high affinity for CCR-6, revealed structural similarities with HBD2, which may account for their common binding to CCR-6 (Perez-Canadillas et al., 2001). HBD1 is similarly chemotactic for cells expressing CCR-6 (Yang et al., 1999). The chemokine activity of CCL20 was approximately 10-fold greater than that of HBD1 and HBD2. The cathelicidin

LL-37 is chemotactic for neutrophils, monocytes, and T cells, but not for dendritic cells (Agerberth et al., 2000; De et al., 2000). This activity is mediated through binding to the formyl-peptide receptor-like 1 receptor (De et al., 2000). It should be noted that, in contrast to the micromolar concentrations needed to kill bacteria, these chemokine activities were observed at nanomolar concentrations.

Other activities have been reported for some defensins and cathelicidins. Certain antimicrobial peptides, including LL-37 (Larrick et al., 1995), have been shown to bind lipopolysaccharide (LPS). The mouse intestinal defensins, cryptdin 2 and 3, promote ion fluxes in epithelial cells (Lencer et al., 1997; Merlin et al., 2001; Yue et al., 2002). Thus antimicrobial peptides have interesting activities beyond killing microbes (Yang et al., 2002). Several of these functions may prove key to the coordination of host defense responses, combining their microbicidal functions with recruitment of immune cells and regulation of ion fluxes.

7.4. ANTIMICROBIAL PEPTIDES OF THE SMALL INTESTINE

The small intestinal epithelium provides a very large luminal absorptive area, which facilitates adequate nutrient absorption, but also increases the opportunity for microorganisms to infect this mucosal surface. Following a meal, the lumen is rich with nutrients, yet to prevent microbial consumption of this aliment, mechanisms must exist to limit the numbers of microbes at this site (Fig. 7.1). There is substantial evidence that antimicrobial peptides play an important role in maintaining a physiological balance to allow the small bowel to efficiently execute its digestive functions in the face of these microbial challenges.

7.4.1. Paneth Cells: A Source of Abundant Antimicrobial Peptides

In most mammals, Paneth cells occupy a position at the base of the small intestinal crypts of Lieberkühn (Fig. 7.4) and are normally found all along the small intestine from the duodenum to the ileum (Porter et al., 2002). Paneth cells have ultrastructural hallmarks of secretory cells, including an extensive endoplasmic reticulum, Golgi network, and secretory granules (Trier, 1963). In response to cholinergic agonists (Satoh et al., 1995) and bacterial stimuli, Paneth cells secrete granules into the crypt lumen (Qu et al., 1996; Ayabe et al., 2000; Putsep et al., 2000). This secretion recently was shown to be modulated by the calcium-activated potassium channel mIKCa1

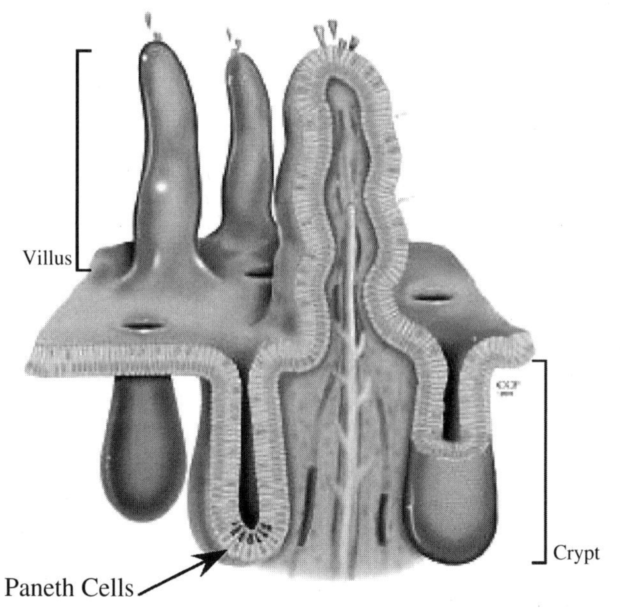

Figure 7.4. Crypt-villus architecture of the small intestinal mucosa. An artist's diagram of the small intestinal mucosa showing the positioning of Paneth cells at the base of the small intestinal crypts. Stem cells reside at the neck of the small intestinal crypt, where they divide and their daughter cells then migrate either toward the villi or deeper toward the base of the crypt. The migration is accompanied by cellular differentiation. The cells at the villus tips ultimately die by apoptosis and are sloughed. The Paneth cells secrete their vesicles into the narrow crypt lumen. These epithelial secretions contain defensins, lysozyme and secretory phospholipase A2, a potent antimicrobial cocktail proposed to defend the crypt, and its vital stem cell population from microbial invasion. (Illustrated by D. Schumick, Department of Medical Illustration, Cleveland Clinic Foundation, © 2001.)

(Ayabe, Wulff et al., 2002). The identification of abundant antimicrobials in Paneth cell granules has provided insights into the immune function of these cells. Paneth cells of rodents and humans produce lysozyme (Peeters and Vantrappen, 1975), secretory phospholipase A_2 ($sPLA_2$) (Senegas-Balas et al., 1984; Harwig et al., 1995) and α-defensins (Selsted et al., 1992; Porter, Liu et al., 1997; Ghosh et al., 2002). Although these cells are key to innate immunity, their ontogeny does not depend on luminal bacteria or dietary antigens. They are detected prenatally during normal human ontogeny (Mallow et al., 1996) and are present in mice, even in those reared under germ-free conditions (Ouellette et al., 1989; Putsep et al., 2000). Although Paneth cell α-defensin expression is largely constitutive, recent studies suggest that gene expression may be increased in the course of pathological processes, including necrotizing enterocolitis (Salzman et al., 1998), inflammatory bowel disease (Cunliffe et al., 2001), and hemorrhagic shock (Condon, 1999).

Murine Paneth cells express at least six α-defensin peptides (cryptdins) (Eisenhauer, Harwig, and Lehrer, 1992; Selsted et al., 1992; Ouellette et al., 1994, 2000; Putsep et al., 2000). Humans express small intestinal defensin genes at levels similar to those in mice (Jones and Bevins, 1992; Mallow et al., 1996; Porter, Liu et al., 1997; Cunliffe et al., 2001; Ghosh et al., 2002). In contrast to the larger number of α-defensin isoforms in mice, only two, HD5 and 6, are found in human Paneth cells (Mallow et al., 1996). Estimated defensin concentrations secreted into the crypt lumen approach high millimolar levels, suggesting they serve as powerful antimicrobial agents at this site (Ayabe et al., 2000; Ghosh et al., 2002). Based on defensin concentrations stored in the ileal mucosa, concentrations of defensins could reach 100 μg/ml or more in the *intestinal* lumen if all of the stored peptide were simultaneously secreted. These estimated levels are sufficient to support a potential antimicrobial role of Paneth cell defensins in the crypt and intestinal lumen.

7.4.2. Proposed Function(s) of Paneth Cell Antimicrobials

The proposed physiological functions of Paneth cell antimicrobial peptides may be grouped into four overlapping categories. First, Paneth cell antimicrobial peptides may protect the epithelial stem cells. The intestinal epithelium is regenerated continually and rapidly throughout the lifetime of mammals, highlighting the vital function of these stem cells. Given the estimated millimolar concentrations of defensins in the crypt lumen, far above their minimal microbicidal concentrations, defensins are very likely to be effective protectors of stem cells. Second, the arsenal of Paneth cell antimicrobials might regulate the numbers of colonizing microbes in the small intestine. Secretions from Paneth cells likely contribute to an antimicrobial milieu that prevents the colonization of the small intestine by microbes in numbers comparable with those found in the adjacent colon. Third, through their selective antimicrobial activity, Paneth cell antimicrobials may shape the composition of the commensal bacteria. Finally, Paneth cell antimicrobials may protect the host from food and waterborne pathogens. Newly colonizing microbes of the small intestine, including pathogens, will be confronted with Paneth cell antimicrobials. Interestingly, parasympathetic (cholinergic) neural activity not only regulates peristalsis and numerous digestive functions throughout the alimentary tract, but it also elicits Paneth cell secretion. It is possible that neurally mediated cholinergic stimulation during oral ingestion induces Paneth cell secretion and thereby equips the intestinal lumen with anticipatory effector molecules, including defensins, that could counter injurious microbes in the food or water. The four proposed functions of Paneth cell secretions should be considered as overlapping rather than as mutually exclusive.

7.4.3. Posttranslational Modification of α-Defensins

As first shown for α-defensins expressed in phagocytic leukocytes, the principal posttranslational modifications of defensin peptides are (1) formation of three intramolecular disulfide bonds and (2) proteolytic processing of the NH_2-terminus to yield the mature peptide (Fig. 7.2). Both modifications are important for antimicrobial activity, as reduction of the disulfide linkages obliterates antimicrobial activity of many defensins and the nature of the NH_2-terminus has significant affects on the bioactivity of the defensin peptides (Valore et al., 1998; Ouellette et al., 2000). In mice, matrilysin (MMP-7), a metalloproteinase expressed in mouse Paneth cells, is essential for cryptdin peptide processing and is indispensable for cryptdins to function in mucosal immunity (Wilson et al., 1999). In vitro, MMP-7 processes pro-cryptdins 1–3 and 6 to active mature peptides by cleaving precursors between residues Ser-58 and Leu-59, precisely at the site identified in vivo. (Selsted et al., 1992; Ouellette et al., 2000; Putsep et al., 2000; Shirafuji et al., 2003).

Ouellette and colleagues recently examined pro-cryptdin processing by MMP-7 in vitro (Shirafuji et al., 2003). Mature cryptdin 4 isolated from the intestinal lumen indicates that final processing takes place at Gly-61. In vitro proteolysis of pro-cryptdin 4 with MMP-7 results in cleavage at Ser-58, unmasking *in vitro* bactericidal activity to the levels comparable with those of mature cryptdin 4 peptide (Shirafuji et al., 2003). Because the cryptdin 4 N-terminus occurs at Gly-61, not at Leu-59, MMP-7 is not sufficient for complete processing of cryptdin 4. The second protease responsible for this final processing step is not yet known. Also, when native pro-cryptdin 4 was subjected to MMP-7 proteolysis, the mature peptide region consistently resisted cleavage and remained intact, but when the tridisulfide array was disrupted by reduction and alkylation, cleavage occurred at three positions of the mature peptide (Shirafuji et al., 2003). This observation indicates that the disulfide bonds are crucial for stability to protease activity, not only for antimicrobial activity.

An L59S mutant of procryptdin 4 was not cleaved in vitro by MMP-7 at the peptide N-terminus. Interestingly, the precursor of cryptdin 4 in C57Black/6 mice contains this L59S substitution (Shirafuji et al., 2003). Mature cryptdin 4 isolated from C57Black/6 mice has Leu-54 as its N-terminus, with no evidence of cleavage at the Ser58–Ser59 cleavage site. Thus polymorphic isoforms of α-defensins resulting from mutations at MMP-7 cleavage sites exist between mouse lines and influence posttranslational processing.

Surprisingly, the processing enzyme for human enteric defensins is not the same as that in mice. Whereas MMP-7 activity is required for α-defensin processing in mice (Wilson et al., 1999), this metalloproteinase is not detected in human Paneth cells. In humans, trypsin was recently identified as the protease responsible for the observed processing of the HD5 (and 6) (Ghosh et al., 2002). The human defensins are cleaved on the C-side of an Arg residue, to generate a 63–94 form of HD5 and a 69–101 form of HD6 (Porter et al., 1998; Ghosh et al., 2002). Trypsin is expressed by human Paneth cells (Ghosh et al., 2002), and current data indicate that Paneth cell trypsin, stored as a zymogen, is activated after secretion, and converts proHD5 into a mature peptide. The protease responsible for activation of Paneth cell trypsinogen is not known and is the focus of active investigation.

Interestingly, the processed lumenal form of mouse cryptdin 4 is cleaved at a site analogous to that of the human defensin HD5, on the C-side of Arg-60 (Fig. 7.2). Given that HD5 is cleaved in vitro by recombinant MMP-7 to a peptide (54–94), there was a question as to whether mouse-to-human differences in Paneth cell defensin processing were reflective of the defensin propeptide primary structure or protease activities in the crypt. Studies of a transgenic model in which HD5 is expressed in the small intestine of mice showed that HD5 was proteolytically processed in the lumen to the same form as that found in humans (aa 63–94) and at a site similar to cryptdin 4 processing (61–92) (Salzman et al., 2003). The mechanisms of this processing are currently under study.

7.4.4. In Vivo Models of Enteric α-Defensin Function

Targeted disruption of the MMP-7 gene impairs enteric host defense. MMP-7 null mice are less effective at clearing orally administered noninvasive *E. coli*, and they succumb more rapidly and to lower doses of virulent *S. typhimurium* (Wilson et al., 1999). A potential mechanism for this effect was suggested by the observation that mice rendered deficient in this defensin-processing enzyme do not produce mature defensins (Wilson et al., 1999) and are deficient in secreted antibacterial activity (Ayabe et al., 2002). It remains a possibility that the loss of other intestinal activities regulated by MMP-7 contributed to the change in host susceptibility. This caveat notwithstanding, these experiments provided the first compelling evidence for the significance of proteolytic processing in defensin biology and the key role of cryptdins in host defense.

In a complementary series of experiments, Salzman et al. developed a transgenic model in which the human defensin HD5 is expressed in mouse Paneth cells (Salzman et al., 2003). In two independently derived lines, using a 2.9-kb HD5 minigene, containing its two exons and 1.4 kb of 5′-flanking sequence, they observed Paneth cell expression. The HD5 mRNA levels in the transgenic mice are comparable with expression of the endogenous mouse cryptdin 4, and the level of transgenic HD5 mRNA is similar to that observed in the human ileal mucosa. In situ hybridization analysis localized HD5 RNA expression to Paneth cells, indicating that the HD5 transgene, under the control of its own promoter, is appropriately expressed in a cell-specific manner. Additionally, as noted in the preceding subsection, the HD5 recovered from the lumen is the same form (63–94) as that detected in the human lumen. The transgenic expression of HD5 provided augmentation of host defense, despite the presence of endogenous cryptdins. The HD5 mice proved to be dramatically resistant to oral challenge with virulent *S. typhimurium*. Following oral challenge, fewer Salmonella bacteria were cultured from the distal small intestine of transgenic mice. Also, compared with wild-type mice, transgenic mice excreted fewer Salmonella bacteria in their feces. The protective effect was detectable very early during the infection (already at 6 h) and was observed with infection only by the oral route (not by the peritoneal route). Taken together, the data demonstrated that Salmonella were more efficiently killed in the transgenic, compared with the wild-type, small intestinal lumen and supported the model of direct lumenal antimicrobial effect of defensins. This is the first report of a gain-of-function transgenic model, which provides evidence for the in vivo function of a defensin peptide.

7.5. ANTIMICROBIAL PEPTIDES OF THE ORAL MUCOSA, ESOPHAGUS, AND STOMACH

The mucosa of the proximal portion of the alimentary tract is faced with both colonizing microflora and variable numbers of microbes in food and water. There is evidence of both constitutive and inducible expression of antimicrobial peptides at these sites. In saliva, which coats the oral cavity and the teeth and permeates ingested food, antimicrobial activity is supplied by several antimicrobial peptides and proteins, including histatins, lysozyme, lactoferrin, and lactoperoxidase (Schenkels, Veerman, and Nieuw Amerongen, 1995). The first evidence of defensin expression in the oral cavity was from Zasloff and co-workers (Schonwetter, Stolzenberg, and Zasloff, 1995), who showed by in situ hybridization that a β-defensin, called a lingual antimicrobial peptide (LAP), is constitutively expressed in the basal layers of the squamous epithelium. LAP mRNA expression was markedly increased near sites of inflammation or injury in the tongue. LAP, isolated from bovine tongue, had antibiotic activity against Gram-positive and Gram-negative bacteria, as well as against *C. albicans* yeast (Schonwetter et al., 1995). More recently the dorsal aspect of the porcine tongue was shown to constitutively express a β-defensin called pBD1 (Zhang et al., 1998; Shi et al., 1999; Zhang et al., 1999). The peptide was highly localized to an approximately 0.1-mm-thick layer in the cornified tips of the filliform papillae. Estimates of the peptide concentration in this layer were in the range 20–100 μg/ml. Although the peptide had microbicidal activity similar to that of LAP, at sublethal concentrations pBD1 showed synergistic activity with porcine cathelicidins against Gram-negative bacteria. This suggested that the cathelicidins from neutrophils attracted to sites of inflammation could enhance the activity of epithelially produced pBD1. Interestingly, defensin expression in the tongue is not limited to mammals, as birds have been shown to express a β-defensin in the tongue also (Zhao et al., 2001).

In the human proximal alimentary tract, the expression of several β-defensin peptides has been documented. In the oral cavity, HBD1 and HBD2 have been observed in the oral mucosa, including the periodontal sulcus epithelium (Krisanaprakornkit et al., 1998; Weinberg, Krisanaprakornkit, and Dale, 1998; Krisanaprakornkit et al., 2000; Dale et al., 2001). HBD1 and HBD2 were also detected in the stomach (O'Neil et al., 2000; Bajaj-Elliott et al., 2002). The mRNA-encoding HBD3 was identified in the tongue, in addition to its prominent expression in skin (Harder et al., 2001).

Recent investigations have investigated mechanisms of HBD2 mRNA regulation in human gingival epithelial cells (Krisanaprakornkit et al., 2000; Krisanaprakornkit, Kimball, and Dale, 2002). When stimulated with a cell wall extract of the periodontal bacterium, *Fusobacterium nucleatum*, human gingival epithelial cells showed NF-κB activation and nuclear translocation of p65, consistent with involvement of NF-κB in HBD2 regulation. However, the HBD2 induction in this model system was not blocked by pretreatment with two inhibitors of the NF-κB pathway, pyrrolidine dithiocarbamate and MG132. Rather, p38 and c-Jun NH(2)-terminal kinase (JNK) pathways appear to mediate HBD2 expression. *F. nucleatum* activated p38 and JNK pathways, and inhibition of p38 and JNK partially blocked HBD2 mRNA induction, and the combination of the two inhibitors completely blocked expression.

As in humans, in mice the proximal alimentary tract also contains several β-defensins. Using Northern blot analysis and RNase protection analyses,

Jia et al. detected mouse β-defensin 4 (mBD4) mRNA expression in tongue and esophagus, as well as in the trachea (Jia et al., 2000). In their examination of 24 tissues, the signals for mBD1, mBD2, and mBD3 were overall much less intense than those for mBD4, with mBD3 weakly detected in the tongue, mBD1 detected in kidney, and no detection of mBD2. When a more sensitive reverse-transcriptase polymerase chain reaction (RT-PCR) method was used, mBD3 mRNA expression was also detected in the esophagus and lung, in addition to the tongue (Bals et al., 1999; Jia et al., 2000). In contrast to the inducible bovine, human, and mouse β-defensins (Diamond, Russell, and Bevins, 1996; Harder, Bartels et al., 1997; Bals et al., 1999), the mBD4 5′-flanking gene sequence had no consensus sequences for NF-κB binding (Jia et al., 2000). Comparing the proximal 5′-flanking region of the mBD3 and mBD4 genes, Jia et al. reported a high degree of similarity except for the omission of 26 bp of sequence in the mBD4 gene that encompassed the NF-κB consensus binding site in the mBD3 gene.

The expression of the α-defensin HD5 has been detected in diseased human esophagus and stomach; the cellular source of this expression was in metaplastic Paneth cells (Shen et al., 2000; Inada et al., 2001). Paneth cell metaplasia has also been observed in the stomach and colon in a variety of disease processes. It is possible that metaplasia of Paneth cells is an adaptive epithelial change that evolved as a host response against microbes but became part of a generalized response to other inflammatory stimuli.

7.6. ANTIMICROBIAL PEPTIDES OF THE COLON

Although the large intestine is in continual contact with a very high lumenal density of colonizing microbes (Simon and Gorbach, 1995), infections and translocation across the colonic mucosa by these bacteria are uncommon (Katouli et al., 1994). This is remarkable, considering the fact that the epithelium that separates the microbe-laden lumen from the body's internal tissues is only a single cell layer thick. How can this organ maintain effective mucosal defense in the face of such a high density of lumenal bacteria? According to one hypothesis, the colon is a complex ecosystem consisting of mutually interacting colonic epithelial cells, resident microflora, and classical immune cells (McCracken and Lorenz, 2001). In a healthy colon the ecosystem is in balance, and this balance might be maintained if epithelial cells provide a strong passive barrier to invasion by colonizing microbes, rather than active secretion of antimicrobials, as seen in the adjacent small intestine. Consistent with this notion, a recent study by Howell et al. identified five antimicrobial polypeptides that normally reside inside cells, and they proposed that these molecules contribute to a barrier function of this epithelium (Howell et al., 2003). The intracellular molecules isolated from normal colonic epithelial cells in this study might appear only extracellularly when the ecosystem is perturbed, perhaps at a time of cellular injury. These investigators suggested that, by not secreting the antimicrobials and keeping them instead as a latent intracellular effector, the colon is able to protect the host from infection and also maintain an abundant microbial flora as part of a balanced ecosystem. This provocative idea will require further, more rigorous examination.

Several studies examining the colonic mucosa have found that the synthesis of defensins and cathelicidins can be induced by inflammation or infection. For example, enteric β-defensin expression in bovine colonic epithelium was elevated during parasitic infection (Tarver et al., 1998). On IL-lα stimulation or infection of those cells with enteroinvasive bacteria, human colonic epithelial cell lines HT-29 and Caco-2 express HBD2 mRNA and protein that is not seen in unstimulated cells (O'Neil et al., 2000). In this study, HBD2 functioned as a NF-κB target gene as blocking NF-κB activation inhibited the rapid induction. Similarly, human fetal intestinal xenografts showed HBD2 upregulated expression in xenografts infected intraluminally with Salmonella (O'Neil et al., 2000). The production of HBD2 in human colonocytes is also induced by bacterial flagellin (Ogushi et al., 2001; Takahashi et al., 2001). Expression of HBD1 was detected in the epithelium of normal and inflamed colon, and in colonic epithelial cell lines (O'Neil et al., 2000), but antimicrobial activity associated with this peptide was not detected in the study of Howell (Howell et al., 2003). By RT-PCR, Fahlgren et al. detected an increase of HBD2, HD5, HD6, and lysozyme expression in the chronic inflammation of ulcerative colitis and Crohn's disease (Fahlgren et al., 2003). Another RT-PCR study by Wehkamp et al. found that HBD2 and HBD3 were increased in ulcerative colitis (Wehkamp et al., 2003).

The expression of LL-37 in human colonocytes appears to be linked to cellular differentiation; mRNA and protein expression was increased in spontaneously differentiating Caco-2 cells and in HCA-7 cells treated with the cell differentiation-inducing agent sodium butyrate (Hase et al., 2002). The virulence of *Shigella* appears to be linked to the ability of bacteria or bacterial DNA to downregulate the expression of both LL-37 and HBD1 (see next section) (Islam et al., 2001).

Other inducible molecules in the colonic epithelium are lysozyme (Cappello et al., 1992) and sPLA2 (Nevalainen, Gronroos, and Kallajoki, 1995; Minami et al., 1997). However, none of these postulated molecules were detected in healthy colonic epithelia. Therefore these studies suggest that, with inflammatory disease, a disruption of the colonic ecosystem, there is an upregulation of antimicrobial factors by colonic epithelial cells that are not normally present in a healthy colon. Further studies of antimicrobial peptide regulation may provide insight into how peptide expression may be modulated to control infection, alter colonization, and/or help resolve inflammation.

7.7. ENTERIC ANTIMICROBIAL PEPTIDES: TARGETS OF BACTERIAL VIRULENCE FACTORS

Mounting evidence supports that many pathogenic microbes have evolved mechanisms to evade killing by antimicrobial peptides and proteins, and these mechanisms serve as critical virulence factors for successful pathogens (Ganz, 2002; Peschel, 2002). Several examples have been identified for enteropathogens. Perhaps the best-understood virulence factors in this regard are in *S. typhimurium*, a Gram-negative pathogen that causes enteritis in humans and a systemic illness in susceptible mouse strains. A Salmonella PhoP-PhoQ regulatory system senses host microenvironments and responds by activating virulence gene transcription (Rakeman and Miller, 1999; Ohl and Miller, 2001). One process, regulated by PhoP-PhoQ, is the covalent modification of the lipid A portion of LPS with aminoarabinose, ethanolamine, and 2-hydroxymyristate. These lipid modifications result in a decrease in negative charge on the bacterial surface and confer an increased resistance to polymyxin and other cationic antimicrobial peptides (Gunn and Miller, 1996; Guo et al., 1998; Gunn et al., 2000). A putative potassium-coupled efflux pump is regulated also by PhoP-PhoQ (Parra-Lopez, Baer, and Groisman, 1993; Parra-Lopez et al., 1994). This efflux pump also confers resistance to cationic antimicrobial peptides. A third virulence factor also regulated by PhoP-PhoQ is the PgtE protease, an outer membrane protease that cleaves α-helical antimicrobial peptides (Guina et al., 2000).

Another intriguing mechanism of resistance to antimicrobial peptides by Gram-negative bacteria was reported by Gudmundsson and colleagues (Islam et al., 2001). *Shigella* sp. were shown to block the expression of HBD1 and LL-37 in epithelial cells, thereby potentially escaping the lethal activity of these microbicidal peptides. Many details on the mechanism of this interesting process remain to be investigated. A similar mechanism of avoiding antimicrobial peptides was reported by Salzman and colleagues, who found that *S. typhimurium* inhibited the expression of α-defensins and Lysozyme in murine Paneth cells in vivo (Salzman et al., 2003).

The ability to resist or avoid the action of cationic antimicrobial peptides is emerging as a widely distributed virulence mechanism of bacterial

Are there additional as yet undescribed antimicrobial effectors in the alimentary tract of mammals?

What phenotypes are associated with mutations in antimicrobial peptide genes (occurring naturally or through genetic manipulation)?

What are the physiological roles of antimicrobial peptides in the digestive tract?

- details on roles in host defense
- possible roles in signaling and electrolyte transport
- crosstalk between the acquired and innate immune systems
- establishment, selection, and maintenance of gastrointestinal flora
- other

What are the key structural determinants of the activity of antimicrobial peptides?

How is the expression of antimicrobial peptides regulated?

How does this defense system vary during development and with aging?

Figure 7.5. Questions for future investigations.

pathogens. These bacterial resistance mechanisms may provide targets for the development of useful therapeutics.

7.8. CONCLUDING REMARKS

Increasing attention has focused on elucidating the contributions of innate immunity to mammalian host defense. Antimicrobial peptides likely contribute to host defense of the alimentary tract through selective antibiotic activity, influencing the composition and limiting the numbers of transient and resident lumenal microbes, and thereby providing a protection that preserves the structural integrity and critical physiological functions of this organ system. Evidence for a key role of antimicrobial peptides in host defense is emerging from recent experiments that assess the impact of ablation and augmentation of antimicrobial peptide production, yet many questions remain (Fig. 7.5). Further clarification of their role in defense of this system may emerge when the impact of genetic mutations, developmental immaturity, chronic inflammation, or concurrent systemic disease on antimicrobial peptide expression and function is linked to susceptibility to disease.

REFERENCES

Agerberth, B., Charo, J., Werr, J., Olsson, B., Idali, F., Lindbom, L., Kiessling, R., Jornvall, H., Wigzell, H., and Gudmundsson, G. H. (2000). The human antimicrobial and chemotactic peptides LL-37 and alpha-defensins are expressed by specific lymphocyte and monocyte populations. *Blood*, 96, 3086–93.

Ayabe, T., Satchell, D. P., Pesendorfer, P., Tanabe, H., Wilson, C. L., Hagen, S. J., and Ouellette, A. J. (2002). Activation of Paneth cell alpha-defensins in mouse small intestine. *Journal of Biological Chemistry*, 277, 5219–28.

Ayabe, T., Satchell, D. P., Wilson, C. L., Parks, W. C., Selsted, M. E., and Ouellette, A. J. (2000). Secretion of microbicidal alpha-defensins by intestinal Paneth cells in response to bacteria. *Nature Immunology*, 1, 113–18.

Ayabe, T., Wulff, H., Darmoul, D., Cahalan, M. D., Chandy, K. G., and Ouellette, A. J. (2002). Modulation of mouse Paneth cell alpha-defensin secretion by mIKCa1, a Ca2+-activated, intermediate conductance potassium channel. *Journal of Biological Chemistry*, 277, 3793–800.

Bajaj-Elliott, M., Fedeli, P., Smith, G. V., Domizio, P., Maher, L., Ali, R. S., Quinn, A. G., and Farthing, M. J. (2002). Modulation of host antimicrobial peptide (beta-defensins 1 and 2) expression during gastritis. *Gut*, 51, 356–61.

Bals, R., Goldman, M. J., and Wilson, J. M. (1998). Mouse beta-defensin 1 is a salt-sensitive antimicrobial peptide present in epithelia of the lung and urogenital tract. *Infection and Immunity*, 66, 1225–32.

Bals, R., Wang, X., Meegalla, R. L., Wattler, S., Weiner, D. J., Nehls, M. C., and Wilson, J. M. (1999). Mouse beta-defensin 3 is an inducible antimicrobial peptide expressed in the epithelia of multiple organs. *Infection and Immunity*, 67, 3542–7.

Bals, R., Wang, X., Wu, Z., Freeman, T., Bafna, V., Zasloff, M., and Wilson, J. M. (1998). Human β-defensin 2 is a salt-sensitive peptide antibiotic expressed in human lung. *Journal of Clinical Investigation*, 102, 874–80.

Bals, R., Wang, X., Zasloff, M., and Wilson, J. M. (1998). The peptide antibiotic LL-37/hCAP-18 is expressed in epithelia of the human lung where it has broad antimicrobial activity at the airway surface. *Proceedings of the National Academy of Sciences USA*, 95, 9541–6.

Bensch, K. W., Raida, M., Mägert, H.-J., Schulz-Knappe, P., and Forssmann, W.-G. (1995). hBD-1: A novel β-defensin from human plasma. *FEBS Letters*, 368, 331–5.

Boman, H. G. (1995). Peptide antibiotics and their role in innate immunity. *Annual Review of Immunology*, 13, 61–92.

Cappello, M., Keshav, S., Prince, C., Jewell, D., and Gordon, S. (1992). Detection of mRNAs for macrophage products in inflammatory bowel disease by in situ hybridisation. *Gut*, 33, 1214–19.

Condon, M. R. (1999). Induction of a rat enteric defensin gene by hemorrhagic shock. *Infection and Immunity*, 67, 4787–93.

Cunliffe, R. N., Rose, F. R., Keyte, J., Abberley, L., Chan, W. C., and Mahida, Y. R. (2001). Human defensin 5 is stored in precursor form in normal Paneth cells and is expressed by some villous epithelial cells and by metaplastic Paneth cells in the colon in inflammatory bowel disease. *Gut*, 48, 176–85.

Dale, B. A., Kimball, J. R., Krisanaprakornkit, S., Roberts, F., Robinovitch, M., O'Neal, R., Valore, E. V., Ganz, T., Anderson, G. M., and Weinberg, A. (2001). Localized antimicrobial peptide expression in human gingiva. *Journal of Periodontal Research*, 36, 285–94.

De, Y., Chen, Q., Schmidt, A. P., Anderson, G. M., Wang, J. M., Wooters, J., Oppenheim, J. J., and Chertov, O. (2000). LL-37, the neutrophil granule- and epithelial cell-derived cathelicidin, utilizes formyl peptide receptor-like 1 (FPRL1) as a receptor to chemoattract human peripheral blood neutrophils, monocytes, and T cells. *Journal of Experimental Medicine*, 192, 1069–74.

Diamond, G., Russell, J. P., and Bevins, C. L. (1996). Inducible expression of an antibiotic peptide gene in lipopolysaccharide-challenged tracheal epithelial cells. *Proceedings of the National Academy of Sciences USA*, 93, 5156–60.

Eisenhauer, P. B., Harwig, S. S. L., and Lehrer, R. I. (1992). Cryptdins: Antimicrobial defensins of the murine small intestine. *Infection and Immunity*, 60, 3556–65.

Fahlgren, A., Hammarstrom, S., Danielsson, A., and Hammarstrom, M. (2003). Increased expression of antimicrobial peptides and lysozyme in colonic epithelial cells of patients with ulcerative colitis. *Clinical and Experimental Immunology*, 131, 90–101.

Frohm, M., Agerberth, B., Ahangari, G., Stahle-Backdahl, M., Liden, S., Wigzell, H., and Gudmundsson, G. H. (1997). The expression of the gene coding for the antibacterial peptide LL-37 is induced in human keratinocytes during inflammatory disorders. *Journal of Biological Chemistry*, 272, 15258–63.

Ganz, T. (2002). Epithelia: Not just physical barriers. *Proceedings of the National Academy of Sciences USA*, 99, 3357–8.

Ghosh, D., Porter, E. M., Shen, B., Lee, S. K., Wilk, D. J., Crabb, J. W., Drazba, J., Yadav, S. Y., Ganz, T., and Bevins, C. L. (2002). Paneth cell trypsin is the processing enzyme for human defensin-5. *Nature Immunology*, 3, 583–90.

Guina, T., Yi, E. C., Wang, H., Hackett, M., and Miller, S. I. (2000). A PhoP-regulated outer membrane protease of *Salmonella enterica* serovar

Typhimurium promotes resistance to alpha-helical antimicrobial peptides. *Journal of Bacteriology*, 182, 4077–86.

Gunn, J. S. and Miller, S. I. (1996). PhoP-PhoQ activates transcription of pmrAB, encoding a two-component regulatory system involved in *Salmonella typhimurium* antimicrobial peptide resistance. *Journal of Bacteriology*, 178, 6857–64.

Gunn, J. S., Ryan, S. S., Ernst, R. K., and Miller, S. I. (2000). Genetic and functional analysis of a PmrA-PmrB-regulated locus necessary for lipopolysaccharide modification, antimicrobial peptide resistance, and oral virulence of *Salmonella enterica* serovar Typhimurium. *Infection and Immunity*, 68, 6139–46.

Guo, L., Lim, K. B., Poduje, C. M., Daniel, M., Gunn, J. S., Hackett, M., and Miller, S. I. (1998). Lipid A acylation and bacterial resistance against vertebrate antimicrobial peptides. *Cell*, 95, 189–98.

Hancock, R. E. W., Falla, T., and Brown, M. (1995). Cationic bactericidal peptides. *Advances in Microbial Physiology*, 37, 135–75.

Harder, J., Bartels, J., Christophers, E., and Schröder, J. M. (1997). A peptide antibiotic from human skin. *Nature (London)*, 387, 861.

Harder, J., Bartels, J., Christophers, E., and Schröder, J. M. (2001). Isolation and characterization of human beta -defensin-3, a novel human inducible peptide antibiotic. *Journal of Biological Chemistry*, 276, 5707–13.

Harder, J., Siebert, R., Zhang, Y., Matthiesen, P., Christophers, E., Schlegelberger, B., and Schröder, J. M. (1997). Mapping of the gene encoding human beta-defensin-2 (DEFB2) to chromosome region 8p22-p23.1. *Genomics*, 46, 472–5.

Harwig, S. S. L., Tan, L., Qu, X.-D., Cho, Y., Eisenhauer, P. B., and Lehrer, R. I. (1995). Bactericidal properties of murine intestinal phospholipase A2. *Journal of Clinical Investigation*, 95, 603–10.

Hase, K., Eckmann, L., Leopard, J. D., Varki, N., and Kagnoff, M. F. (2002). Cell differentiation is a key determinant of cathelicidin LL-37/human cationic antimicrobial protein 18 expression by human colon epithelium. *Infection and Immunity*, 70, 953–63.

Howell, S. J., Wilk, D. J., Yadav, S. Y., and Bevins, C. L. (2003). Antimicrobial polypeptides of the human colonic epithelium. *Peptides*, in press.

Huang, H. W. (1999). Peptide-lipid interactions and mechanisms of antimicrobial peptides. *Novartis Foundation Symposium*, 225, 188–200.

Huttner, K. M., Kozak, C. A., and Bevins, C. L. (1997). The mouse genome encodes a single homolog of the antimicrobial peptide human beta-defensin 1. *FEBS Letters*, 413, 45–49.

Inada, K., Tanaka, H., Nakanishi, H., Tsukamoto, T., Ikehara, Y., Tatematsu, K., Nakamura, S., Porter, E. M., and Tatematsu, M. (2001). Identification of Paneth cells in pyloric glands associated with gastric and intestinal mixed-type intestinal metaplasia of the human stomach. *Virchows Archives*, 439, 14–20.

Islam, D., Bandholtz, L., Nilsson, J., Wigzell, H., Christensson, B., Agerberth, B., and Gudmundsson, G. (2001). Downregulation of bactericidal peptides in enteric infections: A novel immune escape mechanism with bacterial DNA as a potential regulator. *Nature Medicine*, 7, 180–5.

Jia, H.-P., Wowk, S. A., Schutte, B. C., Lee, S. K., Vivado, A., Tack, B. F., Bevins, C. L., and McCray Jr., P. B. (2000). A novel murine β-defensin expressed in tongue, esophagus and trachea. *Journal of Biological Chemistry*, 275, 33314–20.

Jones, D. E. and Bevins, C. L. (1992). Paneth cells of the human small intestine express an antimicrobial peptide gene. *Journal of Biological Chemistry*, 267, 23216–25.

Katouli, M., Bark, T., Ljungqvist, O., Svenberg, T., and Möllby, R. (1994). Composition and diversity of intestinal coliform flora influence bacterial translocation in rats after hemorrhagic stress. *Infection and Immunity*, 62, 4768–74.

Krisanaprakornkit, S., Kimball, J. R., and Dale, B. A. (2002). Regulation of human beta-defensin-2 in gingival epithelial cells: The involvement of mitogen-activated protein kinase pathways, but not the NF-κB transcription factor family. *Journal of Immunology*, 168, 316–24.

Krisanaprakornkit, S., Kimball, J. R., Weinberg, A., Darveau, R. P., Bainbridge, B. W., and Dale, B. A. (2000). Inducible expression of human beta-defensin 2 by *Fusobacterium nucleatum* in oral epithelial cells: Multiple signaling pathways and role of commensal bacteria in innate immunity and the epithelial barrier. *Infection and Immunity*, 68, 2907–15.

Krisanaprakornkit, S., Weinberg, A., Perez, C. N., and Dale, B. A. (1998). Expression of the peptide antibiotic human beta-defensin 1 in cultured gingival epithelial cells and gingival tissue. *Infection and Immunity*, 66, 4222–8.

Larrick, J. W., Hirata, M., Balint, R. F., Lee, J., Zhong, J., and Wright, S. C. (1995). Human CAP18: A novel antimicrobial lipopolysaccharide-binding protein. *Infection and Immunity*, 63, 1291–7.

Lehrer, R. I., Bevins, C. L., and Ganz, T. (1998). Defensins and other antimicrobial peptides. In *Mucosal Immunology*, ed. P. L. Ogra, J. Mestecky, M. E. Lamm, W. M. Strober, and J. Bienstock, pp. 89–99. New York: Academic.

Lehrer, R. I. and Ganz, T. (2002a). Defensins of vertebrate animals. *Current Opinions in Immunology*, 14, 96–102.

Lehrer, R. I. and Ganz, T. (2002b). Cathelicidins: A family of endogenous antimicrobial peptides. *Current Opinions in Hematology*, 9, 18–22.

Lehrer, R. I., Lichtenstein, A. K., and Ganz, T. (1993). Defensins: Antimicrobial and cytotoxic peptides of mammalian cells. *Annual Review of Immunology*, 11, 105–28.

Lencer, W. I., Cheung, G., Strohmeier, G. R., Currie, M. G., Ouellette, A. J., Selsted, M. E., and Madara, J. L. (1997). Induction of epithelial chloride secretion by channel-forming cryptins 2 and 3. *Proceedings of the National Academy of Sciences USA*, 94, 8585–9.

Mallow, E. B., Harris, A., Salzman, N., Russell, J. P., DeBerardinis, J. R., Ruchelli, E., and Bevins, C. L. (1996). Human enteric defensins: Gene structure and developmental expression. *Journal of Biological Chemistry*, 271, 4038–45.

Matsuzaki, K. (1999). Why and how are peptide-lipid interactions utilized for self-defense? Magainins and tachyplesins as archetypes. *Biochimica et Biophysica Acta*, 1462, 1–10.

McCracken, V. and Lorenz, R. (2001). The gastrointestinal ecosystem: A precarious alliance among epithelium, immunity and microbiota. *Cellular Microbiology*, 3, 1–11.

Merlin, D., Yue, G., Lencer, W. I., Selsted, M. E., and Madara, J. L. (2001). Cryptdin-3 induces novel apical conductance(s) in Cl- secretory, including cystic fibrosis, epithelia. *American Journal of Physiology and Cell Physiology*, 280, C296–302.

Minami, T., Shinomura, Y., Miyagawa, J., Tojo, H., Okamoto, M., and Matsuzawa, Y. (1997). Immunohistochemical localization of group II phospholipase A2 in colonic mucosa of patients with inflammatory bowel disease. *American Journal of Gastroenterology*, 92, 289–92.

Morrison, G. M., Davidson, D. J., Kilanowski, F. M., Borthwick, D. W., Crook, K., Maxwell, A. I., Govan, J. R. W., and Dorin, J. R. (1998). Mouse beta defensin-1 is a functional homolog of human beta defensin-1. *Mammalian Genome*, 9, 453–7.

Nevalainen, T. J., Gronroos, J. M., and Kallajoki, M. (1995). Expression of group II phospholipase A2 in the human gastrointestinal tract. *Laboratory Investigation*, 72, 201–8.

O'Neil, D. A., Cole, S. P., Martin-Porter, E., Housley, M. P., Liu, L., Ganz, T., and Kagnoff, M. F. (2000). Regulation of human beta-defensins by gastric epithelial cells in response to infection with *Helicobacter pylori* or stimulation with interleukin-1. *Infection and Immunity*, 68, 5412–15.

Ogushi, K., Wada, A., Niidome, T., Mori, N., Oishi, K., Nagatake, T., Takahashi, A., Asakura, H., Makino, S., Hojo, H., Nakahara, Y., Ohsaki, M., Hatakeyama, T., Aoyagi, H., Kurazono, H., Moss, J., and Hirayama, T. (2001). *Salmonella enteritidis* FliC (flagella filament protein) induces human beta-defensin-2 mRNA production by Caco-2 cells. *Journal of Biological Chemistry*, 276, 30521–6.

Ohl, M. E. and Miller, S. I. (2001). Salmonella: A model for bacterial pathogenesis. *Annual Reviews of Medicine*, 52, 259–74.

Ouellette, A. J., Greco, R. M., James, M., Frederick, D., Naftilan, J., and Fallon, J. T. (1989). Developmental regulation of cryptdin, a corticostatin/defensin precursor mRNA in mouse small intestinal crypt epithelium. *Journal of Cellular Biology*, 108, 1687–95.

Ouellette, A. J., Hsieh, M. M., Nosek, M. T., Cano-Gauci, D. F., Huttner, K. M., Buick, R. N., and Selsted, M. E. (1994). Mouse Paneth cell defensins: Primary structures and antibacterial activities of numerous cryptdin isoforms. *Infection and Immunity*, 62, 5040–7.

Ouellette, A. J., Miller, S. I., Henschen, A. H., and Selsted, M. E. (1992). Purification and primary structure of murine cryptdin-1, a Paneth cell defensin. *FEBS Letters*, 304, 146–8.

Ouellette, A. J., Satchell, D. P., Hsieh, M. M., Hagen, S. J., and Selsted, M. E. (2000). Characterization of luminal Paneth cell alpha-defensins in mouse small intestine: Attenuated antimicrobial activities of peptides with truncated amino termini. *Journal of Biological Chemistry*, 275, 33969–73.

Ouellette, A. J. and Selsted, M. E. (1996). Paneth cell defensins: Endogenous peptide components of intestinal host defense. *FASEB Journal*, 10, 1280–9.

Parra-Lopez, C., Baer, M. T., and Groisman, E. A. (1993). Molecular genetic analysis of a locus required for resistance to antimicrobial peptides in *Salmonella typhimurium*. *EMBO Journal*, 12, 4053–62.

Parra-Lopez, C., Lin, R., Aspedon, A. and Groisman, E. A. (1994). A Salmonella protein that is required for resistance to antimicrobial peptides and transport of potassium. *EMBO Journal*, 13, 3964–72.

Peeters, T. and Vantrappen, G. (1975). The Paneth cell: A source of intestinal lysozyme. *Gut*, 16, 553–8.

Pérez-Cañadillas, J. M., Zaballos, A., Gutierrez, J., Varona, R., Roncal, F., Albar, J. P., Marquez, G., and Bruix, M. (2001). NMR solution structure of murine CCL20/MIP-3alpha, a chemokine that specifically chemoattracts immature dendritic cells and lymphocytes through its highly specific interaction with the beta-chemokine receptor CCR6. *Journal of Biological Chemistry*, 276, 28372–9.

Peschel, A. (2002). How do bacteria resist human antimicrobial peptides? *Trends in Microbiology*, 10, 179–86.

Porter, E., Liu, L., Oren, A., Anton, P., and Ganz, T. (1997). Localization of human intestinal defensin 5 in Paneth cell granules. *Infection and Immunity*, 65, 2389–95.

Porter, E., van Dam, E., Valore, E., and Ganz, T. (1997). Broad spectrum antimicrobial activity of human intestinal defensin 5. *Infection and Immunity*, 65, 2396–401.

Porter, E. M., Bevins, C. L., Ghosh, D., and Ganz, T. (2002). The multifaceted Paneth cell. *Cellular and Molecular Life Sciences*, 59, 156–70.

Porter, E. M., Poles, M. A., Lee, J. S., Naitoh, J., Bevins, C. L., and Ganz, T. (1998). Isolation of human intestinal defensins from ileal neobladder urine. *FEBS Letters*, 434, 272–6.

Putsep, K., Axelsson, L. G., Boman, A., Midtvedt, T., Normark, S., Boman, H. G., and Andersson, M. (2000). Germ-free and colonized mice generate the same products from enteric prodefensins. *Journal of Biological Chemistry*, 275, 40478–82.

Qu, X. D., Lloyd, K. C., Walsh, J. H., and Lehrer, R. I. (1996). Secretion of type II phospholipase A2 and cryptdin by rat small intestinal Paneth cells. *Infection and Immunity*, 64, 5161–5.

Rakeman, J. L. and Miller, S. I. (1999). Salmonella typhimurium recognition of intestinal environments. *Trends in Microbiology*, 7, 221–3.

Salzman, N. H., Chou, M. M., de Jong, H., Liu, L., Porter, E. M., and Paterson, Y. (2003). Enteric Salmonella infection inhibits Paneth cell antimicrobial peptide expression. *Infection and Immunity*, 71, 1109–15.

Salzman, N. H., Polin, R. A., Harris, M. C., Ruchelli, E., Hebra, A., Zirin-Butler, S., Jawad, A., Porter, E. M., and Bevins, C. L. (1998). Enteric defensin expression in necrotizing enterocolitis. *Pediatric Research*, 44, 20–6.

Satoh, Y., Habara, Y., Ono, K., and Kanno, T. (1995). Carbamylcholine- and catecholamine-induced intracellular calcium dynamics of epithelial cells in mouse ileal crypts. *Gastroenterology*, 108, 1345–56.

Schenkels, L. C., Veerman, E. C., and Nieuw Amerongen, A. V. (1995). Biochemical composition of human saliva in relation to other mucosal fluids. *Critical Reviews in Oral Biology and Medicine*, 6, 161–75.

Schonwetter, B. S., Stolzenberg, E. D., and Zasloff, M. A. (1995). Epithelial antibiotics induced at sites of inflammation. *Science*, 267, 1645–8.

Selsted, M. E. and Harwig, S. S. (1989). Determination of the disulfide array in the human defensin HNP-2. A covalently cyclized peptide. *Journal of Biological Chemistry*, 264, 4003–7.

Selsted, M. E., Miller, S. I., Henschen, A. H., and Ouellette, A. J. (1992). Enteric defensins: Antibiotic peptide components of intestine host defense. *Journal of Cell Biology*, 118, 929–36.

Senegas-Balas, F., Balas, D., Verger, R., de Caro, A., Figarella, C., Ferrato, F., Lechene, P., Bertrand, C., and Ribet, A. (1984). Immunohistochemical localization of intestinal phosphlipase A2 in rat Paneth cells. *Histochemistry*, 81, 581–4.

Shai, Y. (1999). Mechanism of the binding, insertion and destabilization of phospholipid bilayer membranes by alpha-helical antimicrobial and cell nonselective membrane-lytic peptides. *Biochimica et Biophysica Acta*, 462, 55–70.

Shen, B., Dumot, J., Goldblum, J. R., Ghosh, D., Skacel, M., Falk, G. W., and Bevins, C. L. (2000). Expression of human defesin-5 in Barrett's esophagus. *American Journal of Gastroenterology*, 95, A81.

Shi, J., Zhang, G., Wu, H., Ross, C., Blecha, F., and Ganz, T. (1999). Porcine epithelial beta-defensin 1 is expressed in the dorsal tongue at antimicrobial concentrations. *Infection and Immunity*, 67, 3121–7.

Shirafuji, Y., Tanabe, H., Satchell, D. P., Henschen-Edman, A., Wilson, C. L., and Ouellette, A. J. (2003). Structural determinants of procryptdin recognition and cleavage by matrix metalloproteinase-7. *Journal of Biological Chemistry*, 278, 7910–19.

Simon, G. L. and Gorbach, S. L. (1995). Normal alimentary tract microflora. In *Infections of the Gastrointestinal Tract*, ed. M. J. Blaser, P. D. Smith, J. I. Ravdin, H. B. Greenberg, and R. L. Guerrant, pp. 53–69. New York, Raven Press.

Singh, P. K., Jia, H. P., Wiles, K., Hesselberth, J., Liu, L., Conway, B. A. D., Greenberg, E. P., Valore, E. V., Welsh, M. J., Ganz, T., Tack, B. F., and McCray Jr., P. B. (1998). Production of beta-defensins by human airway epithelia. *Proceedings of the National Academy of Sciences USA*, 95, 14961–6.

Takahashi, A., Wada, A., Ogushi, K., Maeda, K., Kawahara, T., Mawatari, K., Kurazono, H., Moss, J., Hirayama, T., and Nakaya, Y. (2001). Production of beta-defensin-2 by human colonic epithelial cells induced by *Salmonella enteritidis* flagella filament structural protein. *FEBS Letters*, 508, 484–8.

Tang, Y.-Q. and Selsted, M. E. (1993). Characterization of the disulfide motif in BNBD-12, an antimicrobial beta-defensin peptide from bovine neutrophils. *Journal of Biological Chemistry*, 268, 6649–53.

Tarver, A. P., Clark, D. P., Diamond, G., Cohen, K. M., Erdjument-Bromage, H., Jones, D. E., Sweeney, R., Wines, M., Hwang, S., Tempst, P., and Bevins, C. L. (1998). Enteric β-defensin: Molecular cloning and characterization of a gene with inducible intestinal epithelial expression associated with Cryptospiridium parvum expression. *Infection and Immunity*, 66, 1045–56.

Trier, J. S. (1963). Studies on small intestinal crypt epithelium. I. The fine structure of the crypt epithelium of the proximal small intestine of fasting humans. *Journal of Cellular Biology*, 18, 599–620.

Turner, J., Cho, Y., Dinh, N. N., Waring, A. J., and Lehrer, R. I. (1998). Activities of LL-37, a cathelin-associated antimicrobial peptide of human neutrophils. *Antimicrobial Agents and Chemotherapy*, 42, 2206–14.

Valore, E. V., Park, C. H., Quayle, A. J., Wiles, K. R., McCray Jr., P. B., and Ganz, T. (1998). Human β-defensin-1: An antimicrobial peptide of urogenital tissues. *Journal of Clinical Investigation*, 101, 1633–42.

Wehkamp, J., Harder, J., Weichenthal, M., Mueller, O., Herrlinger, K. R., Fellermann, K., Schroeder, J. M., and Stange, E. F. (2003). Inducible and constitutive beta-defensins are differentially expressed in Crohn's disease and ulcerative colitis. *Inflammatory Bowel Disease*, 9, 215–23.

Weinberg, A., Krisanaprakornkit, S., and Dale, B. A. (1998). Epithelial antimicrobial peptides: review and significance for oral applications. *Critical Reviews in Oral Biology and Medicine*, 9, 399–414.

Wilson, C. L., Ouellette, A. J., Satchell, D. P., Ayabe, T., Lopez-Boado, Y. S., Stratman, J. L., Hultgren, S. J., Matrisian, L. M., and Parks, W. C. (1999). Regulation of intestinal alpha-defensin activation by the metalloproteinase matrilysin in innate host defense. *Science*, 286, 113–17.

Yang, D., Biragyn, A., Kwak, L. W., and Oppenheim, J. J. (2002). Mammalian defensins in immunity: More than just microbicidal. *Trends in Immunology*, 23, 291–6.

Yang, D., Chen, Q., Chertov, O., and Oppenheim, J. J. (2000). Human neutrophil defensins selectively chemoattract naive T and immature dendritic cells. *Journal of Leukocyte Biology*, 68, 9–14.

Yang, D., Chertov, O., Bykovskaia, S. N., Chen, Q., Buffo, M. J., Shogan, J., Anderson, M., Schröder, J. M., Wang, J. M., Howard, O. M. Z., and Oppenheim, J. J. (1999). Beta-defensins: Linking innate and adaptive immunity through dendritic and T cell CCR6. *Science*, 286, 525–8.

Yang, D., Chertov, O., and Oppenheim, J. J. (2001). The role of mammalian antimicrobial peptides and proteins in awakening of innate host defenses and adaptive immunity. *Cellular and Molecular Life Sciences*, 58, 978–89.

Yue, G., Merlin, D., Selsted, M. E., Lencer, W. I., Madara, J. L., and Eaton, D. C. (2002). Cryptdin 3 forms anion selective channels in cytoplasmic membranes of human embryonic kidney cells. *American Journal of Physiology Gastrointestine and Liver Physiology*, 282, G757–65.

Zanetti, M., Gennaro, R., and Romeo, D. (1997). The cathelicidin family of antimicrobial peptide precursors: A component of the oxygen-independent defense mechanisms of neutrophils. *Annals of the New York Academy of Science*, 832, 147–62.

Zanetti, M., Gennaro, R., Skerlavaj, B., Tomasinsig, L., and Circo, R. (2002). Cathelicidin peptides as candidates for a novel class of antimicrobials. *Current Pharmaceutical Design*, 8, 779–93.

Zasloff, M. (2002). Antimicrobial peptides of multicellular organisms. *Nature (London)*, 415, 389–95.

Zhang, G., Hiraiwa, H., Yasue, H., Wu, H., Ross, C. R., Troyer, D., and Blecha, F. (1999). Cloning and characterization of the gene for a new epithelial beta-defensin. Genomic structure, chromosomal localization, and evidence for its constitutive expression. *Journal of Biological Chemistry*, 274, 24031–7.

Zhang, G., Wu, H., Shi, J., Ganz, T., Ross, C. R., and Blecha, F. (1998). Molecular cloning and tissue expression of porcine beta-defensin-1. *FEBS Letters*, 424, 37–40.

Zhao, C., Nguyen, T., Liu, L., Sacco, R. E., Brogden, K. A., and Lehrer, R. I. (2001). Gallinacin-3, an inducible epithelial beta-defensin in the chicken. *Infection and Immunity*, 69, 2684–91.

Zucht, H. D., Grabowski, J., Schrader, M., Liepke, C., Jürgens, M., Schulz-Knappe, P., and Forssmann, W. G. (1998). Human β-defensin-1: A urinary peptide present in variant molecular forms and its putative functional implication. *European Journal of Medical Research*, 3, 315–23.

CHAPTER 8

Antimicrobial peptides suppress microbial infections and sepsis in animal models

Kim A. Brogden, Mark Ackermann, Joseph Zabner,
and Michael J. Welsh

8.1. INTRODUCTION

The presence of antimicrobial peptides in mucosal tissues and secretions at the interface between the host's mucosa and its normal floral commensals or pathogens strongly suggests that these diverse peptides are involved in innate protection against microbial infection. This hypothesis is supported by the high antimicrobial activity many of these peptides have in vitro against Gram-negative bacteria, gram positive bacteria, fungi, and enveloped viruses (Lehrer, Lichtenstein, and Ganz, 1993; Lehrer and Ganz, 1996; Ramanathan et al., 2002; Schutte and McCray, 2002). It is also supported by the observation that individuals deficient in certain antimicrobial peptides [e.g., LL-37, human neutrophil peptides 1–3 (HNP1–HNP3), human β-defensin 2 (HBD2), cathelin-related antimicrobial peptide (CRAMP), etc.] (Ganz et al., 1988; Nizet et al., 2001; Ong et al., 2002; Putsep et al., 2002) or individuals with decreased antimicrobial peptide activity (e.g., cystic fibrosis) (Jiang et al., 1993; Smith et al., 1996; Goldman et al., 1997; Bals et al., 1998b) have chronic microbial infections. Therefore it is logical to assume that natural antimicrobial peptides or synthetic congeners with potent in vitro antimicrobial activity could suppress microbial infections in vivo if given topically, orally, or systemically, similar to conventional antibiotics. These peptides are fast acting and broad spectrum, do not induce bacterial resistance, are easy to synthesize in large quantities, and work in synergy with other antimicrobial agents, characteristics that are highly desirable for pharmacological development.

Although active against a variety of microorganisms in vitro, many peptides cannot be used to prevent or treat microbial infections and sepsis in vivo. Some have undesirable characteristics that include hemolysis and cytotoxicity for host cells of some species and decreased activity in host environments containing high ionic strength conditions or serum. Others have the potential to prevent or treat microbial infections and sepsis. The exact peptide compositions and their prophylactic or therapeutic applications are just now being determined and form the basis of this chapter. Here we present the types of animal models used to assess the efficacy of antimicrobial peptides against microbial infection and sepsis, the characteristics of ideal peptides, the uncertainties of their use, and future areas of study aimed at increasing their in vivo efficacy. We also include a rapidly developing area, which outlines efforts to increase natural endogenous antimicrobial activity in situ.

8.2. TYPES OF ANIMAL MODELS

Since the discovery of lysozyme by Fleming in 1922 (Fleming, 1922), over 750 eucaryotic antimicrobial peptides (Tossi et al., 2004) have been reported. However, only the efficacies of a select few have been assessed against microbial infections and sepsis in small- and large-animal models (Table 8.1). Such studies are providing new information. First, animal models are demonstrating that endogenous expression of mammalian antimicrobial peptides provide defense against bacterial infections, and there are a number of examples. CRAMP-deficient mice developed much larger areas of infection after subcutaneous injection of Group A Strep. (Nizet et al., 2001). Lesion areas increased more rapidly, reached larger maximal size, and persisted longer in CRAMP-deficient mice than in normal littermates, whereas heterozygotes tended to have intermediate-sized lesions. In porcine skin wound chamber models, the application of a specific neutrophil elastase inhibitor blocked the proteolytic activation of protegrins, leading to significantly decreased clearance of bacteria from the wound compared with wounds treated with cathepsin G inhibitor or solvent only (Cole et al., 2001). Administration of 1-μg/mL exogenous protegrin PG-1 4 h after chamber preparation was sufficient to normalize in vivo antimicrobial activity. Killing of *Pseudomonas aeruginosa* by secretions of a xenograft model of human airways was diminished if expression of HBD1 was inhibited in situ with antisense oligonucleotides (Goldman et al., 1997) and overexpression of LL-37/hCAP18, a human cathelicidin antimicrobial peptide, in C57BL/6 mice resulted in augmented protection against *P. aeruginosa* bacterial infection (Bals et al., 1999).

Second, antimicrobial peptides isolated from one species can suppress microbial infections with a natural pathogen in the same species (Brogden et al., 2001; Kalfa et al., 2001b). This suggests that these peptides can act directly as innate effector molecules or indirectly to augment antimicrobial activity of other endogenous antimicrobial proteins and peptides. Third, the

Table 8.1. *Composition and origin of a few antimicrobial peptides used to suppress microbial infections and prevent or treat sepsis in animal models*

Peptide[1]	Composition	Origin	Reference
Buforin II	TRSSRAGLQFPVGRVHRLLRK	Stomach of Asian toad, *Bufo Bufo gargarizans*	(Giacometti et al., 2002; Park et al., 2000)
$CAP18_{109-135}$	FRKSKEKIGKEFKRIVQRIKDFLRNLV	The C-terminal 27-amino-acid fragment of CAP18, a peptide found in rabbit granulocytes	(Kirikae et al., 1998a)
$CAP18_{106-137}$	GDFFRKSKEKIGKEFKRIVQRIKDFLRNLVPR	The C-terminal 32- amino-acid fragment of CAP18, a peptide found in rabbit granulocytes	(Sawa et al., 1998; VanderMeer et al., 1995)
CEMA (also referred to as MBI-28)	KWKLFKKIGIGAVLKVLTTGLPALKLTK	From parts of silk moth cecropin and bee melittin peptides	(Friedrich et al., 1999; Gough et al., 1996; Piers et al., 1994; Piers, Brown, and Hancock, 1993)
CEME (also referred to as MBI-27)	KWKLFKKIGIGAVLKVLTTGLPALIS	From parts of silk moth cecropin and bee melittin peptides	(Friedrich et al., 1999; Gough et al., 1996; Piers et al., 1994; Piers et al., 1993)

(*cont.*)

Table 8.1. *(cont.)*

Peptide[1]	Composition	Origin	Reference
Dermaseptin S4	K_4 S4(1–16), ALWKTLLKKVLKAAAK-amide and K_4-S4(1–13), ALWKTLLKKVLKA-amide	Dermaseptins are a family of peptides expressed in amphibian skin	(Navon-Venezia et al., 2002)
HLD2	TKCFQWQRNMRKVRGPPVSCIKR	The 18- to 40-amino-acid antibacterial region of human lactoferrin	(Haversen et al., 2000)
IB-367	NH_2-RGGLCYCRGRFCVCVGR-$CONH_2$	An analogue of protegrin	(Chen et al., 2000; Loury et al., 1999)
Indolicidin	ILPWKWPWWPWRRG	Cytoplasmic granules of bovine neutrophils	(Giacometti et al., 2002)
KFFKFFKFF	KFFKFFKFF	Synthetic peptide	(Giacometti et al., 2002)
LL-37	LLGDFFRKSKEKIGKEFKRTIVQRIKD FFRNLVPRTES	The C-terminal 37-residue peptide from the unprocessed peptide called LL-37/hCAP18	(Scott et al., 2002)

[1]CAP, cationic antimicrobial proteins; CEMA, positively charged cecropin-bee melittin hybrid peptide; CEME, cecropin-bee melittin hybrid peptide; HLD2, amino acid residues 18 to 40 of the antibacterial region of human lactoferrin.

characteristics of ideal peptides are being identified. With cathelicidins, positive correlations are seen between antimicrobial activity and both the net positive charge of the peptides and the degree of hydrophobicity along the helical axis (Travis et al., 2000). With protegrins, extensive amino acid substitutions are tolerated by the peptide structure, implying that overall structural features such as amphiphilicity, charge, and shape are more important to activity than the presence of specific amino acids (Chen et al., 2000). Finally, studies in small- and large-animal models are identifying the clinical conditions that influence optimal peptide activity. This includes route of administration, times of administration, and concentration of dose.

Peptides with potent antimicrobial activity are of interest in treating microbial infections. Applications include the treatment of burn infections (Steinstraesser et al., 2002), respiratory infections (Kelly et al., 1993; Brogden et al., 2001), skin infections (Nizet et al., 2001), oral cavity infections (Miyasaki et al., 1998; Miyasaki and Lehrer, 1998; Weinberg, Krisanaprakornkit, and Dale, 1998), ocular infections (Haynes, Tighe, and Dua, 1999), and urogenital infections. Efficacy is defined by significant decreases in microbial counts of tissues in treated animals when compared with that of untreated control animals. Peptides with potent antiendotoxin activity are of interest in preventing or treating spepsis. Models of sepsis vary from that induced by direct administration of lipopolysaccharide (LPS) to experimentally induced and natural bacterial peritonitis. The latter is induced by fecal flora from a cecal ligation and puncture model (Giacometti et al., 2002). Efficacy is defined by significant reductions in endotoxin levels and tumor necrosis factor α (TNF-α) concentrations in plasma and significant reductions in bacterial growth in abdominal exudate and plasma of treated groups when compared with those of untreated controls.

Models are also being developed to assess the methods to increase antimicrobial peptide activity in environments that may be inhibiting their activity. Here, osmolytes with low transepithelial permeability are administered to mucosal surfaces, (e.g., the respiratory tract) either before or after experimental challenge to see if their presence can lower the ionic strength conditions of the fluids, a condition that increases antimicrobial activity of endogenous antimicrobial proteins and peptides (Zabner et al., 2000).

8.3. TOXICITY OF ANTIMICROBIAL PEPTIDES IN ANIMALS

By the nature of their structure and composition, many antimicrobial peptides have nonspecific cytotoxic and hemolytic activity toward host cells. Generally, the overall hydrophobicity is a contributing factor (Bessalle et al.,

1993; Dathe et al., 1996; Travis et al., 2000), and efforts to reduce the number of hydrophobic domains or increase the net positive charge on native antimicrobial peptides can result in congeners with increased antibacterial activity and decreased cytotoxic and hemolytic activity (Navon-Venezia et al., 2002). Those peptides with little or no cytotoxic and hemolytic activity toward host cells are further tested for acute toxicity. This can involve examining the peptide administration sites for lesions in nonchallenged control animals of a conventional challenge–treatment study. Lack of significant gross and histopathologic changes usually indicates lack of toxicity. Alternatively, high concentrations of peptides are injected into naive animals to establish toxic or lethal doses. Morbidity, mortality, and pathology are often used as the indicators of toxicity.

8.3.1. Topical Toxicity

Many active antimicrobial peptides, regardless of toxicity, have applications for treatment of cutaneous and localized infections. Topical application also reduces the potential degradation of antimicrobial peptides and proteins or their inactivation by high concentrations of ionic substances or serum components. Cutaneous toxicity is generally low even for the most robust peptides, and topical use has been chosen for various antimicrobial peptides that are currently undergoing clinical phase IIb and III trials. Examples include antimicrobial peptides MBI 594AN and MBI 226 developed by Micrologix Biotech Inc., British Columbia (www.mbiotech.com), for the treatment of acne and central-venous-catheter-related bloodstream infections, respectively (Zasloff, 2002).

8.3.2. Systemic Toxicity

Systemic toxicity is a concern that impedes development of antimicrobial peptides for the treatment of generalized microbial infections and sepsis. With natural peptides used as the template, efforts focus on developing specific congeners that represent the smallest polypeptide segment with the highest activity and the lowest toxicity (Travis et al., 2000; Brogden et al., 2001). Obviously small peptides are going to be easier and more cost effective to synthesize. Larger, more complex peptides may have to be produced by recombinant production methods or by transgenic animals (van Berkel et al., 2002), but this is, as yet, an unproven solution.

Even small alterations in an antimicrobial peptide can dramatically alter its properties. For example, SMAP29, a deduced sheep myeloid antimicrobial peptide of 29 residues, shows little hemolytic activity toward human or sheep

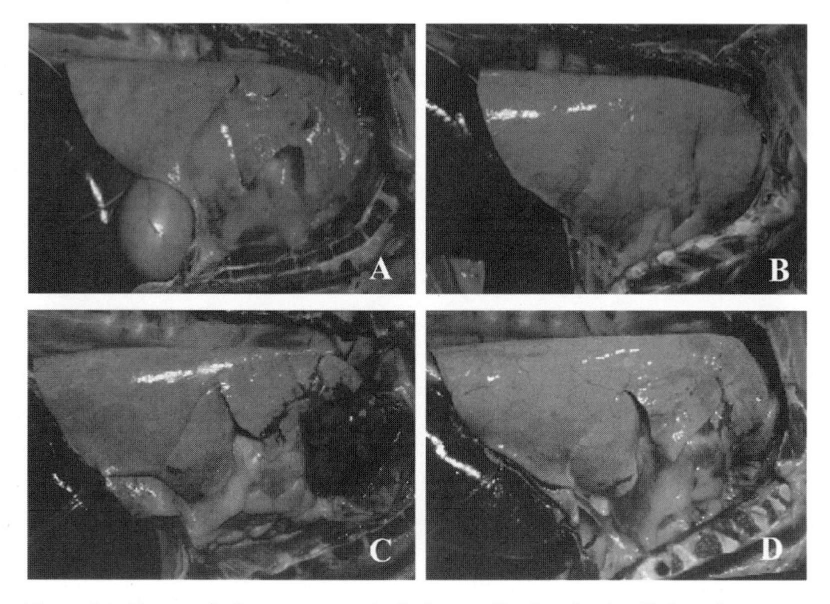

Figure 8.1. No gross lesions were seen in the lungs of lambs after instillation of (A) phosphate buffered saline (PBS) followed by PBS or (B) PBS followed by SMAP29. (C) Consolidation with necrosis and hemorrhage was seen in the anterior part of the right cranial lobe of lambs that received *Mannheimia haemolytica* followed by PBS. Large numbers of neutrophils, cell debris, and proteinaceous exudate were seen. (D) Mild lesions were seen in the anterior part of the right cranial lobe of lambs that received *M. haemolytica* followed by SMAP29. The bronchiolar walls contained moderate numbers of lymphocytes, but there was no acute inflammatory response (Brogden et al., 2001). (See color section.)

erythrocytes (Travis et al., 2000), whereas SMAP28, the isolated peptide with an N-terminal amine, causes hemolysis of human but not sheep erythrocytes (Skerlavaj et al., 1999). Ovispirin-1, a congener of SMAP29, still exhibits considerable in vitro toxicity toward several human epithelial cell lines (Steinstraesser et al., 2002), but novispirin G10, a congener of ovispirin-1, does not induce cytotoxic or hemolytic damage to host cells (Steinstraesser et al., 2002). In rats, the injection site of novispirin G10 did not differ in macroscopic or histologic appearance from the injection site of nontreated controls (Steinstraesser et al., 2002).

Peptides isolated from one animal species are generally not toxic when reintroduced back into the same species and often do not induce any adverse clinical or pathological effects. SMAP29 or anionic peptide (H-DDDDDDD-OH) instilled into ovine lungs did not induce any significant gross pathology (Fig. 8.1B), histopathology (Fig. 8.2), or inflammatory cell filtrates in

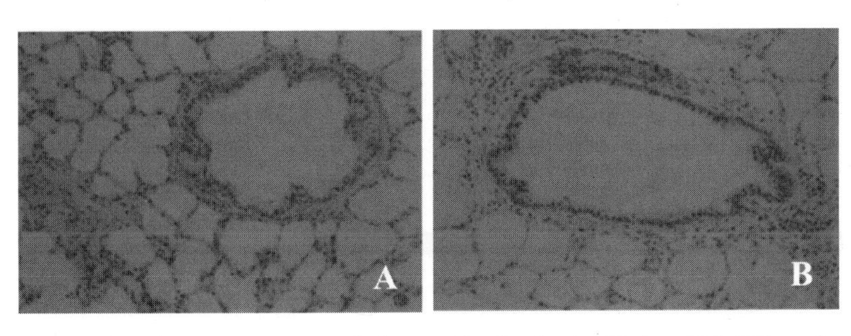

Figure 8.2. No lesions were seen in the lungs of lambs after instillation of (A) PBS followed by PBS or (B) PBS followed by SMAP29. The bronchiolar wall (B) contained minimal-to-mild infiltrates of lymphocytes, but there was no acute inflammatory response in the bronchioles or alveoli (Brogden et al., 2001). (See color section.)

bronchoalveolar lavage fluid (Brogden et al., 2001; Kalfa et al., 2001b). The bronchiolar wall (Fig. 8.2B) contained minimal-to-mild infiltrates of lymphocytes, but there was no acute inflammatory response in the bronchioles or alveoli (Brogden et al., 2001).

Peptides isolated from one animal species may be toxic when introduced into the tissues of another animal species. Derivatives of the cytotoxic amphibian peptide dermaseptin S4 are good examples (Navon-Venezia et al., 2002). Three derivatives were reported with improved toxicity profiles: a 28-residue K_4K_{20}-S4 and two shorter versions, K_4-S4(1–16) and K_4-S4(1–13). At 5.4 mg/kg, mild fur erection was seen in mice minutes after intraperitoneal injection but resolved within 4 h or less. At 11 mg/kg, intravenous injection of K_4K_{20}-S4 caused 30% mortality (Navon-Venezia et al., 2002). No mortality was observed in mice injected with up to 11 mg/kg of K_4-S4(1–16) and K_4-S4(1–13). Intravenous injection of rats with 10 mg/kg K4-S4(1–13) was also tolerated (Navon-Venezia et al., 2002). At 21.7 and 32.6 mg/kg, intravenous injection of K_4-S4(1–16) caused 60% and 100% mortality, respectively (Navon-Venezia et al., 2002).

Cationic antibacterial protein 18 (CAP18) is a LPS-binding protein first isolated from rabbit granulocytes (Larrick et al., 1993; Hirata et al., 1994). $CAP18_{106-137}$, a truncated 32-amino-acid C-terminal peptide of CAP18, induced edema in the respiratory tract of mice, suggesting that it had some pulmonary toxicity (Sawa et al., 1998). Interestingly, $CAP18_{109-135}$, the C-terminal 27 amino-acid fragment thought to be the LPS-neutralizing and antimicrobial domain, did not exhibit any acute toxicity in mice even at doses up to 44 mg/kg of body weight (Kirikae et al., 1998a). Similarly, a 32-amino-acid fragment of CAP18, lacking the five amino acids at the C-terminus, was not toxic for guinea pigs (Tasaka et al., 1996). Intravenous injection of this peptide did not induce lung injury, and there were no significant increases in pulmonary extravascular water, (^{125}I)albumin leakage, or pulmonary histopathology. Additional in vivo studies revealed no renal toxicity at doses up to 20 mg/kg (Tasaka et al., 1996).

Peptides with strong membrane-disintegrating activity may be toxic (Otvos, 2002). Cecropin, for example, is deadly to mice at a dose of 100 mg/kg (Moore, Devine, and Bibby, 1994), and PR-39 rapidly enters microvascular endothelial cells (Chan et al., 2001).

Short, proline-rich peptides, like drosocin, appear to be completely nontoxic to healthy animals when given intravenously in two 100 mg/kg doses (Hoffmann et al., 1999). Similarly, pyrrhocoricin does not lyse sheep erythrocytes and is nontoxic for mammalian COS cells up to 256 and 50 μM, respectively (Cudic et al., 1999; Otvos et al., 2000a; Otvos, 2002). In vivo, pyrrhocoricin and analogs were not toxic up to the maximum applied dose of 50 mg/kg.

Lysozyme appears to have little toxicity. In mice transgenic for rat lysozyme, rat lysozyme cDNA was expressed in distal respiratory epithelial cells, resulting in a 2- or 4-fold increase in protein in bronchoalveolar lavage fluid with a 6.6-fold to 17-fold increase in lysozyme enzymatic activity (Akinbi et al., 2000). Lung structure and cellular composition of bronchoalveolar lavage fluid were not altered.

8.3.3. Oral Toxicity

Lactoferrins have little toxicity and are very safe, especially when administered orally. Bovine lactoferrin, a by-product from cheese whey or skim milk, is commercially available as a health food in Japan (Iwasa et al., 2002) and has been used in a wide variety of products since it was first added to infant formula in 1986 (Tomita et al., 2002). The oral toxicity of purified lactoferrin was judged to be extremely low, and 2000 (mg/kg)/day did not cause any adverse effects in rats of both sexes (Yamauchi et al., 2000b; Tomita et al., 2002). Similarly, 434-mg/kg lactoferrin was not toxic for mice (Zagulski et al., 1989). From these results, lactoferrin was considered to be a highly safe food additive in supplemented infant formula, milk, skim milk, yogurt, chewing gum, nutritional supplements, skin care cosmetics, therapeutic diet for relief of inflammation in dogs and cats, and in aquaculture feed (Tomita et al., 2002).

8.4. ANTIMICROBIAL PEPTIDES FOR THE TREATMENT OF INFECTION

The efficacy of cathelicidins, bactericidal permeability-increasing protein, lactoferrin, and other antimicrobial peptides have been assessed against microbial infections in small- and large-animal models (Table 8.2).

8.4.1. Cathelicidins

Cathelicidins are generally protective in homologous cutaneous and systemic models of infection. SMAP29 is a potent antimicrobial cathelicidin from sheep (Skerlavaj et al., 1999; Travis et al., 2000; Brogden et al., 2001; Kalfa et al., 2001a). In lambs infected with the ovine respiratory pathogen *Mannheimia haemolytica*, a single bronchial instillation of 0.5-mg SMAP29 reduced the concentration of bacteria in the bronchial alveolar lavage fluid and consolidated pulmonary tissues (Brogden et al., 2001). The lesion was significantly reduced, and the bronchiolar walls contained moderate numbers of lymphocytes (Figure 8.1D). CRAMP, an important murine cathelicidin, protected mice against necrotic skin infection caused by Group A Strep. (Nizet et al., 2001). However, prevention or treatment of infections with cathelicidins in heterologous animal models has not always been fruitful. CAP18-derived peptides are antimicrobial for Gram-negative bacteria and Gram-positive bacteria in vitro (Larrick et al., 1993; Travis et al., 2000; Brogden et al., 2001), but they appear to have no antibacterial activity in vivo. For example, intratracheal instillation of $CAP18_{106-137}$ mixed with *P. aeruginosa* significantly reduced pulmonary injury but did not reduce the number of bacteria in the lungs of infected mice and did not improve the survival of the infected mice (Sawa et al., 1998). Similarly, injection of $CAP18_{109-135}$ into *P. aeruginosa*-infected mice did not cause a significant decrease in the number of viable bacteria in their blood (Kirikae et al., 1998a). Furthermore, injection of $CAP18_{109-135}$ together with ceftazidime did not decrease the number of viable infecting bacteria that was present after ceftazidime treatment alone (Kirikae et al., 1998a).

8.4.2. Bactericidal Permeability-Increasing Protein

Bactericidal permeability-increasing protein (BPI) is a potent antibacterial protein stored in the primary granules of human polymorphonuclear leukocytes and released into the phagocytic vacuole during phagocytosis (Elsbach and Weiss, 1993; Kelly et al., 1993; Elsbach, Weiss, and Levy, 1994).

Table 8.2. *Examples of antimicrobial peptides used to suppress microbial infections and prevent or treat sepsis in animal models*

Peptide[1]	Dose[2]	Model	Results	Reference
BPI	4 mg/kg every 2 h for 4 doses	Pulmonary infection induced by intratracheal instillation of *Escherichia coli* in mice	$rBPI_{23}$ reduced mortality when administered immediately (95% survival) or 2 h (65% survival) after infection	(Kelly et al., 1993)
Buforin II	Single intraperitoneal or intravenous injection of 1 mg/kg	Sepsis induced by intraperitoneal administration of LPS, induction of *E. coli* peritonitis, and cecal ligation and puncture	Treatment resulted in significant reductions in plasma endotoxin levels, TNF-α levels, and bacterial growth in abdominal exudate and plasma	(Giacometti et al., 2002)
$CAP18_{109-135}$	Single intraperitoneal injection of 4.4 mg/kg	Sepsis induced by intraperitoneal injection of 100 ng of *Salmonella enterica* serovar Minnesota LPS in mice	Injection of $CAP18_{109-135}$ protected 100% of D-GaIN-sensitized mice from death caused by LPS	(Kirikae et al., 1998a)

(cont.)

Table 8.2. (*cont.*)

Peptide[1]	Dose[2]	Model	Results	Reference
$CAP18_{106-137}$	Single intratracheal instillation of 0.07–0.285 mg/kg	Pulmonary infection induced by intratracheal instillation of *P. aeruginosa* in mice	Concentration of *P. aeruginosa* was not significantly decreased when $CAP18_{106-137}$ was added to the inoculum	(Sawa et al., 1998)
$CAP18_{106-137}$	Intravenous infusion of 4 (mg/kg)/h for 4 h	Sepsis induced in pigs by intravenous infusion of 3 (μg/kg)/h of *E. coli* LPS	Treatment blocked LPS-induced increases in 6-keto-prostaglandin F1 alpha and TNF-α and prevented changes in cardiac output, arterial PO_2, phagocyte activation, and leukocyte count	(VanderMeer et al., 1995)
CEMA (also referred to as MBI-28)	Single intraperitoneal injection of 7.7 mg/kg within 10 min of LPS of *P. aeruginosa* or at 2 and 14 h after injection of *P. aeruginosa*	Infection induced by intraperitoneal injection of *P. aeruginosa* in neutropenic mice and endotoxic shock was induced by intraperitoneal injection of 20 μg of *E. coli* 0111:B4 LPS in galactosamine-sensitized mice	Significantly protected mice against an LD_{90} dose of *P. aeruginosa* strain M2 and protected mice from a lethal LPS challenge	(Gough et al., 1996)

CEME (also referred to as MBI-27)	Single intraperitoneal injection of 7.7 mg/kg within 10 min of LPS or *P. aeruginosa* or at 2 and 14 h after injection of *P. aeruginosa*	Infection induced by intraperitoneal injection of *P. aeruginosa* in neutropenic mice and endotoxic shock was induced by intraperitoneal injection of 20 μg of *E. coli* 0111:B4 LPS in galactosamine-sensitized mice	Significantly protected mice against an LD_{90} dose of *P. aeruginosa* strain M2 and protected mice from a lethal LPS challenge	(Gough et al., 1996)
CEME	200-μg CEME was delivered continuously by miniosmotic pumps placed in the peritoneal cavity	12 days after pump implantation, juvenile coho salmon received intraperitoneal injections of *Vibrio anguillarum* at a dose that would kill 50 to 90% of the population	Salmon receiving 200 μg of CEME per day survived longer and had significantly lower accumulated mortalities (13%) than the control groups (50%–58%)	(Jia et al., 2000)
Dermaseptin S4 analogs K_4-S4(1–16) and K_4-S4(1–13)	Single intraperitoneal injection of 4.5 mg/kg	Sepsis induced by intraperitoneal injection of 10^6 CFU *P. aeruginosa*	Mice receiving K_4-S4(1–16) and K_4-S4(1–13) had 18% and 36% mortality, respectively. Mice in the control group had 75% mortality	(Navon-Venezia et al., 2002)

(cont.)

Table 8.2. (*cont.*)

Peptide[1]	Dose[2]	Model	Results	Reference
HLD2, a fragment of human lactoferrin	Oral administration of 500 μg	Mouse model of urinary tract infection induced by *E. coli*	HLD2 was comparable with human lactoferrin in reducing the number of *E. coli* in the kidneys of infected mice	(Haversen et al., 2000)
IB-367, an analogue of protegrin	0.5 or 2.0 mg/mL of IB-367 applied topically to the buccal mucosa 5 or 6 times per day starting 6 to 8 h before abrasion	Hamster oral microflora reduction model	Mucositis scores were significantly lower in hamsters given formulations	(Loury et al., 1999)
Indolicidin	Single intraperitoneal or intravenous injection of 1 mg/kg	Sepsis induced by intraperitoneal administration of LPS, induction of *E. coli* peritonitis, and cecal ligation and puncture	Treatment resulted in significant reductions in plasma endotoxin levels, TNF-α levels, and bacterial growth in abdominal exudate and plasma	(Giacometti et al., 2002)

KFFKFFKFF	Single intraperitoneal or intravenous injection of 1 mg/kg	Sepsis induced by intraperitoneal administration of LPS, induction of *E. coli* peritonitis, and cecal ligation and puncture	Treatment resulted in significant reductions in plasma endotoxin levels, TNF-α levels, and bacterial growth in abdominal exudate and plasma	(Giacometti et al., 2002)
Novispirin G10	Single intradermal injection of 1, 3, or 6 mg/kg	Infected, partial-thickness burn in rats	CFU/gm of burned tissue were significantly lower for the novispirin G10 treated rats	(Steinstraesser et al., 2002)
Pleurocidin amide (a C-terminally amidated form of the natural flounder peptide)	Pleurocidin amide was delivered continuously by miniosmotic pumps placed in the peritoneal cavity	12 days after pump implantation, juvenile coho salmon received intraperitoneal injections of *V. anguillarum* at a dose that would kill 50%–90% of the population	Salmon receiving pleurocidin amide at 250 μg per day also survived longer and had significantly lower accumulated mortalities (5%) than the control groups (67%–75%)	(Jia et al., 2000)

(cont.)

Table 8.2. *(cont.)*

Peptide[1]	Dose[2]	Model	Results	Reference
Polyphemusin I	Single intraperitoneal injection of 7.7 mg/kg within 10–30 min of LPS or *P. aeruginosa*	Infection induced by intraperitoneal injection of *P. aeruginosa* in neutropenic mice and endotoxic shock was induced by intraperitoneal injection of 0.3 μg of *E. coli* 0111:B4 LPS in galactosamine-sensitized mice	Polyphemusin I, which has high in vitro antimicrobial activity, did not provide a high level of protection in mice against infection with *P. aeruginosa* strain M2 and did not seem to be highly protective against endotoxic shock	(Zhang et al., 2000)
Structural variants (PV5, PV7, and PV8) of polyphemusin I	Single intraperitoneal injection of 7.7 mg/kg within 30 min of LPS or *P. aeruginosa*	Infection induced by intraperitoneal injection of *P. aeruginosa* in neutropenic mice and endotoxic shock was induced by intraperitoneal injection of 0.3 μg of *E. coli* 0111:B4 LPS in galactosamine-sensitized mice	Structural variants (PV5, PV7, and PV8) of polyphemusin I, which have low in vitro antimicrobial activities, had high levels of protection in mice against infection with *P. aeruginosa* strain M2 and were protective against endotoxic shock	(Zhang et al., 2000)

[1] BPI, bactericidal permeability-increasing protein.
[2] Dose is based on a 25-g mouse.

In mice with *Escherichia coli* pneumonia, $rBPI_{23}$, a recombinant amino-terminal fragment of BPI, administered every 2 h significantly reduced overall mortality rates (Kelly et al., 1993). Treatment of mice with $rBPI_{23}$ at 2 h after infection (65% survival) was less effective than treatment with $rBPI_{23}$ immediately after infection (95% survival). Xoma LLC (Berkeley, California) is assessing the commerical potential of various modified recombinant BPI derivatives, including $rBPI_{21}$, and Neuprex, an injectable formulation of $rBPI_{21}$. These peptides are being used to treat acne, Gram-positive and Gram-negative bacterial infections, Crohn's disease, and severe meningococcal sepsis (Levin et al., 2000).

8.4.3. Lactoferrin

Lactoferrin is a dynamic, multifunctional immunoregulatory protein (Baveye et al., 1999; Vorland, 1999) associated with host defense at mucosal surfaces through its antibacterial properties (Tomita et al., 1994; Dionysius and Milne, 1997; Haversen et al., 2000; Tomita et al., 2002) and ability to prevent biofilm development (Singh et al., 2002). The antimicrobial domain of lactoferrin, called lactoferricin, is near the N-terminus of lactoferrin and consists of a loop of 18 amino acids (Bellamy et al., 1992).

In the gastrointestinal tract, lactoferrin suppresses the movement or translocation of organisms from the intestines into local lymph nodes (Teraguchi et al., 1995a) and suppresses the proliferation of enteric pathogens (Teraguchi et al., 1995b). For example, mice that were fed bovine milk had a high incidence of enteric bacteria in the mesenteric lymph nodes, which could be significantly reduced if the milk diet was supplemented with bovine lactoferrin (Teraguchi et al., 1995a). Similarly, neonatal rats orally fed recombinant human lactoferrin were protected from systemic infection (Edde et al., 2001). Treated rats had less bacteremia and lower disease severity scores after intestinal infection with *E. coli*. Lactoferrin also reduces chronic inflammation and intractable stomatitis in feline immunodeficiency virus-positive and feline immunodeficiency virus-negative cats (Sato et al., 1996).

Lactoferrin is excreted into the urinary tract 2 h after feeding (Haversen et al., 2000), making it a potential therapeutic agent for urinary tract infections. Intravenous administration of 1 mg of human lactoferrin, bovine lactoferrin, or bovine lactoferrin hydrolysate to mice infected intravenously with staphylococci reduced kidney infections 30%–50% (Bhimani, Vendrov, and Furmanski, 1999). Viable bacterial counts in the kidneys decreased 5-fold to 12-fold. Feeding mice 2% bovine lactoferrin in drinking water also reduced staphylococcal kidney infections by 40%–60% and viable bacterial counts by 5-fold to 12-fold (Bhimani et al., 1999). Similarly, bovine lactoferrin, human lactoferrin, or synthetic peptide sequences based on the antibacterial region of human lactoferrin were given orally to female mice 30 min after the instillation of *E. coli* into the urinary bladder (Haversen et al., 2000) and partially prevented experimentally induced urinary tract infections. The inflammatory response, characterized by increased neutrophil numbers and interleukin-6 concentrations in the urine, was also reduced in the human-lactoferrin-treated mice. Overall, human lactoferrin was more effective than bovine lactoferrin as an anti-infectious and anti-inflammatory agent. The antibacterial region, comprising residues 18–40 of the human lactoferrin molecule, is perhaps the active antibacterial site (Haversen et al., 2000).

Oral lactoferrin administration is also effective against bacterial and fungal systemic infections. Bovine and human lactoferrin were shown to protect mice (Zagulski et al., 1989) and rabbits (Zagulski et al., 1986) against bacterial infections. The mortality rate in lactoferrin-pretreated mice was only 11%–16% and the mortality rate in untreated, *E. coli*-infected mice was 92–96% (Zagulski et al., 1989). Similarly, administration of lactoferrin improved the dermatological symptoms of *Trichophyton mentagrophytes* infection in guinea pigs (Wakabayashi et al., 2000), *Candida albicans* infection in mice (Abe et al., 2000; Tomita et al., 2002), *Toxoplasma gondii* in mice (Isamida et al., 1998), and *Trichophyton rubrum* and *T. mentagrophytes* infection in humans (Yamauchi et al., 2000a). Patients treated with 600 or 2000 mg of lactoferrin had lower dermatological lesion scores, and patients with moderate vesicular or interdigital tinea pedis had significantly decreased dermatological lesion scores in the lactoferrin-treated groups.

8.4.4. Proline-, Arginine-, and Lysine-Rich Peptides

Pyrrhocoricin, a 20-amino-acid proline-rich peptide from the European sapsucking bug *Pyrrhocoris apterus* (Otvos, 2002), protected mice against *E. coli* infection at intravenous doses of 10 or 25 mg/kg given 1 and 5 h after infection. However, higher doses (e.g., 50 mg/kg) were toxic to compromised animals. Similarly, drosocin was also toxic to infected animals, possibly because of the inhibition of a mammalian chaperonine, which is produced during infection (Otvos et al., 2000b). Analogs of pyrrhocoricin, in which Val1 is replaced with 1-amino-cyclohexane-carboxylic acid and Asn20 is replaced with β-acetyl-diamino-propionic acid, also protects mice in a 10–50-mg/kg dose range but without any toxic effects (Otvos et al., 2000a). Intranasal administration of a single 20-mg/kg dose of pyrrhocoricin dimer was found to reduce the concentration of *Haemophilus influenzae* in the bronchoalveolar lavage of infected mice (Otvos, 2002).

8.4.5. Protegrin

Protegrin has recently emerged as a potential antimicrobial agent for the prevention of polymicrobial infections that exacerbate oral mucositis (Chen et al., 2000). Oral mucositis was induced in hamsters by intraperitoneal injection of 5-fluorouracil followed by superficial abrasion of the buccal mucosa (Sonis et al., 1990). In this model, mucositis scores were significantly lower in hamsters given formulations containing 0.5 or 2.0 mg/mL protegrin IB-367 than in placebo-treated controls (Loury et al., 1999; Chen et al., 2000). In a human Phase III clinical trial conducted by IntraBiotics Pharmaceuticals, Inc., Palo Alto, California, 509 patients undergoing high-dose chemotherapy for the treatment of cancer were treated with iseganan hydrochloride or a placebo. 43% of patients treated with iseganan did not develop severe oral mucositis versus 37% of patients treated with placebo, a difference that did not reach statistical significance ($p = .18$). These results indicated that iseganan did not meet its primary end point of reducing oral mucositis.

8.4.6. Lysozyme

Lysozyme is a 14-kd cationic enzyme secreted by macrophages (Goldstein, 1983) and found in the primary azurophil and secondary specific granules of neutrophils (Ganz, Selsted, and Lehrer, 1986). It is specific for bacterial cell wall peptidoglycan and cleaves the glycosidic bond between the two major repeating units *N*-acetyl muramic acid and *N*-acetylglucosamine (Chipman and Sharon, 1969; Holtje, 1996; Kirby, 2001). The ability of lysozyme to prevent infection was recently demonstrated in transgenic mice deficient in lysozyme M (lys M -/-) after challenge with the nonpathogenic, lysozyme-sensitive bacterium *Micrococcus luteus* (Ganz et al., 2003). 24 h after subcutaneous injection of *M. luteus*, the injection sites in all mice were swollen and red, but were much more prominent and persistent in lys M -/- mice. At necropsy 2 to 5 days after injection, the lesions appeared inflamed and purulent in lys M -/- mice, but not in control mice. Interestingly, in other studies, lesions induced by purified *M. luteus* peptidoglycan were also much larger and more inflamed in lys M-/- mice than those in control mice (Ganz et al., 2003). The ability of lysozyme to prevent infection was also shown in mice transgenic for rat lysozyme under the control of a lung-specific promoter from human surfactant protein C (Akinbi et al., 2000). These mice expressed rat lysozyme cDNA in distal respiratory epithelial cells and produced about 2-fold or 4-fold increase in protein in bronchoalveolar lavage fluid with 6.6-fold and 17-fold as much lysozyme enzymatic activity as control mice. Increased production of lysozyme in respiratory epithelial cells of transgenic mice enhanced bacterial killing in the lung in vivo and was associated with decreased systemic dissemination of pathogen and increased survival following infection (Akinbi et al., 2000). Killing of group B streptococci was significantly enhanced (2-fold and 3-fold) in transgenic mice at 6 h postinfection and was accompanied by a decrease in systemic dissemination of pathogen. Killing of *P. aeruginosa* was also significantly enhanced (5-fold and 30-fold) in transgenic mice. *P. aeruginosa* is generally not lysozyme sensitive, so this mechanism may involve stimulation on other innate immune mechanisms.

8.4.7. Miscellaneous Peptides

Anionic peptides are small antimicrobial peptides containing homopolymeric regions of aspartic acid (Brogden et al., 1996) found in respiratory epithelium and mucosal secretions of ruminants (Brogden, Ackermann, and Huttner, 1998) and humans (Brogden et al., 1999). Intrapulmonary administration of H-DDDDDDD-OH (5.0 mg) before infection with *M. haemolytica*, a respiratory pathogen of sheep, did not prevent pulmonary inflammation or reduce the concentration of organisms in infected lung tissues (Kalfa et al., 2001b). However, administration of H-DDDDDDD-OH (5.0 mg) after *M. haemolytica* infection substantially reduced pulmonary inflammation and the concentration of organisms in infected lung tissues (Kalfa et al., 2001b).

Two 26- to 28- amino-acid α-helical peptides, MBI-27 (also referred to as cecropin-bee melittin hybrid peptide called CEME) and MBI-28 (also referred to as CEME's more positively charged derivative called CEMA), derived from parts of silk moth cecropin and bee melittin peptides, are active against Gram-negative bacteria (Piers, Brown, and Hancock, 1994; Piers and Hancock, 1994; Friedrich et al., 1999;) with minimum inhibitory concentrations (MICs) ranging from 1 to 16 μg/ml (Gough, Hancock, and Kelly, 1996). In a neutropenic mouse model, both MBI-27 and MBI-28 significantly protected mice against an LD_{90} dose of *P. aeruginosa* strain M2 (Gough et al., 1996).

Polyphemusins and tachyplesins are 18- and 17-residue β-sheet antimicrobial peptides from *Limulus polyphemus* and *Tachypleus* spp., respectively. These peptides inhibit the growth of bacteria, fungi, and viruses (Ohto et al., 1992; Zhang et al., 2000). Polyphemusin I and three structural variants (PV5, PV7, and PV8) were given as a single dose of 200 μg 30 min after challenge of cyclophosphamide-induced neutropenic mice with an LD_{100} dose of *P. aeruginosa* strain M2 (Zhang et al., 2000). Polyphemusin I, which has the best in vitro antimicrobial activity, did not provide the highest level of protection in mice against infection with *P. aeruginosa*. The variants, which have lower in vitro antimicrobial activities, had higher levels of protection in mice against infection with *P. aeruginosa* (Zhang et al., 2000). It appears that simple antimicrobial activity may not enough to induce protection and other factors may be involved, such as in vivo stability or synergy with other innate host defenses.

Dermaseptin S4 derivatives K_4K_{20}-S4, K_4-S4(1–16), and K_4-S4(1–13) were shown to be antibacterial in a *P. aeruginosa*-induced peritonitis model of mice. A single intraperitoneal injection of 4.5-mg/kg K_4-S4(1–16) or K_4 S4(1–13) protected mice and only 18% and 36% mortality was seen, respectively. Naive mice in the vehicle control group exhibited 75% mortality (Navon-Venezia et al., 2002). In vivo bactericidal activity was confirmed in neutropenic mice, in which intraperitoneal administration of K_4-S4(1–16) reduced the number of viable colony forming units (CFU) in a dose-dependent manner by >3 log units within 1 h of exposure.

8.5. ANTIMICROBIAL PEPTIDE MODELS FOR THE TREATMENT OF SEPSIS

Severe sepsis and septic shock are major causes of morbidity and mortality in neutropenic individuals, hospitalized patients, and all immunocompromised subjects (Giacometti et al., 2002). Sepsis, which often follows antimicrobial therapy, results in the release of endotoxins from the outer membrane of Gram-negative bacteria (Kadurugamuwa and Beveridge, 1997; Kirikae et al., 1998b). LPSs, components of endotoxin, activate host effector cells through stimulation of receptors on their surface (Rietschel et al., 1996). These target cells then secrete large quantities of the inflammatory cytokines, TNF, interleukin-1, interleukin-6, interleukin-8, platelet-activating factor, arachidonic acid metabolites, erythropoietin, and endothelin (Morrison and Ryan, 1987; Dinarello, 1996). These cytokines then induce systemic shock followed by death (Vincent, 1996). Many antimicrobial peptides bind to the LPS in the outer membrane of gram negative bacteria. The interaction of antimicrobial peptides with LPSs may go beyond their simple binding and inactivation. CEMA, a synthetic α-helical peptide, selectively modulates the transcriptional response of macrophages to LPS and can alter gene expression in macrophages. CEMA blocked LPS-induced gene expression in the RAW 264.7 macrophage cell line (Scott et al., 2000). The ability of LPS to induce the expression of >40 genes was strongly inhibited by CEMA, whereas LPS-induced expression of another 16 genes was relatively unaffected. In addition, CEMA itself induced the expression of a distinct set of 35 genes, including genes involved in cell adhesion and apoptosis (Scott et al., 2000). Similarly, LL-37 may contribute to the immune response by limiting the damage caused by bacterial products and by recruiting immune cells to the site of infection. LL-37 was found to directly upregulate 29 genes and downregulate another 20 genes (Scott et al., 2002). LL-37 upregulated the expression of chemokines in macrophages and the mouse lung (monocyte chemoattractant protein 1), human A549 epithelial cells (IL-8), and whole human blood (monocyte chemoattractant protein 1 and IL-8), without stimulating the proinflammatory cytokine TNFα. LL-37 also upregulated the chemokine receptors CXCR-4, CCR2, and IL-8RB (Scott et al., 2002).

8.5.1. Cathelicidins

The only human cathelicidin was isolated from the bone marrow (Agerberth et al., 1995; Larrick et al., 1995). The C-terminal 37-residue peptide encoded by this gene is LL-37, whereas the unprocessed peptide is referred to as LL-37/hCAP18 (Kirikae et al., 1998a; Bals et al., 1999). The C-terminal 27-amino-acid fragment of CAP18, and the internal fragment of LL-37, is a LPS-neutralizing and antimicrobial domain (Hirata, Wright, and Larrick, 1996; Kirikae et al., 1998a). This peptide, called $CAP18_{109-135}$, prevented ceftazidime-induced endotoxic shock in mice with septicemia (Kirikae et al., 1998a). Injection of $CAP18_{109-135}$ protected mice injected with LPS or ceftazidime-induced endotoxin from death and lowered their TNF levels in serum in a dose-dependent manner. Treatment with ceftazidime caused death of D-(+)-galactosamine-sensitized *P. aeruginosa*-infected mice within 48 hr, whereas injection with $CAP18_{109-135}$ rescued the mice from death (Kirikae et al., 1998a); endotoxin levels in plasma and TNF production by liver tissues were decreased but the numbers of viable infecting bacteria in their blood were not decreased significantly and remained at the levels in ceftazidime-treated mice. In guinea pigs, a 32-amino-acid fragment of CAP18, lacking five amino acids at the C-terminus, administered at 0 and 10 min after intravenous injection of 0.02 mg/kg endotoxin attenuated lung injury, whereas administration of peptide at 60 min after intravenous injection of endotoxin did not (Tasaka et al., 1996). Significantly lower lung wet-to-dry ratios, (^{125}I) albumin leakage in lung tissue and bronchoalveolar lavage fluids, and inflammatory cell infiltrates were seen.

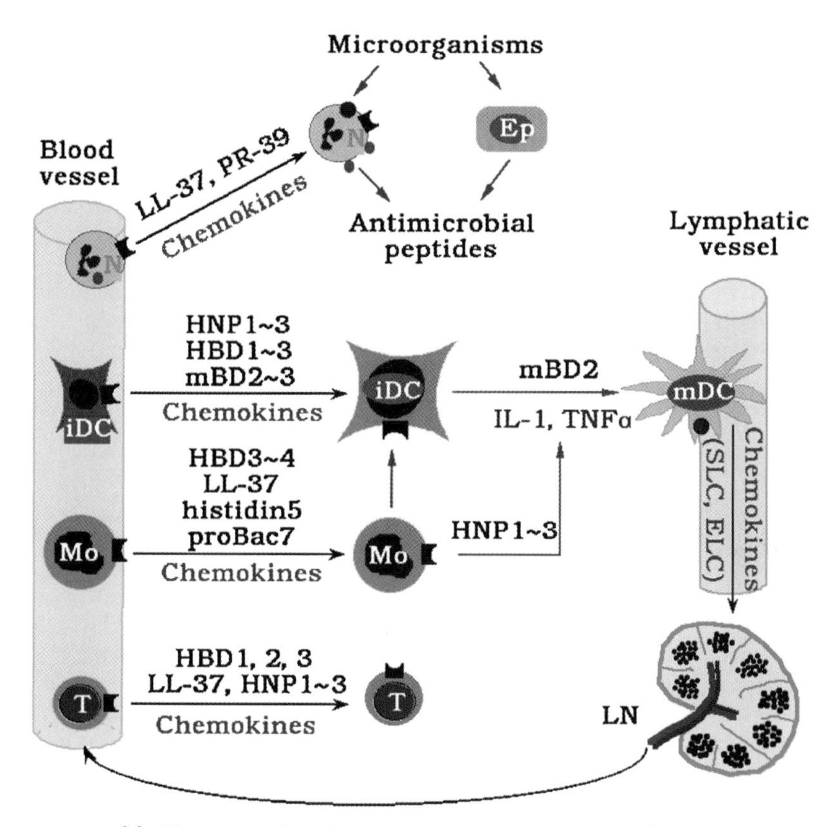

Figure 3.2. Schematic illustration of the potential mechanisms by which mammalian antimicrobial peptides enhance host adaptive antimicrobial immunity. In infected tissue, antimicrobial peptides are abundant because of the production predominantly by epithelial cells (Ep, including keratinocytes) and infiltrating neutrophils (N). Antimicrobial peptides form chemotactic gradients and potentially participate in the recruitment of both DC precursors (Mo) and iDCs to sites of infection. In addition, antimicrobial peptides promote the maturation of iDCs to mDCs directly or indirectly through inducing the production of TNFα and IL-1. Participation of antimicrobial peptides in the recruitment and maturation of DCs certainly contributes to promote antigen capture and presentation, thereby enhancing the induction of adaptive antimicrobial immune response. Furthermore, antimicrobial peptides may also contribute to the effector phase of adaptive antimicrobial immunity by facilitating the recruitment of effector T cells to infected tissues. Chemokines have similar roles, particularly in the recruitment of leukocytes to infected tissue. SLC, secondary lymphoid tissue chemokine; ELC, EBI1-ligand chemokine; LN, lymph node.

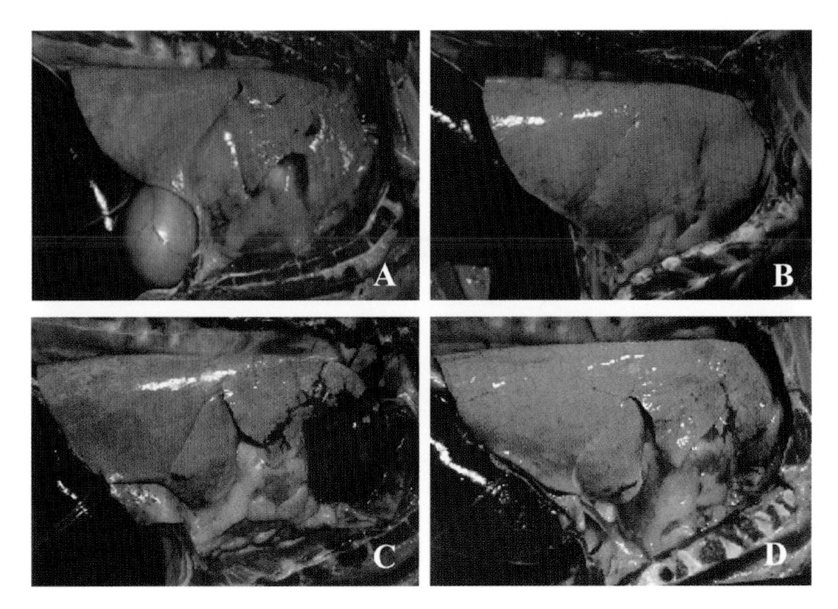

Figure 8.1. No gross lesions were seen in the lungs of lambs after instillation of (A) phosphate buffered saline (PBS) followed by PBS or (B) PBS followed by SMAP29. (C) Consolidation with necrosis and hemorrhage was seen in the anterior part of the right cranial lobe of lambs that received *Mannheimia haemolytica* followed by PBS. Large numbers of neutrophils, cell debris, and proteinaceous exudate were seen. (D) Mild lesions were seen in the anterior part of the right cranial lobe of lambs that received *M. haemolytica* followed by SMAP29. The bronchiolar walls contained moderate numbers of lymphocytes, but there was no acute inflammatory response (Brogden et al., 2001).

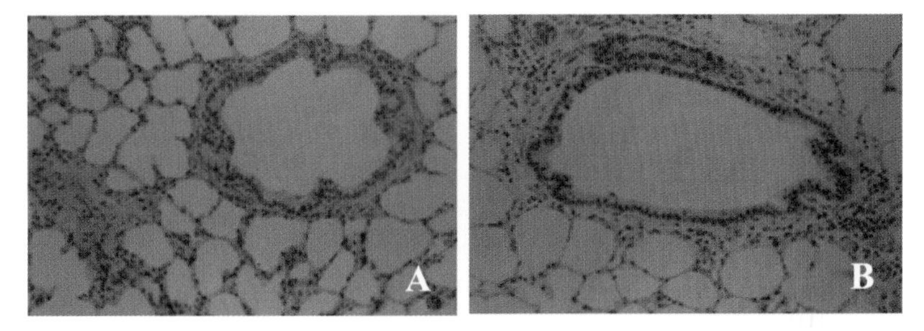

Figure 8.2. No lesions were seen in the lungs of lambs after instillation of (A) PBS followed by PBS or (B) PBS followed by SMAP29. The bronchiolar wall (B) contained minimal-to-mild infiltrates of lymphocytes, but there was no acute inflammatory response in the bronchioles or alveoli (Brogden et al., 2001).

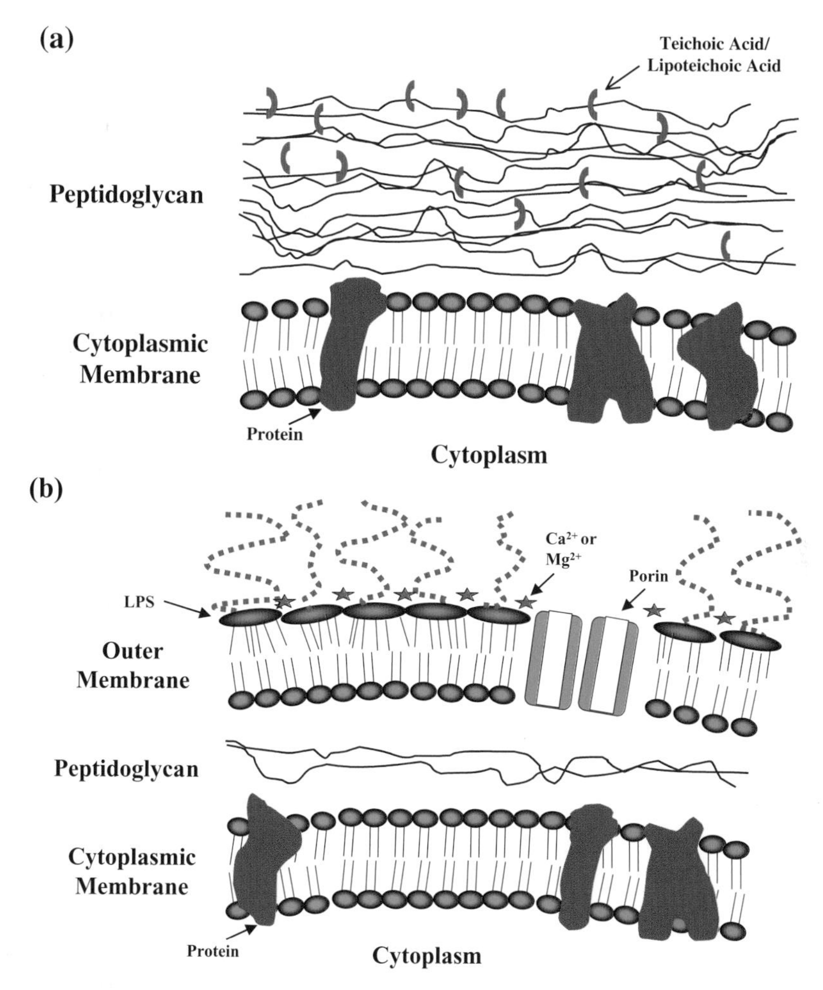

Figure 9.1. Schematic representation of the structure of the (a) Gram positive and (b) Gram negative cell envelopes.

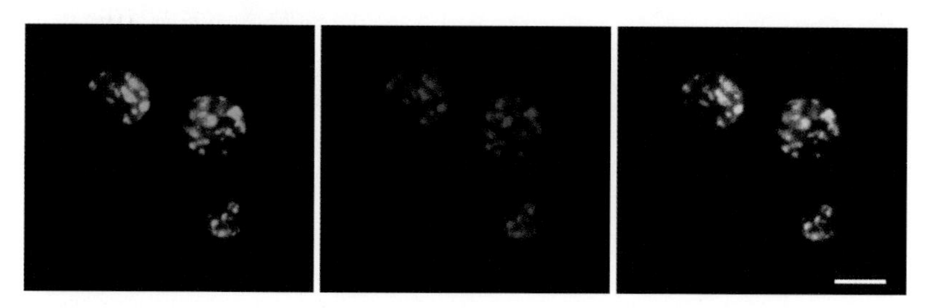

Figure 10.3. Mitochondrial targeting of histatin 5 in *C. albicans*. Left, FITC-histatin 5; middle, mitotracker orange; right, colocalization. Bar: 5 μm [From Helmerhorst et al. (1999a).]

In pigs, treatment with $CAP18_{106–137}$ neutralized many of the deleterious effects of LPS (VanderMeer et al., 1995). $CAP18_{106–137}$ [4 (mg/kg)/h for 4 h] blocked LPS-induced increases in plasma levels of 6-keto-prostaglandin F1-α and TNF-α and prevented LPS-induced changes in cardiac output, arterial PO_2, phagocyte activation, and peripheral leukocyte counts.

8.5.2. Bactericidal Permeability-Increasing Protein

BPI also binds LPS and neutralizes many of its effects in vitro and in vivo. A recombinant 23-kDa NH_2-terminal fragment of BPI (BPI23) reduces acute lung injury in endotoxemic pigs (Vandermeer et al., 1994). In pigs given 250 μg/kg LPS intravenously over 50 min, treatment with BPI23 (3-mg/kg bolus and 3 mg/kg over 60 min) significantly ameliorated LPS-induced hypoxemia, functional upregulation of opsonin receptors on circulating phagocytes, and alveolitis but had no effect on the elaboration of TNF-α or thromboxane A2. It was thought that the salutory effects of BPI23 on acute lung injury in endotoxemic pigs may be mediated, at least in part, by inhibition of direct activation of phagocytes by LPS (VanderMeer et al., 1994).

8.5.3. Lactoferrin

Lactoferrin also protects against endotoxin-induced septic shock (Lee et al., 1998). Piglets that were fed lactoferrin before challenge with intravenous *E. coli* LPS had a significant decrease in mortality and an overall increase in wellness. In vitro, a flow cytometric assay system demonstrated that LPS binding to porcine monocytes was inhibited by lactoferrin in a dose-dependent fashion, suggesting that it may inhibit LPS binding in vivo to monocytes and macrophages and, in turn, prevent induction of monocyte- and macrophage-derived inflammatory cytokines (Lee et al., 1998).

8.5.4. Proline-, Arginine-, and Lysine-Rich Peptides

Buforin II, indolicidin, and KFFKFFKFF, three cationic antimicrobial peptides with broad-spectrum antibacterial activities and outer-membrane permeability-increasing properties (Vaara and Porro, 1996; Falla and Hancock, 1997; Park et al., 2000; Giacometti et al., 2002) were assessed in models of rats in which septic shock was induced by intraperitoneal administration of *E. coli* LPS, by induction of peritonitis with 10^{10} CFU of *E. coli*, and by cecal ligation and puncture (Giacometti et al., 2002). Treatment of rats with 1 mg/kg of each peptide resulted in significant reductions in plasma endotoxin and TNF-α concentrations in all three models of sepsis and significant reductions in bacterial growth in abdominal exudate and plasma of rats with *E. coli* peritonitis and in cecal ligation and puncture models (Giacometti et al., 2002). Survival rates were 73.4%, 60.0%, and 46.7% for rats with *E. coli* peritonitis treated with Buforin II, indolicidin, and KFFKFFKFF, respectively, and 66.7%, 46.7%, and 33.3% for rats with cecal ligation and puncture treated with Buforin II, indolicidin, and KFFKFFKFF, respectively (Giacometti et al., 2002).

MBI-27 and MBI-28 also protect galactosamine-sensitized mice from lethal LPS challenges. Both peptides bind to LPSs (Piers et al., 1994; Piers and Hancock, 1994) and prevent its ability to induce a TNF response in both a macrophage tissue culture cell line and galactosamine-sensitized mice (Gough, Hancock, and Kelly, 1996). In the same endotoxemia model, polyphemusin I and three structural variants (PV5, PV7, and PV8) were given as a single dose of 200 μg 30 min after challenge of galactosamine-sensitized mice with *E. coli* 0111:B4 (Zhang et al., 2000). Polyphemusin I, which has the best in vitro antimicrobial activity, did not provide the highest level of protection in mice against endotoxemia induced with *E. coli* 0111:B4 LPS. The variants, which have lower in vitro antimicrobial activities, had higher levels of protection in mice against endotoxemia induced with *E. coli* 0111:B4 LPS (Zhang et al., 2000).

8.6. ACTIVATION/STIMULATION OF ANTIMICROBIAL PEPTIDE ACTIVITY IN SITU

An attractive alternative approach to using antimicrobials to prevent or treat infection and sepsis in animals would be to directly induce antimicrobial peptide production in situ or to stimulate increased antimicrobial peptide activity for those molecules already present but inactive or in an inhibiting environment. This approach has the added advantage in that molecules are (1) present in the host at the site of their natural production, (2) present in the appropriate physiologic concentrations for antimicrobial activity, and (3) in the presence of other complementary antimicrobial substances and innate clearance mechanisms. An added advantage is the reduced side effects and acute toxicity to the host.

This approach also increases the efficacy of antimicrobial peptides whose activity is dependent on the ionic strength of their microenvironment. This includes most antimicrobial peptides found in the thin layer of airway surface liquid (ASL), including lysozyme, lactoferrin, secretory leukoproteinase inhibitor, HBDs, secretory phospholipase A2, and the cathelicidin LL-37.

For example, in cystic fibrosis, the presence of NaCl in the mucosal milieu may directly abrogate the optimal activity of antimicrobial peptides and proteins already present in these secretions, perhaps explaining the presence of *P. aeruginosa* associated with chronic respiratory inflammation in patients with cystic fibrosis. Therefore attempts to reduce the salt content of ASL could increase the activity of endogenous antimicrobial substances.

Sugars, in general, are mucolytic and have been used to assist mucous clearance by cough and physiotherapy. Dextrans are good examples (Feng et al., 1998; Sudo, Boyd, and King, 2000). A promising osmolyte for lowering the ionic strength of ASL is the five-carbon sugar, xylitol (Zabner et al., 2000). It has a low transepithelial permeability and is poorly metabolized by many bacterial species. As a preventative agent, xylitol pretreatment increases killing of instilled bacteria. Mice that received xylitol pretreatment for up to 3 h before inoculation with *P. aeruginosa* had a significant decrease in the number of live bacteria as compared with those that were pretreated with saline, suggesting that xylitol can have a long-lasting effect on ASL salt concentration and bacterial killing.

As a therapeutic agent, xylitol enhances the bacterial killing in the lungs of normal mice, enhances bacterial killing in the ASL of mouse lungs, and decreases pulmonary inflammation in mice following bacteria instillation. *P. aeruginosa* in 20-μl 300-mOsm xylitol instilled intranasally in mice are more readily cleared by 18 h than *P. aeruginosa* in 300-mOsm saline. As early as 2 min following instillations, mice receiving *P. aeruginosa* in xylitol had very few live bacteria in bronchoalveolar lavage fluid as compared with mice that received *P. aeruginosa* in saline. The ability of xylitol to accelerate the disappearance of live bacteria is consistent with bacterial killing by innate bactericidal substances (Coonrod, 1986), which act very quickly. As a result, pulmonary inflammation is minimized; there are significantly fewer neutrophils in the xylitol-treated mice compared with the saline group, and concentrations of the CXC chemokines KC and MIP-2 in bronchoalveolar lavage fluid 24 h after bacterial inoculation were almost normal. These results suggest that enhancing bacterial killing by ASL may result in less inflammation.

Similarly, xylitol enhances the bacterial killing in the lungs of sheep. *M. haemolytica* in 1-ml 300-mOsm xylitol instilled intratracheally to lambs are more readily cleared by 20 min and 24 h than *M. haemolytica* in 1-ml 300-mOsm saline. The effect of xylitol in this model appeared to be dependent on the concentration of *M. haemolytica* instilled; preliminary data with a small number of animals per group suggest that xylitol has a significant effect on airway sterility at 20 min and 24 h when the bacterial inocula is 10^7 CFU or less.

8.7. CONDITIONS AFFECTING PEPTIDE EFFICACY IN VIVO

Some antimicrobial peptides that are very active against microorganisms in vitro may simply not work in vivo or may be too toxic to the host (Table 8.3). These peptides may be (1) poorly retained and rapidly eliminated by the host, (2) inactivated by high concentrations of ionic substances at the site of peptide activity, (3) inactivated by host proteases, (4) neutralized by serum components, or (5) acutely toxic for host cells and tissues. Antimicrobial peptides are small molecules and, when given systemically, are likely to be rapidly secreted from the host. In mice, $rBPI_{23}$ was administered with *E. coli* pneumonia every 2 h to maintain adequate tissue levels because of rapid blood clearance (Kelly et al., 1993).

The activity of antimicrobial peptides is also diminished in high-ionic-strength solutions (Goldman et al., 1997; Bals, Goldman, and Wilson, 1998a; Bals et al., 1998b; Garcia et al., 2001) or solutions containing other cations. The addition of calcium and magnesium inhibits the potency of novispirin G10, an effect that could be reversed by the addition of citrate, a chelator of these cations (Steinstraesser et al., 2002). Because the normal concentrations of magnesium and calcium in rat wound fluid are 0.8–2 mM, the peptide may not be as effective in vivo. Surprisingly, novispirin G10 was highly effective in reducing the number of bacteria in the wound, presumably because of its rapid effect and considerable potency (Steinstraesser et al., 2002). Divalent cations also reduce the antimicrobial activity of lactoferricin (Bellamy et al., 1993), protegrins (Cho et al., 1998), α-defensins (Lehrer et al., 1988), and β-defensins (Tomita et al., 2000). The mechanism of this effect is likely due to the ability of divalent cations to inhibit the self-promoted uptake of cationic antimicrobial peptides by bacterial membranes (Hancock, 1997). This self-promoted uptake hypothesis states that cationic compounds interact at sites, on the outer-membrane surface, at which divalent cations crossbridge adjacent LPS molecules (Hancock et al., 1991). This causes a destabilization of the outer membrane that is proposed to permit uptake of the interacting molecule and/or of other molecules in the environment of the cell.

Cationic antimicrobial peptides, particularly defensins, also bind to and are inactivated by blood proteins in what is thought to be a mechanism to regulate their nonspecific activity in inflammatory processes. Defensins bind to alpha 1-proteinase inhibitor (Panyutich et al., 1995), 1-antichymotrypsin (Panyutich et al., 1995), alpha 2-antiplasmin (Panyutich et al., 1995), antithrombin III (Panyutich et al., 1995), complexes of activated C1 complement (Panyutich et al., 1994), and activated alpha 2-macroglobulin (Panyutich and Ganz, 1991). Likewise, drosocin fails to protect mice from systemic *E. coli*

Table 8.3. *Characteristics of antimicrobial peptides that may limit their use as therapeutic agents*

Peptide	Concern	Reference
BPI	$rBPI_{23}$ was administered every 2 h to maintain adequate tissue levels because of rapid blood clearance	(Kelly et al., 1993)
$CAP18_{109-135}$	Plasma components inhibited the in vitro antibacterial activity of $CAP18_{109-135}$, perhaps explaining why this fragment was not effective against infection with *P. aeruginosa*	(Kirikae et al., 1998a)
$CAP18_{106-137}$	$CAP18_{106-137}$ was thought to bind to negatively charged tissues, thus causing the lung edema seen in mice	(Sawa et al., 1998)
Drosocin	25–100 mg/kg dose fails to protect mice from *E. coli* infection, possibly because of low stability of the native glycopeptide in animal blood	(Hoffmann et al., 1999)
HLD2, a fragment of human lactoferrin	Low antimicrobial activity of HLD2 in undiluted mouse urine was probably an effect of the salt concentration, as bactericidal activity was inhibited when the NaCl concentration was increased from 0.035 to 0.14 M in test solutions	(Haversen et al., 2000)
Novispirin G10	Calcium and magnesium inhibited the antimicrobial potency of Novispirin G10	(Steinstraesser et al., 2002)

infection in the 25–100-mg/kg dose range, possibly because of the low stability of the native glycopeptide in animal blood (Hoffmann et al., 1999). The peptide was significantly more stable in insect hemolymph. In contrast, LL-37 protected in the transgenic mice studies of Bals et al. (Bals et al., 1999).

8.8. FUTURE AREAS OF STUDY

The current use of antibiotics as growth promotors in animal feeds (Witte, 1998; Wegener et al., 1999; Witte et al., 2000) and the extensive use of antibiotics to treat human or animal infections (Gaynes, 1997; Gaynes and Monnet, 1997; Diekema, Brueggemann, and Doern, 2000) are thought to be the underlying cause of an alarming increase in antibiotic resistance among bacterial and fungal pathogens, a highly debated and controversial issue (Marwick, 1999; Gorbach, 2001). Bacterial resistance to many classes of antibiotics is becoming a major clinical problem (Arthur, Reynolds, and Courvalin, 1996; Jacoby, 1996). Therefore the search continues for new classes of antibiotics. Synthetic congeners of natural antimicrobial peptides are good candidates; they are active in vivo, fast acting, and broad spectrum, and do not induce bacterial resistance. However, only a select few of the vast numbers of antimicrobial peptides currently known are being assessed in animal models of microbial infections and sepsis.

8.8.1. Identification and Development of New Molecules

New molecular tools may help identify additional antimicrobial peptides, particularly with the annotation of mammalian genomes. One approach is to use HMMER, a computational search tool based on hidden Markov models, in combination with BLAST. This strategy recently identified 28 new HBDs and 43 new mouse β-defensin genes in five syntenic chromosomal regions (Schutte et al., 2002). Preliminary analysis indicated that at least 26 of the predicted genes were transcribed. These results demonstrated the value of a genome-wide search strategy to identify genes with conserved structural motifs (Schutte et al., 2002).

Structure–activity relationship studies on currently described antimicrobial peptides will also help identify those peptides with increased activity and decreased toxicity in animal models of infection and sepsis. Synthetically engineered antimicrobial peptides can be prepared in large numbers, resulting in the formation of families of related molecules (Travis et al., 2000; Brogden et al., 2001; Kalfa et al., 2001a). For example, an extensive structure–activity relationship study was conducted on several hundred protegrin analogs (Chen et al., 2000). A large number of amino acid substitutions were tolerated by the protegrin structure, implying that overall structural features such as amphiphilicity, charge, and shape were more important to activity than the presence of specific amino acids. Future studies will continue to assess the structure–activity relationship of synthetic peptides by using conformationally defined combinatorial libraries (Blondelle et al., 1996; Hong et al., 1998). Peptides can be rapidly screened by their circular dichroism spectra, antimicrobial activity, and cytotoxicity.

8.8.2. New Methods of Administration

Antimicrobial peptides are small molecules and, when given systemically, are likely to be rapidly secreted from the host. Research in this area will identify new methods of administration for better retention. For example, Demegen (Pittsburgh, Pennsylvania) is developing a synthetic antimicrobial peptide (called D2A21) in a gel formulation as a wound-healing product to treat infected burns and wounds. In vitro *P. aeruginosa is* killed within 30 min by 1–5-μM D2A21. In an infected in vivo rodent burn model, D2A21 demonstrated significant antibacterial activity against *P. aeruginosa*. After three days, there was a fourfold decrease in subeschar bacterial growth and zero eschar bacterial growth in the D2A21-treated group versus the control group, which had 9.5×10^7 eschar colony-forming units. In this same model, D2A21 demonstrated significant improvement in percent survival (85% versus 0% in the control group, $p < .001$).

REFERENCES

Abe, S., Okutomi, T., Tansho, S., Ishibashi, H., Wakabayashi, H., Teraguchi, S., Hayasawa, H., and Yamaguchi, H. (2000). Augmentation by lactoferrin of host defense against Candida infection in mice. In *Lactoferrin: Structure, Function, and Applications*, ed. Shimazaki, K. Tsuda, H., Tomita, M., Kuwata, T., and Perraudin, J. P. pp. 195–201. Amsterdam: Elsevier Science.

Agerberth, B., Gunne, H., Odeberg, J., Kogner, P., Boman, H. G., and Gudmundsson, G. H. (1995). FALL-39, a putative human peptide antibiotic, is cysteine-free and expressed in bone marrow and testis. *Proceedings of the National Academy of Science USA*, 92, 195–9.

Akinbi, H. T., Epaud, R., Bhatt, H., and Weaver, T. E. (2000). Bacterial killing is enhanced by expression of lysozyme in the lungs of transgenic mice. *Journal of Immunology*, 165, 5760–6.

Arthur, M., Reynolds, P., and Courvalin, P. (1996). Glycopeptide resistance in enterococci. *Trends in Microbiology*, 4, 401–7.

Bals, R., Goldman, M. J., and Wilson, J. M. (1998a). Mouse β-defensin 1 is a salt-sensitive antimicrobial peptide present in epithelia of the lung and urogenital tract. *Infection and Immunity*, 66, 1225–32.

Bals, R., Wang, X. R., Wu, Z. R., Freeman, T., Bafna, V., Zasloff, M., and Wilson, J. M. (1998b). Human beta-defensin 2 is a salt-sensitive peptide antibiotic expressed in human lung. *Journal of Clinical Investigation*, 102, 874–80.

Bals, R., Weiner, D. J., Moscioni, A. D., Meegalla, R. L., and Wilson, J. M. (1999). Augmentation of innate host defense by expression of a cathelicidin antimicrobial peptide. *Infection and Immunity*, 67, 6084–9.

Baveye, S., Elass, E., Mazurier, J., Spik, G., and Legrand, D. (1999). Lactoferrin: A multifunctional glycoprotein involved in the modulation of the inflammatory process. *Clinical Chemistry and Laboratory Medicine* 37, 281–6.

Bellamy, W., Takase, M., Yamauchi, K., Wakabayashi, H., Kawase, K., and Tomita, M. (1992). Identification of the bactericidal domain of lactoferrin. *Biochimica et Biophysica Acta*, 1121, 130–6.

Bellamy, W. R., Wakabayashi, H., Takase, M., Kawase, K., Shimamura, S., and Tomita, M. (1993). Role of cell-binding in the antibacterial mechanism of lactoferricin B. *Journal of Applied Bacteriology*, 75, 478–84.

Bessalle, R., Gorea, A., Shalit, I., Metzger, J. W., Dass, C., Desiderio, D. M., and Fridkin, M. (1993). Structure-function studies of amphiphilic antibacterial peptides. *Journal of Medicinal Chemistry*, 36, 1203–9.

Bhimani, R. S., Vendrov, Y., and Furmanski, P. (1999). Influence of lactoferrin feeding and injection against systemic staphylococcal infections in mice. *Journal of Applied Microbiology*, 86, 135–44.

Blondelle, S. E., Takahashi, E., Houghten, R. A., and Perez-Paya, E. (1996). Rapid identification of compounds with enhanced antimicrobial activiey by using conformationally defined combinatorial libraries. *Biochemical Journal*, 313, 141–7.

Brogden, K. A., Ackermann, M., and Huttner, K. M. (1998). Detection of anionic antimicrobial peptides in ovine bronchoalveolar lavage fluid and respiratory epithelium. *Infection and Immunity*, 66, 5948–54.

Brogden, K. A., Ackermann, M. R., McCray, Jr., P. B., and Huttner, K. M. (1999). Differences in the concentrations of small, anionic, antimicrobial peptides in bronchoalveolar lavage fluid and in respiratory epithelia of patients with and without cystic fibrosis. *Infection and Immunity*, 67, 4256–9.

Brogden, K. A., De Lucca, A. J., Bland, J., and Elliott, S. (1996). Isolation of an ovine pulmonary surfactant-associated anionic peptide bactericidal for *Pasteurella haemolytica*. *Proceedings of the National Academy of Science USA*, 93, 412–6.

Brogden, K. A., Kalfa, V. C., Ackermann, M. R., Palmquist, D. E., McCray, Jr., P. B., and Tack, B. F. (2001). The ovine cathelicidin SMAP29 kills ovine respiratory pathogens in vitro and in an ovine model of pulmonary infection. *Antimicrobial Agents and Chemotherapy*, 45, 331–4.

Chan, Y. R., Zanetti, M., Gennaro, R., and Gallo, R. L. (2001). Anti-microbial activity and cell binding are controlled by sequence determinants in the anti-microbial peptide PR-39. *Journal of Investigative Dermatology*, 116, 230–5.

Chen, J., Falla, T. J., Liu, H., Hurst, M. A., Fujii, C. A., Mosca, D. A., Embree, J. R., Loury, D. J., Radel, P. A., Cheng Chang, C., Gu, L., and Fiddes, J. C. (2000). Development of protegrins for the treatment and prevention of oral mucositis: Structure-activity relationships of synthetic protegrin analogues. *Biopolymers*, 55, 88–98.

Chipman, D. M. and Sharon, N. (1969). Mechanism of lysozyme action. *Science*, 165, 454–65.

Cho, Y., Turner, J. S., Dinh, N. N., and Lehrer, R. I. (1998). Activity of protegrins against yeast-phase *Candida albicans*. *Infection and Immunity*, 66, 2486–93.

Cole, A. M., Shi, J., Ceccarelli, A., Kim, Y. H., Park, A., and Ganz, T. (2001). Inhibition of neutrophil elastase prevents cathelicidin activation and impairs clearance of bacteria from wounds. *Blood*, 97, 297–304.

Coonrod, J. D. (1986). The role of extracellular bactericidal factors in pulmonary host defense. *Seminars in Respiratory Infections*, 1, 118–29.

Cudic, M., Bulet, P., Hoffmann, R., Craik, D. J., and Otvos Jr., L. (1999). Chemical synthesis, antibacterial activity and conformation of diptericin, an 82-mer peptide originally isolated from insects. *European Journal of Biochemistry*, 266, 549–58.

Dathe, M., Schumann, M., Wieprecht, T., Winkler, A., Beyermann, M., Krause, E., Matsuzaki, K., Murase, O., and Bienert, M. (1996). Peptide helicity and membrane surface charge modulate the balance of electrostatic and hydrophobic interactions with lipid bilayers and biological membranes. *Biochemistry*, 35, 12612–22.

Diekema, D. J., Brueggemann, A. B., and Doern, G. V. (2000). Antimicrobial-drug use and changes in resistance in *Streptococcus pneumoniae*. *Emerging Infectious Diseases*, 6, 552–6.

Dinarello, C. A. (1996). Cytokines as mediators in the pathogenesis of septic shock. *Current Topics in Microbiology and Immunology*, 216, 133–65.

Dionysius, D. A. and Milne, J. M. (1997). Antibacterial peptides of bovine lactoferrin: purification and characterization. *Journal of Dairy Science*, 80, 667–74.

Edde, L., Hipolito, R. B., Hwang, F. F., Headon, D. R., Shalwitz, R. A., and Sherman, M. P. (2001). Lactoferrin protects neonatal rats from gut-related systemic infection. *American Journal of Physiology Gastrointestinal and Liver Physiology*, 281, G1140–50.

Elsbach, P. and Weiss, J. (1993). Bactericidal/permeability increasing protein and host defense against Gram-negative bacteria and endotoxin. *Current Opinion in Immunology*, 5, 103–7.

Elsbach, P., Weiss, J., and Levy, O. (1994). Integration of antimicrobial host defenses: Role of the bactericidal/permeability-increasing protein. *Trends in Microbiology*, 2, 324–8.

Falla, T. J. and Hancock, R. E. (1997). Improved activity of a synthetic indolicidin analog. *Antimicrobial Agents and Chemotherapy*, 41, 771–5.

Feng, W., Garrett, H., Speert, D. P., and King, M. (1998). Improved clearability of cystic fibrosis sputum with dextran treatment in vitro. *American Journal of Respiratory and Critical Care Medicine*, 157, 710–4.

Fleming, A. (1922). On a remarkable bacteriolytic element found in tissues and secretions. *Proceedings of the Royal Society of London. Series B. Biological Sciences*, 93, 306–17.

Friedrich, C., Scott, M. G., Karunaratne, N., Yan, H., and Hancock, R. E. (1999). Salt-resistant alpha-helical cationic antimicrobial peptides. *Antimicrobial Agents and Chemotherapy*, 43, 1542–8.

Ganz, T., Gabayan, V., Liao, H. I., Liu, L., Oren, A., Graf, T., and Cole, A. M. (2003). Increased inflammation in lysozyme M-deficient mice in response to *Micrococcus luteus* and its peptidoglycan. *Blood*, 101, 2388–92.

Ganz, T., Metcalf, J. A., Gallin, J. I., Boxer, L. A., and Lehrer, R. I. (1988). Microbicidal/cytotoxic proteins of neutrophils are deficient in two disorders: Chediak–Higashi syndrome and "specific" granule deficiency. *Journal of Clinical Investigation*, 82, 552–6.

Ganz, T., Selsted, M. E., and Lehrer, R. I. (1986). Antimicrobial activity of phagocyte granule proteins. *Seminars in Respiratory Infections*, 1, 107–17.

Garcia, J. R., Krause, A., Schulz, S., Rodriguez-Jimenez, F. J., Kluver, E., Adermann, K., Forssmann, U., Frimpong-Boateng, A., Bals, R., and Forssmann, W. G. (2001). Human β-defensin 4: A novel inducible peptide with a specific salt-sensitive spectrum of antimicrobial activity. *Federation of American Societies for Experimental Biology Journal*, 15, 1819–21.

Gaynes, R. (1997). The impact of antimicrobial use on the emergence of antimicrobial-resistant bacteria in hospitals. *Infectious Disease Clinic of North America*, 11, 757–65.

Gaynes, R. and Monnet, D. (1997). The contribution of antibiotic use on the frequency of antibiotic resistance in hospitals. *CIBA Foundation Symposium*, 207, 47–56.

Giacometti, A., Cirioni, O., Ghiselli, R., Mocchegiani, F., Del Prete, M. S., Viticchi, C., Kamysz, W., E, L. E., Saba, V., and Scalise, G. (2002). Potential therapeutic role of cationic peptides in three experimental models of septic shock. *Antimicrobial Agents and Chemotherapy*, 46, 2132–6.

Goldman, M. J., Anderson, G. M., Stolzenberg, E. D., Kari, U. P., Zasloff, M., and Wilson, J. M. (1997). Human beta-defensin-1 is a salt-sensitive antibiotic in lung that is inactivated in cystic fibrosis. *Cell*, 88, 553–60.

Goldstein, E. (1983). Hydrolytic enzymes of alveolar macrophages. *Reviews of Infectious Diseases*, 5, 1078–92.

Gorbach, S. L. (2001). Antimicrobial use in animal feed – time to stop. *New England Journal of Medicine*, 345, 1202–3.

Gough, M., Hancock, R. E. W., and Kelly, N. M. (1996). Antiendotoxin activity of cationic peptide antimicrobial agents. *Infection and Immunity*, 64, 4922–7.

Hancock, R. E., Farmer, S. W., Li, Z. S., and Poole, K. (1991). Interaction of aminoglycosides with the outer membranes and purified lipopolysaccharide and OmpF porin of *Escherichia coli*. *Antimicrobial Agents and Chemotherapy*, 35, 1309–14.

Hancock, R. E. W. (1997). Antibacterial peptides and the outer membranes of gram-negative bacilli. *Journal of Medical Microbiology*, 46, 1–3.

Haversen, L. A., Engberg, I., Baltzer, L., Dolphin, G., Hanson, L. A., and Mattsby-Baltzer, I. (2000). Human lactoferrin and peptides derived from a surface-exposed helical region reduce experimental *Escherichia coli* urinary tract infection in mice. *Infection and Immunity*, 68, 5816–23.

Haynes, R. J., Tighe, P. J., and Dua, H. S. (1999). Antimicrobial defensin peptides of the human ocular surface. *British Journal of Ophthalmology*, 83, 737–41.

Hirata, M., Shimomura, Y., Yoshida, M., Morgan, J. G., Palings, I., Wilson, D., Yen, M. H., Wright, S. C., and Larrick, J. W. (1994). Characterization of a rabbit cationic protein (CAP18) with lipopolysaccharide-inhibitory activity. *Infection and Immunity*, 62, 1421–6.

Hirata, M., Wright, S. C., and Larrick, J. W. (1996). Endotoxin-neutralizing proteins for sepsis and endotoxin shock. In *Shock*, ed. K. Okada and H. Ogata, pp. 109–15. Amsterdam: Elsevier Science.

Hoffmann, R., Bulet, P., Urge, L., and Otvos Jr., L. (1999). Range of activity and metabolic stability of synthetic antibacterial glycopeptides from insects. *Biochimica et Biophysica Acta*, 1426, 459–67.

Holtje, J. V. (1996). Lysozyme substrates. *Exs*, 75, 105–10.

Hong, S. Y., Oh, J. E., Kwon, M. Y., Choi, M. J., Lee, J. H., Lee, B. L., Moon, H. M., and Lee, K. H. (1998). Identification and characterization of novel antimicrobial decapeptides generated by combinatorial chemistry. *Antimicrobial Agents and Chemotherapy*, 42, 2534–41.

Isamida, T., Tanaka, T., Omata, Y., Yamauchi, K., Shimazaki, K., and Saito, A. (1998). Protective effect of lactoferricin against *Toxoplasma gondii* infection in mice. *Journal of Veterinary Medical Science,* 60, 241–4.

Iwasa, M., Kaito, M., Ikoma, J., Takeo, M., Imoto, I., Adachi, Y., Yamauchi, K., Koizumi, R., and Teraguchi, S. (2002). Lactoferrin inhibits hepatitis C virus viremia in chronic hepatitis C patients with high viral loads and HCV genotype 1b. *American Journal of Gastroenterology,* 97, 766–7.

Jacoby, G. A. (1996). Antimicrobial-resistant pathogens in the 1990s. *Annual Review of Medicine,* 47, 169–79.

Jia, X., Patrzykat, A., Devlin, R. H., Ackerman, P. A., Iwama, G. K., and Hancock, R. E. (2000). Antimicrobial peptides protect coho salmon from *Vibrio anguillarum* infections. *Applied and Environmental Microbiology,* 66, 1928–32.

Jiang, C., Finkbeiner, W. E., Widdicombe, J. H., McCray Jr., P. B., and Miller, S. S. (1993). Altered fluid transport across airway epithelium in cystic fibrosis. *Science,* 262, 424–7.

Kadurugamuwa, J. L. and Beveridge, T. J. (1997). Natural release of virulence factors in membrane vesicles by *Pseudomonas aeruginosa* and the effect of aminoglycoside antibiotics on their release. *Journal of Antimicrobial Chemotherapy,* 40, 615–21.

Kalfa, V. C., Jia, H. P., Kunkle, R. A., McCray Jr., P. B., Tack, B. F., and Brogden, K. A. (2001a). Congeners of SMAP29 kill ovine pathogens and induce ultrastructural damage in bacterial cells. *Antimicrobial Agents* and *Chemotherapy,* 45, 3256–61.

Kalfa, V. C., Palmquist, D., Ackermann, M. R., and Brogden, K. A. (2001b). Suppression of *Mannheimia (Pasteurella) haemolytica* serovar 1 infection in lambs by intrapulmonary administration of ovine antimicrobial anionic peptide. *International Journal of Antimicrobial Agents,* 17, 505–10.

Kelly, C. J., Cech, A. C., Argenteanu, M., Gallagher, H., Shou, J., Minnard, E., and Daly, J. M. (1993). Role of bactericidal permeability-increasing protein in the treatment of gramnegative pneumonia. *Surgery,* 114, 140–6.

Kirby, A. J. (2001). The lysozyme mechanism sorted – after 50 years. *Nature Structural Biology,* 8, 737–9.

Kirikae, T., Hirata, M., Yamasu, H., Kirikae, F., Tamura, H., Kayama, F., Nakatsuka, K., Yokochi, T., and Nakano, M. (1998a). Protective effects of a human 18-kilodalton cationic antimicrobial protein (CAP18)-derived peptide against murine endotoxemia. *Infection and Immunity,* 66, 1861–8.

Kirikae, T., Kirikae, F., Saito, S., Tominaga, K., Tamura, H., Uemura, Y., Yokochi, T., and Nakano, M. (1998b). Biological characterization of endotoxins released from antibiotic-treated *Pseudomonas aeruginosa* and *Escherichia coli. Antimicrobial Agents and Chemotherapy,* 42, 1015–21.

Larrick, J. W., Hirata, M., Balint, R. F., Lee, J., Zhong, J., and Wright, S. C. (1995). Human CAP18: A novel antimicrobial lipopolysaccharide-binding protein. *Infection and Immunity*, 63, 1291–7.

Larrick, J. W., Hirata, M., Shimomoura, Y., Yoshida, M., Zheng, H., Zhong, J., and Wright, S. C. (1993). Antimicrobial activity of rabbit CAP18-derived peptides. *Antimicrobial Agents and Chemotherapy*, 37, 2534–9.

Lee, W. J., Farmer, J. L., Hilty, M., and Kim, Y. B. (1998). The protective effects of lactoferrin feeding against endotoxin lethal shock in germfree piglets. *Infection and Immunity*, 66, 1421–6.

Lehrer, R. I. and Ganz, T. (1996). Endogenous vertebrate antibiotics. Defensins, protegrins, and other cysteine-rich antimicrobial peptides. *Annals of the New York Academy of Sciences*, 797, 228–39.

Lehrer, R. I., Ganz, T., Selsted, M. E., Babior, B. M., and Curnutte, J. T. (1988). Neutrophils and host defense. *Annals of Internal Medicine*, 109, 127–42.

Lehrer, R. I., Lichtenstein, A. K., and Ganz, T. (1993). Defensins: Antimicrobial and cytotoxic peptides of mammalian cells. *Annual Review of Immunology*, 11, 105–28.

Levin, M., Quint, P. A., Goldstein, B., Barton, P., Bradley, J. S., Shemie, S. D., Yeh, T., Kim, S. S., Cafaro, D. P., Scannon, P. J., and Giroir, B. P. (2000). Recombinant bactericidal/permeability-increasing protein (rBPI21) as adjunctive treatment for children with severe meningococcal sepsis: a randomised trial. rBPI21 Meningococcal Sepsis Study Group. *Lancet*, 356, 961–7.

Loury, D., Embree, J. R., Steinberg, D. A., Sonis, S. T., and Fiddes, J. C. (1999). Effect of local application of the antimicrobial peptide IB-367 on the incidence and severity of oral mucositis in hamsters. *Oral Surgery, Oral Medicine, Oral Pathology, Oral Radiology and Endodontics*, 87, 544–51.

Marwick, C. (1999). Animal feed antibiotic use raises drug resistance fear. *Journal of the American Medical Association*, 282, 120–2.

Miyasaki, K. T., Iofel, R., Oren, A., Huynh, T., and Lehrer, R. I. (1998). Killing of *Fusobacterium nucleatum, Porphyromonas gingivalis* and *Prevotella intermedia* by protegrins. *Journal of Periodontal Research*, 33, 91–8.

Miyasaki, K. T. and Lehrer, R. I. (1998). Beta-sheet antibiotic peptides as potential dental therapeutics. *International Journal of Antimicrobial Agents*, 9, 269–80.

Moore, A. J., Devine, D. A., and Bibby, M. C. (1994). Preliminary experimental anticancer activity of cecropins. *Peptide Research*, 7, 265–9.

Morrison, D. C. and Ryan, J. L. (1987). Endotoxins and disease mechanisms. *Annual Review of Medicine*, 38, 417–32.

Navon-Venezia, S., Feder, R., Gaidukov, L., Carmeli, Y., and Mor, A. (2002). Antibacterial properties of dermaseptin S4 derivatives with in vivo activity. *Antimicrobial Agents and Chemotherapy*, 46, 689–94.

Nizet, V., Ohtake, T., Lauth, X., Trowbridge, J., Rudisill, J., Dorschner, R. A., Pestonjamasp, V., Piraino, J., Huttner, K., and Gallo, R. L. (2001). Innate antimicrobial peptide protects the skin from invasive bacterial infection. *Nature (London)*, 414, 454–7.

Ohto, M., Ito, H., Masuda, K., Tanaka, S., Arakawa, Y., Wacharotayankun, R., and Kato, N. (1992). Mechanisms of antibacterial action of tachyplesins and polyphemusins, a group of antimicrobial peptides isolated from horseshoe crab hemocytes. *Antimicrobial Agents and Chemotherapy*, 36, 1460–5.

Ong, P. Y., Ohtake, T., Brandt, C., Strickland, I., Boguniewicz, M., Ganz, T., Gallo, R. L., and Leung, D. Y. (2002). Endogenous antimicrobial peptides and skin infections in atopic dermatitis. *New England Journal of Medicine*, 347, 1151–60.

Otvos Jr., L. (2002). The short proline-rich antibacterial peptide family. *Cellular and Molecular Life Sciences*, 59, 1138–50.

Otvos Jr., L., Bokonyi, K., Varga, I., Otvos, B. I., Hoffmann, R., Ertl, H. C., Wade, J. D., McManus, A. M., Craik, D. J., and Bulet, P. (2000a). Insect peptides with improved protease-resistance protect mice against bacterial infection. *Protein Science*, 9, 742–9.

Otvos Jr., L., O. I., Rogers, M. E., Consolvo, P. J., Condie, B. A., Lovas, S., Bulet, P., and Blaszczyk-Thurin, M. (2000b). Interaction between heat shock proteins and antimicrobial peptides. *Biochemistry*, 39, 14150–9.

Panyutich, A. and Ganz, T. (1991). Activated alpha2-macroglobulin is a principal defensin-binding protein. *American Journal of Respiratory Cell and Molecular Biology*, 5, 101–6.

Panyutich, A. V., Hiemstra, P. S., van Wetering, S., and Ganz, T. (1995). Human neutrophil defensin and serpins form complexes and inactivate each other. *American Journal of Respiratory Cell and Molecular Biology*, 12, 351–7.

Panyutich, A. V., Szold, O., Poon, P. H., Tseng, Y., and Ganz, T. (1994). Identification of defensin binding to C1 complement. *Federation of European Biological Sciences Letters*, 356, 169–73.

Park, C. B., Yi, K. S., Matsuzaki, K., Kim, M. S., and Kim, S. C. (2000). Structure-activity analysis of buforin II, a histone H2A-derived antimicrobial peptide: The proline hinge is responsible for the cell-penetrating ability of buforin II. *Proceedings of the National Academy of Science USA*, 97, 8245–50.

Piers, K. L., Brown, M. H., and Hancock, R. E. (1994). Improvement of outer membrane-permeabilizing and lipopolysaccharide-binding activities of an antimicrobial cationic peptide by C-terminal modification. *Antimicrobial Agents and Chemotherapy*, 38, 2311–6.

Piers, K. L., Brown, M. H., and Hancock, R. E. W. (1993). Recombinant DNA procedures for producing small antimicrobial cationic peptides in bacteria. *Gene*, 134, 7–13.

Piers, K. L. and Hancock, R. E. W. (1994). The interaction of a recombinant cecropin/melittin hybrid peptide with the outer membrane of *Pseudomonas aeruginosa*. *Molecular Microbiology*, 12, 951–8.

Putsep, K., Carlsson, G., Boman, H., and Andersson, M. (2002). Deficiency of antibacterial peptides in patients with morbus Kostmann: An observation study. *Lancet*, 360, 1144–49.

Ramanathan, B., Davis, E. G., Ross, C. R., and Blecha, F. (2002). Cathelicidins: Microbicidal activity, mechanisms of action, and roles in innate immunity. *Microbes and Infection*, 4, 361–72.

Rietschel, E. T., Brade, H., Holst, O., Brade, L., Muller-Loennies, S., Mamat, U., Zahringer, U., Beckmann, F., Seydel, U., Brandenburg, K., Ulmer, A. J., Mattern, T., Heine, H., Schletter, J., Loppnow, H., Schonbeck, U., Flad, H. D., Hauschildt, S., Schade, U. F., Di Padova, F., Kusumoto, S., and Schumann, R. R. (1996). Bacterial endotoxin: Chemical constitution, biological recognition, host response, and immunological detoxification. *Current Topics in Microbiology and Immunology*, 216, 39–81.

Sato, R., Inanami, O., Tanaka, Y., Takase, M., and Naito, Y. (1996). Oral administration of bovine lactoferrin for treatment of intractable stomatitis in feline immunodeficiency virus (FIV)-positive and FIV-negative cats. *American Journal of Veterinary Research*, 57, 1443–6.

Sawa, T., Kurahashi, K., Ohara, M., Gropper, M. A., Doshi, V., Larrick, J. W., and Wiener-Kronish, J. P. (1998). Evaluation of antimicrobial and lipopolysaccharide-neutralizing effects of a synthetic CAP 18 fragment against *Pseudomonas aeruginosa* in a mouse model. *Antimicrobial Agents and Chemotherapy*, 42, 3269–75.

Schutte, B. C. and McCray Jr., P. B. (2002). β-defensins in lung host defense. *Annual Review of Physiology*, 64, 709–48.

Schutte, B. C., Mitros, J. P., Bartlett, J. A., Walters, J. D., Jia, H. P., Welsh, M. J., Casavant, T. L., and McCray Jr., P. B. (2002). Discovery of five conserved β-defensin gene clusters using a computational search strategy. *Proceedings of the National Academy of Science USA*, 99, 2129–33.

Scott, M. G., Davidson, D. J., Gold, M. R., Bowdish, D., and Hancock, R. E. (2002). The human antimicrobial peptide LL-37 is a multifunctional modulator of innate immune responses. *Journal of Immunology*, 169, 3883–91.

Scott, M. G., Rosenberger, C. M., Gold, M. R., Finlay, B. B., and Hancock, R. E. (2000). An alphahelical cationic antimicrobial peptide selectively modulates macrophage responses to lipopolysaccharide and directly alters macrophage gene expression. *Journal of Immunology*, 165, 3358–65.

Singh, P. K., Parsek, M. R., Greenberg, E. P., and Welsh, M. J. (2002). A component of innate immunity prevents bacterial biofilm development. *Nature (London)*, 417, 552–5.

Skerlavaj, B., Benincasa, M., Risso, A., Zanetti, M., and Gennaro, R. (1999). SMAP-29: A potent antibacterial and antifungal peptide from sheep leukocytes. *Federation of European Biological Sciences Letters*, 463, 58–62.

Smith, J. J., Travis, S. M., Greenberg, E. P., and Welsh, M. J. (1996). Cystic fibrosis airway epithelia fail to kill bacteria because of abnormal airway surface fluid. *Cell*, 85, 229–36.

Sonis, S. T., Tracey, C., Shklar, G., Jenson, J., and Florine, D. (1990). An animal model for mucositis induced by cancer chemotherapy. *Oral Surgery, Oral Medicine, and Oral Pathology*, 69, 437–43.

Steinstraesser, L., Tack, B. F., Waring, A. J., Hong, T., Boo, L. M., Fan, M. H., Remick, D. I., Su, G. L., Lehrer, R. I., and Wang, S. C. (2002). Activity of novispirin G10 against *Pseudomonas aeruginosa* in vitro and in infected burns. *Antimicrobial Agents and Chemotherapy*, 46, 1837–44.

Sudo, E., Boyd, W. A., and King, M. (2000). Effects of dextran sulfate on tracheal mucociliary velocity in dogs. *Journal of Aerosol Medicine*, 13, 87–96.

Tasaka, S., Ishizaka, A., Urano, T., Sayama, K., Sakamaki, F., Nakamura, H., Terashima, T., Waki, Y., Soejima, K., Nakamura, M., Matsubara, H., Fujishima, S., Kanazawa, M., and Larrick, J. W. (1996). A derivative of cationic antimicrobial protein attenuates lung injury by suppressing cell adhesion. *American Journal of Respiratory Cell and Molecular Biology*, 15, 738–44.

Teraguchi, S., Shin, K., Ogata, T., Kingaku, M., Kaino, A., Miyauchi, H., Fukuwatari, Y., and Shimamura, S. (1995a). Orally administered bovine lactoferrin inhibits bacterial translocation in mice fed bovine milk. *Applied and Environmental Microbiology*, 61, 4131–4.

Teraguchi, S., Shin, K., Ozawa, K., Nakamura, S., Fukuwatari, Y., Tsuyuki, S., Namihira, H., and Shimamura, S. (1995b). Bacteriostatic effect of orally administered bovine lactoferrin on proliferation of Clostridium species in the gut of mice fed bovine milk. *Applied and Environmental Microbiology*, 61, 501–6.

Tomita, M., Takase, M., Wakabayashi, H., and Bellamy, W. (1994). Antimicrobial peptides of lactoferrin. *Advances in Experimental Medicine and Biology*, 357, 209–18.

Tomita, M., Wakabayashi, H., Yamauchi, K., Teraguchi, S., and Hayasawa, H. (2002). Bovine lactoferrin and lactoferricin derived from milk: Production and applications. *Biochemistry and Cell Biology-Biochimie et Biologie Cellulaire*, 80, 109–12.

Tomita, T., Hitomi, S., Nagase, T., Matsui, H., Matsuse, T., Kimura, S., and Ouchi, Y. (2000). Effect of ions on antibacterial activity of human beta defensin 2. *Microbial Immunology*, 44, 749–54.

Tossi, A., Mitaritonna, N., Tarantino, C., Giangaspero, A., Sandri, L., and Winterstein, K. (2002). Antimicrobial Sequences Database http://www.bbcm.univ.trieste.it/~tossi/pag5.htm

Travis, S. M., Anderson, N. N., Forsyth, W. R., Espiritu, C., Conway, B. D., Greenberg, E. P., McCray Jr., P. B., Lehrer, R. I., Welsh, M. J., and Tack, B. F. (2000). Bactericidal activity of mammalian cathelicidin-derived peptides. *Infection and Immunity*, 68, 2748–55.

Vaara, M. and Porro, M. (1996). Group of peptides that act synergistically with hydrophobic antibiotics against Gram-negative enteric bacteria. *Antimicrobial Agents and Chemotherapy*, 40, 1801–5.

van Berkel, P. H., Welling, M. M., Geerts, M., van Veen, H. A., Ravensbergen, B., Salaheddine, M., Pauwels, E. K., Pieper, F., Nuijens, J. H., and Nibbering, P. H. (2002). Large scale production of recombinant human lactoferrin in the milk of transgenic cows. *Nature Biotechnology*, 20, 484–7.

VanderMeer, T. J., Menconi, M. J., O'Sullivan, B. P., Larkin, V. A., Wang, H., Kradin, R. L., and Fink, M. P. (1994). Bactericidal/permeability-increasing protein ameliorates acute lung injury in porcine endotoxemia. *Journal of Applied Physiology*, 76, 2006–14.

VanderMeer, T. J., Menconi, M. J., Zhuang, J., Wang, H., Murtaugh, R., Bouza, C., Stevens, P., and Fink, M. P. (1995). Protective effects of a novel 32-amino acid C-terminal fragment of CAP18 in endotoxemic pigs. *Surgery*, 117, 656–62.

Vincent, J. L. (1996). Definition and pathogenesis of septic shock. *Current Topics in Microbiology and Immunology*, 216, 1–13.

Vorland, L. H. (1999). Lactoferrin: a multifunctional glycoprotein. *Acta Pathologica Microbiologica et Immunologica Scandinavica*, 107, 971–81.

Wakabayashi, H., Uchida, K., Yamauchi, K., Teraguchi, S., Hayasawa, H., and Yamaguchi, H. (2000). Lactoferrin given in food facilitates dermatophytosis cure in guinea pig models. *Journal of Antimicrobial Chemotherapy*, 46, 595–602.

Wegener, H. C., Aarestrup, F. M., Jensen, L. B., Hammerum, A. M., and Bager, F. (1999). Use of antimicrobial growth promoters in food animals and *Enterococcus faecium* resistance to therapeutic antimicrobial drugs in Europe. *Emerging Infectious Diseases*, 5, 329–35.

Weinberg, A., Krisanaprakornkit, S., and Dale, B. A. (1998). Epithelial antimicrobial peptides: Review and significance for oral applications. *Critical Reviews in Oral Biology and Medicine*, 9, 399–414.

Witte, W. (1998). Medical consequences of antibiotic use in agriculture. *Science*, 279, 996–7.

Witte, W., Tschape, H., Klare, I., and Werner, G. (2000). Antibiotics in Animal Feed. *Acta Veterinaria Scandinavica*, 93 (Suppl.), 37–45.

Yamauchi, K., Hiruma, M., Yamazaki, N., Wakabayashi, H., Kuwata, H., Teraguchi, S., Hayasawa, H., Suegara, N., and Yamaguchi, H. (2000a). Oral administration of bovine lactoferrin for treatment of tinea pedis. A placebo-controlled, double-blind study. *Mycoses*, 43, 197–202.

Yamauchi, K., Toida, T., Nishimura, S., Nagano, E., Kusuoka, O., Teraguchi, S., Hayasawa, H., Shimamura, S., and Tomita, M. (2000b). 13-week oral repeated administration toxicity study of bovine lactoferrin in rats. *Food and Chemical Toxicology*, 38, 503–12.

Zabner, J., Seiler, M. P., Launspach, J. L., Karp, P. H., Kearney, W. R., Look, D. C., Smith, J. J., and Welsh, M. J. (2000). The osmolyte xylitol reduces the salt concentration of airway surface liquid and may enhance bacterial killing. *Proceedings of the National Academy of Science USA*, 97, 11614–9.

Zagulski, T., Jedra, M., Jarzabek, Z., and Zagulska, A. (1986). Protective effect of lactoferrin during a systemic experimental infection of rabbits with *Escherichia coli*. *Animal Science Papers and Reports*, 1, 59–74.

Zagulski, T., Lipinski, P., Zagulska, A., Broniek, S., and Jarzabek, Z. (1989). Lactoferrin can protect mice against a lethal dose of *Escherichia coli* in experimental infection in vivo. *British Journal of Experimental Pathology*, 70, 697–704.

Zasloff, M. (2002). Antimicrobial peptides of multicellular organisms. *Nature (London)*, 415, 389–95.

Zhang, L., Scott, M. G., Yan, H., Mayer, L. D., and Hancock, R. E. (2000). Interaction of polyphemusin I and structural analogs with bacterial membranes, lipopolysaccharide, and lipid monolayers. *Biochemistry*, 39, 14504–14.

CHAPTER 9

Bacterial structure and physiology: Influence on susceptibility to cationic antimicrobial peptides

Robert E. W. Hancock

9.1. INTRODUCTION

The importance of antimicrobial peptides in nature has only begun to be appreciated in the past decade. They are now clearly established as nature's antibiotics (Hancock, 1998) because cationic amphiphilic peptides with antimicrobial activity have been found in virtually all species of life from microbes to humans. In addition, it has recently become clear that they have a profound role in regulating other mechanisms of innate immunity (Gudmundsson and Agerberth, 1999; Hancock and Diamond, 2000), as well as a possible role in regulating the transition from innate to adaptive immunity (Lillard et al., 1999).

Natural cationic peptides with these properties range in size from twelve to hundreds of amino acids (Tossi, 2004). However, I am restricting discussion here to those peptides, both natural and synthetic, that range in size from 6 to 35 residues, as these have been best studied. Furthermore my own recent studies have indicated that the antimicrobial and innate immunity-enhancing properties of cationic peptides are clearly separable, and only the antibiotic activities are discussed here. It should be noted that many natural peptides have extremely weak antibacterial activity. This may indicate that other activities are more important. Conversely, it is possible that synergy between the multiple peptides found at most body locations in animals, and/or very high concentrations, such as those found in the azurophilic granules of neutrophils or at inflammatory sites, contribute to their in vivo role as antibiotic substances (Matsuzaki et al., 1998; Gudmundsson and Agerberth, 1999; Hancock and Diamond, 2000; Yan and Hancock, 2001).

Antibiotic cationic peptides have enormous chemical and sequence diversity. The most conserved themes are the presence of two or more (up to 9) net positive charges, contributed by the basic amino acids arginine and lysine, as well as around 50% hydrophobic residues, and an ability to fold into an amphiphilic structure (often in the presence of membranes) that leads to spatial separation of charged/polar and hydrophobic domains. Enormous heterogeneity is provided by variations in length, secondary structure (α-helical, β-sheet, β-turn loop, or extended), and sequence, even within a secondary structure class. This is to some extent reflected by heterogeneity in their mechanism of action on, and interaction with, bacteria, although in this chapter I try to emphasize thematic similarities.

Given the dearth of new antibiotics, the cationic antimicrobial peptides have engendered considerable excitement as prospective novel antibiotic agents. Despite the dampening of this excitement by recent clinical failures (Magainin's MSI-78 and Intrabiotics's IB-367), commercial development continues. Thus understanding the interaction of peptides with bacteria is a key element in understanding structure–activity relationships.

9.2. THE BACTERIAL ENVELOPE

9.2.1. Overview

There are two major classes of bacteria, Gram negative and Gram positive, according to their ability to retain crystal violet stain. These classes correspond to differences in envelope composition, and this in turn has had a substantial impact on antibiotic development. Thus, although antibiotics with broad-spectrum activity (against both types of bacteria) exist, it has proven, generally speaking, easier to make antibiotics with activity against Gram positive than against Gram negative bacteria. In contrast, cationic antimicrobial peptides with broad-spectrum activity or selectivity for Gram negative bacteria outnumber those with exclusive activity against Gram positive bacteria.

These properties can largely be explained by the properties of the envelopes of these two classes of bacteria (schematically diagrammed in Fig. 9.1). The cell envelope of Gram positive bacteria consists of a cytoplasmic membrane surrounded by a relatively thick (15–80-nm) layer of peptidoglycan. Gram negative bacteria have two cell envelope membranes, a cytoplasmic membrane and an outer membrane, separated by the periplasm, which contains a relatively thinner (~2-nm) peptidoglycan layer. Further details of these structures are described in the next subsection. However, many conventional antibiotics have weak activity against Gram negative bacteria because of a

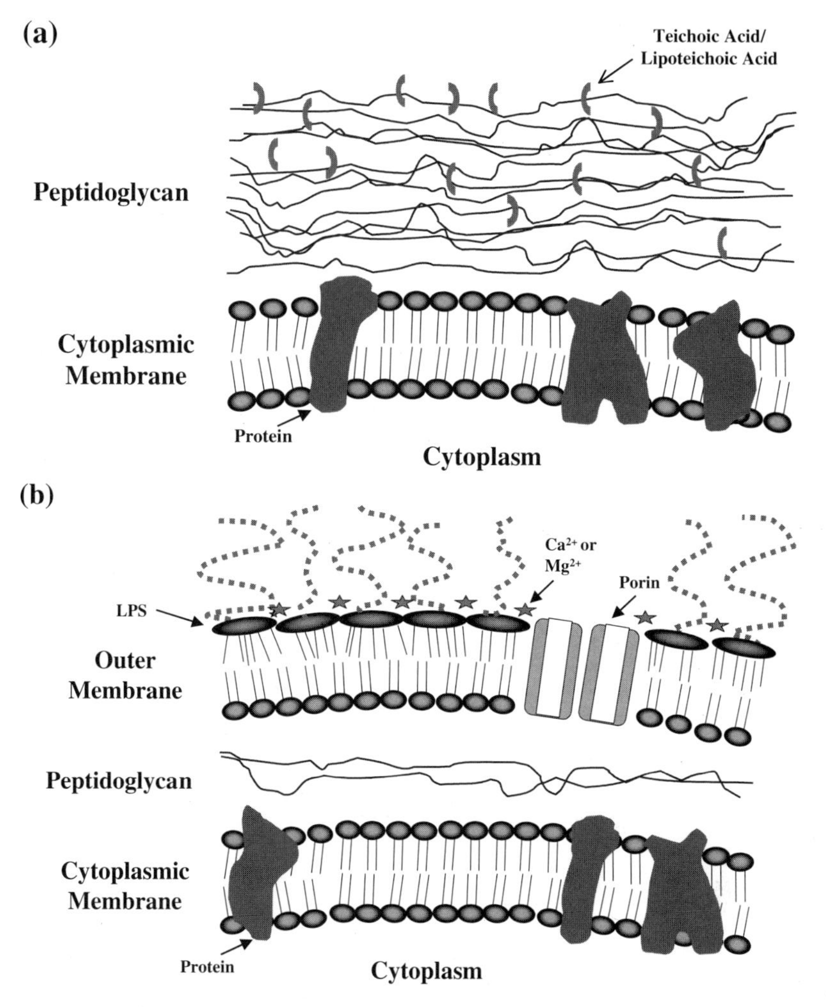

Figure 9.1. Schematic representation of the structure of the (a) Gram positive and (b) Gram negative cell envelopes. (See color section.)

combination of restricted uptake through the channels of porin proteins in the outer membrane and active efflux from the cell (Hancock, 1997). Conversely, cationic antimicrobial peptides demonstrate enhanced uptake into gram negative bacteria because of their utilization of an alternative antibiotic uptake mechanism, termed self-promoted uptake (Hancock and Chapple, 1999), which probably serves to focus peptide to a discrete region of the cytoplasmic membrane.

9.2.2. Gram Negative Bacterial Outer Membrane

Outer-membrane structure has been previous reviewed in some detail (Hancock, Egli, and Karunaratne, 1994), and only an overview is presented here. The outer membrane is an asymmetric bilayer with the inner monolayer containing lipids, usually phospholipids, and the outer monolayer containing predominantly an unusual glycolipid called lipopolysaccharide (LPS). A modest number of protein species (around 50–150) are found in the outer membrane, the most important of which are the porins, which form water-filled channels across the membrane. One class of these proteins, the non-specific porins, determines the general ability of the outer membrane to take up some small molecules, restrict the uptake of others (e.g., conventional antibiotics) based on their size relative to the channel diameter, and exclude larger molecules, e.g., peptides larger than around 5 amino acids. This has led researchers to discuss the outer membrane as a molecular sieve, which is in fact incorrect as this conceptually suggests a surface with holes in an impervious material. Indeed, it is clear that the "fabric" of the outer membrane is permeable to both hydrophobic substances (which may subsequently be rapidly effluxed out; Zgurskaya and Nikaido, 2000), and to large amphipathic polycations (including cationic peptides) (Hancock and Chapple, 1999). The key property of the outer membrane that permits uptake of these polycations is conferred by the surface LPS. LPS is a highly negatively charged glycolipid that comprises a membrane-inserted, conserved lipidic molecule, Lipid A (usually fatty-acylated diglucosamine phosphate) and attached oligosaccharides (termed the Rough core and O-antigen). The strong negative charge, which is due to attached phosphate residues and the negatively charged sugar 2-keto-3-deoxy-octulosonate, is partly neutralized by divalent cations, predominantly Ca^{2+} and Mg^{2+}. Polycationic molecules such as polymyxins, aminoglycosides, and cationic antimicrobial peptides interact with polyanionic LPS molecules, resulting in the displacement of divalent cations. This in turn causes localized disruption of the outer membrane (visualized as large blebs; Fig. 9.2). Uptake of additional molecules of the permeabilizing polycation has been proposed to occur through these lesions, giving rise to the term self-promoted uptake.

9.2.3. Influence of the Outer Membrane on Peptide Susceptibility

It seems reasonable to predict, based on the self-promoted uptake mechanism, that cationic antimicrobial peptides will be found in higher

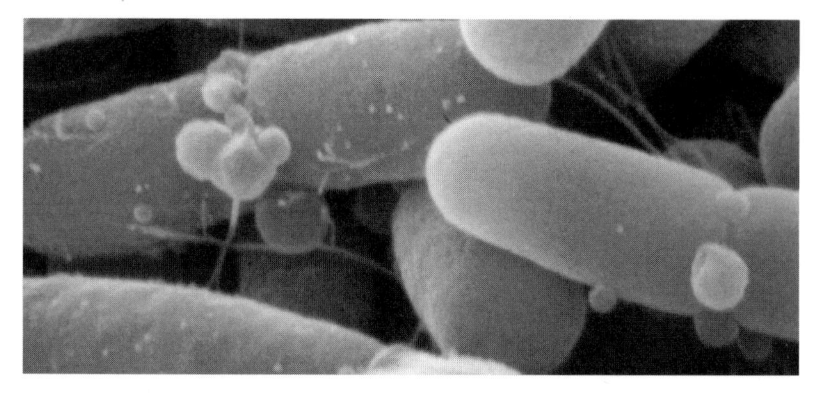

Figure 9.2. Effect of addition of the synthetic cecropin:melittin derivative antimicrobial peptide CEMA on *Escherichia coli* as observed by scanning electron microscopy. Peptide was added at three times the MIC (6.4 μg/ml) for 30 min. Compared with control cells that exhibited a smooth surface, peptide-treated cells demonstrated the development of large blebs.

concentration at regions of the cytoplasmic membrane underlying the outer-membrane blebs described in the preceding subsection. However, in my opinion there is no convincing evidence to show that peptide interaction with the outer membrane is itself sufficient to cause bacterial cell death. To address this issue a series of peptides was produced based on the amphipathic α-helical structural theme with zero, one, or two proline residues (Zhang, Benz, and Hancock, 1999). Each of the peptides had rather similar outer-membrane interaction, as revealed by binding affinity for LPS [dansyl polymyxin (DPX) displacement assay] and ability to permeabilize the outer membrane to the hydrophobic probe NPN. However, they differed up to 16-fold in minimum inhibitory concentration (MIC) for *Escherichia coli* and *Pseudomonas aeruginosa*. Nevertheless, because there is no direct assay for peptide translocation across the outer membrane, only limited conclusions can be made.

The first step in self-promoted uptake involves the binding of peptides to polyanionic LPS and displacement of divalent cations. From this, one would predict that added cations would antagonize self-promoted uptake and, if this were merely an ionic event, that divalent cations would be only 3-fold to 4-fold better at antagonism than monovalent cations. Indeed, it has been experimentally observed that added cations are antagonistic, but that the divalent cations Mg^{2+} and Ca^{2+} are generally greater than 20-fold more antagonistic than monovalent cations Na^{+} and K^{+} (Friedrich et al., 1999). This is more in keeping with the existence of a specific divalent cation-binding site than a general ionic interaction.

Tight binding to LPS would presumably involve both charge:charge interactions between the positive charges on the peptide and negative charges on the LPS, and hydrophobic interactions between the hydrophobic residues of the peptide and the fatty-acyl region of Lipid A. When purified LPS is arrayed in monolayers, the insertion of peptide into the monolayers is relatively resistant to antagonism by divalent cations (Zhang et al., 2000). Indeed, theoretically high ionic strengths should drive hydrophobic interactions. Thus the extent of cation antagonism for a given peptide will probably reflect antagonism of the charge:charge interaction and promotion of hydrophobic insertion. Indeed, even related peptides can differ in their ability to resist antagonism. It should also be reemphasized that generally 1-mM Mg^{2+} or Ca^{2+} is as antagonistic as 100-mM Na^+ or K^+. Despite the presence of 3-mM divalent cations in human blood, divalent cation antagonism is rarely studied, and, in the peptide literature, salt resistance is a euphemism for resistance to physiological saline.

The importance of the outer membrane in determining susceptibility to cationic peptides is revealed through the studies of cells grown in low Mg^{2+} medium and investigation of mutants in the PhoPQ and PmrAB two-component regulators that influence the LPS composition of cells (see Chapter 12 by Tomayo, Portillo, and Gunn; Zhou et al., 2001). Intriguingly, it was recently demonstrated that, in *P. aeruginosa*, cationic peptides regulate the expression of the two-component regulator PmrAB and, through this, regulate increased intrinsic resistance to the peptides themselves, presumably by PmrAB-mediated alterations in LPS composition (McPhee, Lewenza, and Hancock, 2003, submitted). Indeed, the activity of individual peptides against *P. aeruginosa* is roughly inversely related to their ability to induce PmrAB.

9.2.4. Peptidoglycan

The peptidoglycan of bacteria is often referred to by the term cell wall. It generally comprises polymers of the alternating disaccharides *N*-acetyl glucosamine and *N*-acetyl muramic acid, with the latter covalently linked to a reasonably conserved tetrapeptide. The size of the polymers varies, and individual polymers are covalently cross-linked through their peptide side chains by virtue of a peptide or pentaglycine bridge. Overall, this gives the peptidoglycan a meshlike structure and the term "peptidoglycan gel" has been used to describe its general nature. Generally speaking, Gram positive bacteria have a much thicker peptidoglycan layer than Gram negative bacteria and also differ in the nature of the third amino acid in the side chain (lysine or diaminopimelate) and in the nature of the interpolymer cross link. The Gram positive bacterial peptidoglycan tends also to be covalently decorated with other polymers such as teichoic and lipoteichoic acid. Few studies have addressed the general permeability of the peptidoglycan. However, because such molecules as DNA (during transformation) are able to cross the cell wall, it can be assumed that the peptidoglycan gel contains at least some larger passageways.

9.2.5. Influence of Peptidoglycan on Cationic Peptide Susceptibility

Few studies have addressed the issue of how the peptidoglycan influences susceptibility to cationic peptides. It has been demonstrated that lipoteichoic acid (Scott, Gold, and Hancock, 1999) and teichoic acid (Vorland et al., 1999) can bind cationic peptides, and the former, especially, is conceptually like Gram negative LPS. However, Scott, Gold, and Hancock (1999) were unable to find a correlation between lipoteichoic acid binding ability and MIC for a range of peptides. In contrast, Peschel et al. (1999) utilized mutants in the *dlt* operon (involved in D-alanyl esterification of teichoic acid) to demonstrate a link between teichoic acid structure (leading to altered surface charge) and peptide susceptibility.

Some peptides demonstrate effects on the peptidoglycan of Gram positive bacteria that can be visualized by electron microscopy as fraying or thinning of the cell wall (e.g., Friedrich et al., 2000) (Fig. 9.3). Other studies have reported cell lysis as a by-product of action of certain peptides or even as the actual mechanism of action (Ginsburg, 2001), although in my experience, such extreme lysis of cells is relatively unusual at the minimal bactericidal concentration. Such observations are consistent with the induction of autolytic enzymes (which are naturally involved in peptidoglycan metabolism during e.g., cell division) and/or inhibition of peptidoglycan biosynthesis. Thus the impact of cationic peptides on peptidoglycan biosynthesis deserves more attention.

9.2.6. Cytoplasmic Membrane

The cytoplasmic membrane of bacteria differs from eukaryotic cell plasma membranes in two important respects. Both are fluid mosaic membranes comprising a lipid bilayer, but bacterial cytoplasmic membranes tend to have a much higher negative surface charge and carry a much larger

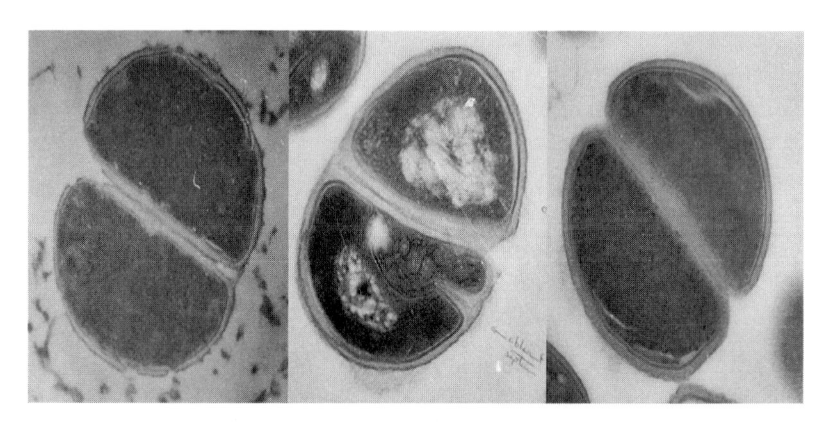

Figure 9.3. Effect of addition of antimicrobial peptides to the bacterium *Staphylococcus epidermidis* as observed with thin-section transmission electron microscopy. Electron micrographs are in order of (right) untreated, (middle) Bac2A-NH_2-treated, and (left) CP29-treated. All peptides were at concentrations of 10-fold the MIC.

membrane electrical potential gradient across them (oriented internal negative). These features combine to explain the relatively higher intrinsic susceptibility of bacteria to peptides and the relatively greater resistance of eukaryotic cells.

9.2.7. Influence of the Cytoplasmic Membrane on Peptide Activity

It is well accepted that most, if not all, cationic antimicrobial peptides interact with the cytoplasmic membrane. Indeed, this topic was recently reviewed in great detail (Hancock and Rozek, 2002), and other chapters in this book also discuss this topic in detail.

Interaction with the cytoplasmic membrane probably involves four factors, although the following discussion is largely based on model membrane studies. It has been clearly demonstrated that lipid composition is an important determinant of the ability of peptides to interact with membranes (e.g., Matsuzaki et al., 1995). Thus many peptides will insert into only the membranes of monolayers and bilayers containing negatively charged lipids (e.g., phosphatidyl glycerol and cardiolipin – both found in the bacterial cytoplasmic membrane). *E. coli*, as a typical Gram negative bacterium, contains around 20% anionic lipids, whereas most studies report a lower level of anionic lipids in the Gram positive bacterium *Staphylococcus aureus*. It is not known if this influences the relative susceptibility of these two organisms.

However, Peschel et al. (2001) demonstrated that *S. aureus* mutants defective in the *mprF* gene that modifies the anionic lipid phosphatidylglycerol with lysine and thereby reduces the net negative charge exhibited increased sensitivity to host peptides and decreased virulence in a mouse model.

The insertion of lipids into membranes should be affected by two other properties, namely the fatty-acid composition (which varies from bacterium to bacterium and according to growth conditions) and the temperature. These will influence the ability of membrane to undergo a phase transition, and it is generally held that phase-transition behavior influences the permeability characteristics of membranes (de Kruijff, 1997). The fourth cytoplasmic membrane property that probably influences peptide susceptibility is the size of the membrane potential gradient. Energized bacteria carry a large membrane potential gradient that can exceed –140 mV and is oriented internal negative, thus tending to elecrophorese peptides into the membrane. In one study (Falla, Karunaratne, and Hancock, 1996), depolarizing of the *E. coli* membrane led to resistance to the bovine peptide, indolicidin. However, in my experience this is not universally true, and the depolarizing compound carbonyl cyanide *m*-chlorophenol hydrazone can have effects ranging from antagonism to synergism, depending on the bacterium and the peptide (Friedrich, 2001).

9.3. BACTERIAL PHYSIOLOGY

9.3.1. Mode of Action

The actual mechanism of action of some peptides has been ascribed to the effects on the bacterial cytoplasmic membrane (e.g., Oren and Shai, 1998), whereas other peptides are known to cross the membrane and possibly act on cytoplasmic targets. Without reiterating recently made arguments (Hancock and Rozek, 2002), it is my opinion that at high enough concentrations virtually all peptides will destroy membrane integrity as revealed by complete depolarization (i.e., loss of the membrane potential gradient) of the cytoplasmic membrane. However, at their minimal effective concentrations, many cationic peptides do not destroy membrane potential and probably traverse the membrane to act in the cytoplasm (next subsection).

Friedrich et al. (2000) previously proposed that peptides have several potential targets in bacteria, including (but not limited to) cytoplasmic membrane integrity, peptidoglycan metabolism including activation of autolysis, cell division, macromolecular synthesis, and specific enzymes. In this multitarget model any given peptide might have a dominant target in a particular bacterium. However, this would not necessarily imply that a related peptide would have the same major target or that the given peptide would have the same target in a different bacterial species. Furthermore, it is well established that it is difficult to make bacteria resistant to peptides (Steinberg et al., 1997), arguing against a specific target. This is consistent with the concept that removal of any given target by mutation would still leave remaining another target that might be inhibited at a somewhat higher peptide concentration, resulting in little change in susceptibility.

It has clearly been established that some peptides have internal targets (reviewed in Devine and Hancock, 2002) and inhibit cells at concentrations below those required for seeing substantial membrane effects. However, it is, generally speaking, more difficult to study cytoplasmic targets than membrane activity and, as discussed previously, most cationic amphipathic peptides will permeabilize (depolarize) bacterial membranes at high enough concentrations. For this reason it is possible that far more peptides have cytoplasmic targets than those mentioned in Subsection 9.3.3.

9.3.2. Growth Phase/Rate

Relatively few studies have examined the influence of bacterial growth rate and growth phase on peptide susceptibility. Miyasaki and colleagues (1990) examined the activity of rabbit defensin NP-1 against a range of oral Gram negative facultative bacteria and reported that the bactericidal activity was not dependent on bacterial growth. Many other authors have followed up on these studies and demonstrated killing by a variety of peptides in buffer in the absence of an added carbon source. Koo, Yeaman, and Bayer (1996) examined the influence of a wide range of growth conditions on the bactericidal activity of thrombin-induced platelet microbicidal protein (tPMP-1). They observed more rapid *S. aureus* killing activity against log phase than stationary phase cells, although both were killed. This mirrors the unpublished observations of me and my colleagues (Monisha Scott and I) with both *P. aeruginosa* and *S. aureus*, for which we have observed a longer lag time to killing of stationary phase cells and somewhat slower killing. Conversely, Steinberg et al. (1997) observed very rapid of killing of both logarithmic and stationary phase methicillin-resistant *S. aureus* (MRSA) and *P. aeruginosa* by the protegrin derivative PG-1 applied at 1-fold to 2.5-fold the MIC. In contrast, it has been demonstrated for a wide variety of other antimicrobials that there is virtually no activity on stationary phase bacteria (see Steinberg et al., 1997, for data on norfloxacin, vancomycin, and gentamicin at 16- to 32-fold the MIC).

9.3.3. Interaction with Macromolecular Synthesis

It has been clearly demonstrated that fluorescein-tagged buforin, a histone-related peptide from the frog stomach, can translocate into *E. coli* (Park, Kim, and Kim, 1998). This same study indicated that fluorescein-tagged magainin II lodged in the cytoplasmic membrane, although another study that used labeled antibodies to unmodified magainin II demonstrated that both magainin II and lactoferricin, a lactoferrin-derived peptide, were able to enter the bacterial cytoplasm (Haukland et al., 2001). Differences between these studies could be explained by use of a fluorescein tag, as the former study demonstrated that modest changes in buforin sequence were sufficient to impede translocation into *E. coli* (Kobayashi et al., 2000).

Translocation across membranes has also been demonstrated by use of liposomes, indicating that an electrical potential gradient across the cytoplasmic membrane may not be a prerequisite for translocation. Peptides that have been demonstrated to translocate across liposome bilayers include four α-helical peptides, including buforin, and the β-hairpin peptide polyphemusin (Kobayashi et al., 2000; Zhang, Rozek, and Hancock, 2001).

Other evidence of translocation across membranes comes from studies demonstrating the inhibition of macromolecular synthesis, and/or cell killing, at concentrations that do not cause cell depolarization. Thus the peptides buforin (Park et al., 1998) and P-Der (Patrykzat et al., 2002) in separate studies inhibited RNA synthesis without depolarizing cells. Recently, the insect antimicrobial peptide, cecropin A, that had been thought to kill by interaction with and destruction of the cytoplasmic membrane was shown at sublethal concentrations to influence the expression of 26 different genes, implying that it found its way into the cytoplasm even at these sublethal concentrations (Hong et al., 2003). Similarly Oh et al. (2000) have indicated that cationic antimicrobial peptides appear to induce selective transcription of *micF* and *osmY* in *E. coli* and have proposed that this is related to the mechanism of antibacterial action of these agents. Several other studies have indicted that different peptides target macromolecular synthesis (reviewed in Devine and Hancock, 2002).

9.3.4. Other Cytoplasmic Targets

It should be noted that depolarization of cell membranes and inhibition of macromolecular synthesis do not necessarily lead to cell killing, and there are known inhibitors of these targets that are bacteriostatic rather than bactericidal, such as the cationic antimicrobial peptides. Indeed, for several peptides it has been suggested that, although they depolarize cytoplasmic membranes, their actual lethal targets are in the cytoplasm and remain unknown. The work of Otvos and collaborators (Kragol et al., 2001) has clearly pointed to the cytoplasmic chaperone and heat-shock protein DnaK as one possible target. Conversely, Barker et al. (2000) indicated that lethality that was due to a fragment of the cationic protein, BPI (bactericidal permeability-increasing protein), was due to disruption of the respiratory chain. In fact, it is not hard to make a case for any anionic or hydrophobic molecule as a potential target.

9.4. KEY QUESTIONS

We now know a lot about cationic peptides, but there is a lot left to know. Here are some of the key questions that should be addressed over the next few years:

1. Is it the antibacterial activity or the immunity-regulating activity of peptides that represents their principal biological function?
2. What is the mechanism of synergy of combinations of peptides?
3. Do membranes represent the actual target for most peptides or are they merely a barrier that must be crossed on the way to anionic cytoplasmic targets? How is peptide activity influenced by the structure of this membrane?
4. What role do anionic macromolecules in the cell wall of Gram positive bacteria play?
5. Does binding of peptides to anionic molecules in the cell envelope impede or enhance uptake and subsequent activity?
6. Which conditions in vivo, for example, temperature, cation nature and concentration, presence of polyanions, and presence of lipidic materials, are most influential on the activity of peptides?

We have come a long way in the past decade since research on cationic antimicrobial peptides became more intensive, but we have a long way to go.

ACKNOWLEDGMENTS

I would like to acknowledge the Canadian Cystic Fibrosis Foundation and the Canadian Bacterial Diseases Network that have sponsored my own research on these topics. I am currently supported by the Canada Research Chair program.

REFERENCES

Barker, H. C., Kinsella, N., Jaspe, A., Friedrich, T., and O'Connor, C. D. (2000). Formate protects stationary-phase *Escherichia coli* and *Salmonella* cells from killing by a cationic antimicrobial peptide. *Molecular Microbiology*, 35, 1518–29.

de Kruijff, B. (1997). Lipid polymorphism and biomembrane function. *Current Opinions in Chemical Biology*, 1, 564–9.

Devine, D. A. and Hancock, R. E. W. (2002). Cationic peptides: Distribution and mechanisms of resistance. *Current Pharmaceutical Design*, 8, 99–110.

Falla, T. J., Karunaratne, D. N., and Hancock, R. E. W. (1996). Mode of action of the antimicrobial peptide indolicidin. *Journal of Biological Chemistry*, 271, 19298–303.

Friedrich, C. L. (2001). Structure/Function and Mode of Action of Antimicrobial Cationic Peptides on Gram positive Bacteria, Ph.D. thesis. Vancouver: University of British Columbia.

Friedrich, C., Scott, M. G., Karunaratne, N., Yan, H., and Hancock, R. E. W. (1999). Salt-resistant alpha-helical cationic antimicrobial peptides. *Antimicrobial Agents and Chemotherapy*, 43, 1542–8.

Friedrich, C. L., Moyles, D., Beveridge, T. J., and Hancock, R. E. W. (2000). Antibacterial action of structurally diverse cationic peptides on Gram-positive bacteria. *Antimicrobial Agents and Chemotherapy*, 44, 2086–92.

Ginsburg, I. (2001). Cationic peptides from leukocytes might kill bacteria by activating their autolytic enzymes causing bacteriolysis: Why are publications proposing this concept never acknowledged? *Blood*, 97, 2530–1.

Gudmundsson, G. H. and Agerberth, B. (1999). Neutrophil antibacterial peptides, multifunctional effector molecules in the mammalian immune system. *Journal of Immunological Methods*, 232, 45–54.

Hancock, R. E. W. (1997). The bacterial outer membrane as a drug barrier. *Trends in Microbiology*, 5, 37–42.

Hancock, R. E. W. (1998). The therapeutic potential of cationic peptides. *Expert Opinion in Investigational Diseases*, 7, 167–74.

Hancock, R. E. W. and Chapple, D. S. (1999). Peptide antibiotics. *Antimicrobial Agents and Chemotherapy*, 43, 1317–23.

Hancock, R. E. W. and Diamond, G. (2000). The role of cationic antimicrobial peptides in innate host defences. *Trends in Microbiology*, 8, 402–10.

Hancock, R. E. W., Egli, C., and Karunaratne, N. (1994). Molecular organization and structural role of outer membrane macromolecules. In *Bacterial Cell Envelope*, ed. J. M. Ghuysen and R. Hakenbeck, pp. 263–279. Amsterdam: Elsevier Science.

Hancock, R. E. W. and Rozek, A. (2002). Role of membranes in the activities of antimicrobial cationic peptides. *FEMS Microbiology Letters*, 206, 143–9.

Haukland, H. H., Ulvatne, H., Sandvik, K., and Vorland, L. H. (2001). The antimicrobial peptides lactoferricin B and magainin 2 cross over the bacterial cytoplasmic membrane and reside in the cytoplasm. *FEBS Letters*, 508, 389–93.

Hong R. W., Schepetov, M., Weiser, J. N., and Axelsen, P. H. (2003). Transcriptional profile of the *Escherichia coli* response to the antimicrobial insect peptide cecropin A. *Antimicrobial Agents and Chemotherapy*, 47, 1–6.

Kobayashi, S., Takeshima, K., Park, C. B., Kim, S. C., and Matsuzaki, K. (2000). Interactions of the novel antimicrobial peptide buforin 2 with lipid bilayers: Proline as a translocation promoting factor. *Biochemistry*, 29, 8648–54.

Koo, S. P., Yeaman, M. R., and Bayer, A. S. (1996). Staphylocidal action of thrombin-induced platelet microbicidal protein is influenced by microenvironment and target cell growth phase. *Infection and Immunity*, 64, 3758–64.

Kragol, G., Lovas, S., Varadi, G., Condie, B. A., Hoffmann, R., and Otvos Jr., L. (2001). The antibacterial peptide pyrrhocoricin inhibits the ATPase actions of DnaK and prevents chaperone-assisted protein folding. *Biochemistry*, 40, 3016–26.

Lillard, J. W., Boyaka, P. N., Chertov, O., Oppenheim, J. J., and McGhee, J. R. (1999). Mechanisms for induction of acquired host immunity by neutrophil peptide defensins. *Proceedings of the National Academy of Sciences USA*, 96, 651–6.

Matsuzaki, K., Mitani, Y., Akada, K. Y., Murase, O., Yoneyama, S., Zasloff, M., and Miyajima, K. (1998). Mechanism of synergism between antimicrobial peptides magainin 2 and PGLa. *Biochemistry*, 37, 15144–53.

Matsuzaki, K., Sugishita, K., Fujii, N., and Miyajima, K. (1995). Molecular basis for membrane selectivity of an antimicrobial peptide, magainin 2. *Biochemistry*, 34, 3423–9.

McPhee, J. B., Lewenza, S., and Hancock, R. E. W. (2003). Cationic antimicrobial peptides activate a two-component regulatory system, PmrA–PmrB, that regulates resistance to polymyx B and cationic antimicrobial peptides in *Pseudomonas* aeruginosa. *Molecular Microbiology*, 50, 205–17.

Miyasaki, K. T., Bodeau, A. L., Selsted, M. E., Ganz, T., and Lehrer, R. I. (1990). Killing of oral, Gram negative, facultative bacteria by the rabbit defensin, NP-1. *Oral Microbiology and Immunology*, 5, 315–9.

Oh, J. T., Cajal, Y., Skowronska, E. M., Belkin, S., Chen, J., Van Dyk, T. K., Sasser, M., and Jain, M. K. (2000). Cationic peptide antimicrobials induce selective transcription of *micF* and *osmY* in *Escherichia coli*. *Biochimica et Biophysica Acta*, 1463, 43–54.

Oren, Z. and Shai, Y. (1998). Mode of action of linear amphipathic alpha-helical antimicrobial peptides. *Biopolymers*, 47, 451–63.

Park, C. B., Kim, H. S., and Kim, S. C. (1998). Mechanism of action of the antimicrobial peptide buforin II: Buforin II kills microorganisms by penetrating the cell membrane and inhibiting cellular functions. *Biochemical Biophysical Research Communications*, 244, 253–7.

Patrykzat, A., Friedrich, C. L., Zhang, L., Mendoza, V., and Hancock, R. E. W. (2002). Sub-lethal concentrations of pleurocidin-derived antimicrobial peptides inhibit macromolecular synthesis in *Escherichia coli*. *Antimicrobial Agents and Chemotherapy*, 46, 605–14.

Peschel, A., Jack, R. W., Otto, M., Collins, L. V., Staubitz, P., Nicholson, G., Kalbacher, H., Nieuwenhuizen, W. F., Jung, G., Tarkowski, A., van Kessel, K. P., and van Strijp, J. A. (2001). *Staphylococcus aureus* resistance to human defensins and evasion of neutrophil killing via the novel virulence factor MprF is based on modification of membrane lipids with l-lysine. *Journal of Experimental Medicine*, 193, 1067–76.

Peschel, A., Otto, M., Jack, R. W., Kalbacher, H., Jung, G., and Götz, F. (1999). Inactivation of the *dlt* operon in *Staphylococcus aureus* confers sensitivity to defensins, protegrins, and other antimicrobial peptides. *Journal of Biological Chemistry*, 274, 8405–10.

Scott, M. G., Gold, M. R., and Hancock, R. E. W. (1999). Interaction of cationic peptides with lipoteichoic acid and Gram positive bacteria. *Infection and Immunity*, 67, 6445–53.

Steinberg, D. A., Hurst, M. A., Fujii, C. A., Kung, A. H., Ho, J. F., Cheng, F. C., Loury, D. J., and Fiddles, J. C. (1997). Protegrin-1: A broad-spectrum, rapidly microbicidal peptide with in vivo activity. *Antimicrobial Agents and Chemotherapy*, 41, 1738–42.

Tossi, A. (2004). Antimicrobial sequences database. www.bbcm.univ.trieste.it/~tossi/pag1.htm

Vorland, L. H., Ulvatne, H., Rekdal, O., and Svendsen, J. S. (1999). Initial binding sites of antimicrobial peptides in *Staphylococcus aureus* and *Escherichia coli*. *Scandinavian Journal of Infectious Diseases*, 31, 467–73.

Yan, H. and Hancock, R. E. W. (2001). Synergistic interactions between mammalian antimicrobial defense peptides. *Antimicrobial Agents and Chemotherapy*, 45, 1558–60.

Zgurskaya, H. I. and Nikaido, H. (2000). Multidrug resistance mechanisms: Drug efflux across two membranes. *Molecular Microbiology*, 37, 219–25.

Zhang, L., Rozek, A., and Hancock, R. E. W. (2001). Interaction of cationic antimicrobial peptides with model membranes. *Journal of Biological Chemistry*, 276, 35714–22.

Zhang, L., Scott, M. G., Yan, H., Mayer, L. D., and Hancock, R. E. W. (2000). Interaction of polyphemusin I and structural analogs with bacterial membranes, lipopolysaccharide and lipid monolayers. *Biochemistry*, 39, 14504–14.

Zhang, L., Benz, R., and Hancock, R. E. W. (1999). Influence of proline residues on the antibacterial and synergistic activities of α-helical peptides. *Biochemistry*, 38, 8102–11.

Zhou, Z., Ribeiro, A. A., Lin, S., Cotter, R. J., Miller, S. I., and Raetz, C. R. (2001). Lipid A modifications in polymyxin-resistant Salmonella typhimurium: PMRA-dependent 4-amino-4-deoxy-L-arabinose, and phosphoethanolamine incorporation. *Journal of Biological Chemistry*, 276, 43111–21.

CHAPTER 10

The antifungal mechanisms of antimicrobial peptides

Eva J. Helmerhorst and Frank G. Oppenheim

10.1. INTRODUCTION

Antimicrobial peptides are potentialy of vital importance in the innate host defense against fungal and bacterial infections in a wide variety of species. These peptides are mostly cationic in nature and frequently display antibacterial as well as antifungal activities. Highly active peptides, such as mammalian α-defensins, are sequestered in the primary granules of phagocytic cells and exert their antifungal or antibacterial activities intracellularly after the granules fuse with the phagosome containing the ingested microorganism. Other peptides are secreted by phagocytic, epithelial, pituitary or salivary gland cells and are expected to act without harm to the host in an environment surrounded by host tissue cells in complex biological fluids such as blood, interstitial fluid, or saliva.

Mammalian antimicrobial peptides with established antifungal properties are listed in Table 10.1. For most of the peptides listed, there are considerable data available on in vitro antifungal activities but very limited data on their in vivo efficacies. Most *in vitro* antifungal studies are conducted with *Candida albicans*, as this fungus is the most frequently isolated species from clinical specimens. *C. albicans* is an opportunistic pathogen that causes mucocutaneous or disseminated candidiasis in immunocompromised hosts. In those situations, the activities of innate host-defense factors such as antimicrobial peptides to control or combat fungal infections may be of eminent importance.

The in vitro susceptibilities of *C. albicans* and other medically important fungi like *Cryptococcus* and *Aspergillus* species to antifungal peptides have been established in a variety of growth inhibition and killing assays. As will be described in subsequent sections, the activity of antimicrobial peptides

Table 10.1. *Mammalian antimicrobial peptides and proteins with antifungal activity*

Family	Peptide(s)[1]	Source	Cell type	Reference[2]
α-defensins	HNP 1, 2, 3	Human	Leukocytes	Lehrer et al., 1985; Newman et al., 2000
	NP 1, 2, 3a, 3b, 4, 5	Rabbit	Leukocytes	Patterson-Delafield et al., 1981; Selsted et al., 1985; Levitz et al., 1986
β-defensins	HBD	Human	Epithelial cells	Schröder and Harder, 1999; Harder et al., 2001
	PBD	Pig	Epithelial cells	Shi et al., 1999
	TAP	Cow	Tracheal epithelial cells	Diamond et al., 1991
	LAP	Cow	Lingual epithelial cells	Schonwetter, Stolzenberg, and Zasloff 1995
θ-defensins	RTD	Monkey	Leukocytes	Tang et al., 1999
Cathelicidins	CRAMP-18	Mouse	Leukocytes	Shin et al., 2000
	PMAP-23	Pig	Leukocytes	Lee et al., 2001
	SMAP-29	Sheep	Leukocytes	Mahoney et al., 1995
	eCATH 1, 2, 3	Horse	Leukocytes	Skerlavaj et al., 2001
	Bactenecin 5	Cow	Leukocytes	Raj and Edgerton, 1995
	Tritrpticin	Pig	Leukocytes	Lawyer et al., 1996; Yang et al., 2002
	Indolicidin	Cow	Leukocytes	Ösapay et al., 2000
	Protegrin 1, 2, 3, 4, 5	Pig	Leukocytes	Cho et al., 1998
Histatins	Histatin 1, 3, 5	Human	Salivary gland cells	Pollock et al., 1984; Oppenheim et al., 1988; Helmerhorst et al., 1999b
	m-histatin 1	Monkey	Salivary gland cells	Xu et al., 1990

Other	α-MSH	Cow	Pituitary gland cells	Cutuli et al., 2000
	Seminalplasmin	Cow	Prostate cells	Scheit and Bhargava, 1985
	Lactoferrin	Cow	Leukocytes	Kirkpatrick et al., 1971; Lupetti et al., 2000
	Lactoferricin	Cow	Leukocytes	Bellamy et al., 1993
	NK-lysin	Human	Leukocytes	Andrä and Leippe, 1999
	Cecropins	Pig	Intestinal epithelial cells	Andrä, Berninghausen, and Leippe, 2001
	BPI	Human	Leukocytes	Newman et al., 2000
	Vasostatin	Cow	Chromaffin cells	Lugardon et al., 2000
	Chromofungin	Cow	Chromaffin cells	Lugardon et al., 2001
	MUC7 fragment	Human	Salivary gland cells	Gururaja et al., 1999; Liu et al., 2000
	BSA	Cow	Hepatic cells	Olson et al., 1977
	hBNP-32	Human	Brain cells	Krause et al., 2001

[1] HNP, human neutrophil peptide; HBD, human β-defensin; TAP, tracheal antimicrobial peptide; LAP, lingual antimicrobial peptide; CRAMP, cathelin-related antimicrobial peptide; PMAP, pig myeloid antimicrobial peptide; SMAP, sheep myeloid antimicrobial peptide; eCATH, equine cathelicidin; α-MSH, α-melanocyte stimulating hormone; NK, natural killer; BPI, bactericidal permeability-inducing protein; MUC, mucin; BSA, bovine serum albumin; hBNP, human "brain-type" natriuretic peptide

[2] Reference in which activity against a particular fungus was first described

is strongly influenced by the composition of the media in which the antifungal assays are conducted. Therefore, the number of antimicrobial peptides for which antifungal activities have been described may either represent an underestimation or an overestimation of peptides that will exhibit true antifungal activity in vivo. Considering the complex composition of the body fluid in which antifungal peptides are present in vivo, the ultimate extent to which fungal cell killing occurs is likely the result of the cooperative destructive effects of several components that act on the fungus in similar or different ways. More and more studies focus on potential synergistic interactions (McCafferty et al., 1999; Nagaoka et al., 2000; Yan and Hancock, 2001). To understand or predict synergism, it is imperative to decipher the antifungal mechanisms of individual components participating in fungicidal events. For some of the peptides in Table 10.1, considerable insight has been obtained on their modes of action. In the present review, four antifungal mechanistic models are proposed that take into account all current evidence gathered from studies that focus on the molecular events that occur on exposure of fungal cells to mammalian antifungal peptides.

10.2. PEPTIDE–FUNGUS INTERACTION

10.2.1. Electrostatic Interactions

The initial interaction between antifungal peptides and the fungal target is believed to be electrostatic in nature because the cell surface of fungi such as *C. albicans* is negatively charged (Klotz, 1994) and the majority of antifungal peptides are cationic in nature. Binding and activity of many cationic antifungal peptides in general are weak and antagonized by the presence of ions that disturb the low-affinity electrostatic interactions of, for example, histatin 3 (Xu et al., 1999), histatin 5 (Xu et al., 1991, Helmerhorst et al., 1997), NP1 (Patterson-Delafield et al., 1981, Lehrer et al., 1985, Selsted et al., 1985) sheep myeloid antimicrobial peptide 29 (SMAP-29) (Lee et al., 2002), and insect antifungal proteins tenecin 3 (Kim et al., 2001) and AFP (Iijima, Kurata, and Natori, 1993). Unlike these antifungal peptides, protegrin-1, a cathelicidin peptide from porcine neutrophils, retains its antimicrobial activity in the presence of 100-mM NaCl, but killing occurs at a faster rate in a low-salt buffer, indicating that electrostatic interactions are involved in the activity of this peptide as well (Cho et al., 1998). Further evidence for the importance of electrostatic interactions between cationic peptides and the fungal cell comes from studies in which either the charge of the peptide or the target cell was modified. Introduction of lysine residues in the mammalian cathelin-related antimicrobial peptide CRAMP-18 that increases peptide charge increases antifungal activity against *C. albicans* and *Aspergillus fumigatus* (Shin et al., 2000), whereas reduction in the cationicity of peptides derived from lactoferrin (Lupetti et al., 2000) or histatin 5 (Tsai, Raj, and Bobek, 1996) greatly diminishes anticandidal activity. Furthermore, N- or C-terminal elongations of a model peptide, Lys-Lys-Val-Val-Phe-Lys-Val-Lys-Phe-Lys-amide, with lysine residues notably increased antifungal activity but not antibacterial activity, suggesting that charge:charge interactions are particularly important for antifungal properties (Hong, Park, and Lee, 2001).

The interactions of peptides with dye-containing liposome vesicles of varying phospholipid contents indicate that the use of zwitterionic phospholipids such as phosphatidyl choline (PC) or phosphatidyl ethanolamine (PE) eliminate dye leakage caused by defensins (Hristova, Selsted, and White, 1997). In contrast, incorporation of the negatively charged phospholipid cardiolipin [diphosphatidylglycerol (diPG)] greatly enhances lysis by NP1, and the effect is even greater than that anticipated from charged interactions alone (Hristova et al., 1997). These studies demonstrate that alteration of charged moieties on either the peptide or in the plasma membrane that increase electrostatic interactions are paralleled by biological effects.

10.2.2. Hydrophobic Interactions

Although electrostatic interactions play an important role in the initial interaction between the peptide and the fungal cell, hydrophobic interactions are involved as well. For example, if cationic residues in histatin 5 are replaced with hydrophobic leucine and isoleucine residues, no decrease in antifungal activity is observed (Tsai et al., 1996), suggesting that a decrease in peptide charge can be compensated for by an increase in peptide hydrophobicity. The importance of hydrophobic residues for anticandidal activity was also demonstrated in the porcine cathelicidin peptide, tritrpticin. This peptide contains three sequential tryptophan residues, and substitution of these for alanine residues abolishes anticandidal activity, whereas substitition of the tryptophan residues for hydrophobic phenylalanine residues does not (Yang et al., 2002). Therefore interactions between peptide hydrophobic amino acid side chains and hydrophobic *Candida* cell-wall constituents such as hydrophobic proteins with low levels of glycosylation (Fukazawa and Kagaya, 1997) are likely to occur in the recognition of peptides by yeast cells.

10.2.3. Peptide-Related Interaction Characteristics

Antifungal peptides display a large variation in their primary amino acid sequences and overall amino acid composition (Table 10.2). Nevertheless, at least two antifungal peptides from insect origin, AFP and tenecin 3 from *Sarcophaga peregrina* and *Tenebrio molitor* larva, respectively, and histatins from human saliva share an enrichment in the relatively uncommon amino acid histidine. In both AFP and tenecin 3, approximately 1-in-5 residues is a histidine residue, and, in histatin 5, more than 1-in-3 residues is a histidine. As is subsequently discussed, tenecin 3 and histatins show some similarities in their mechanisms of antifungal action.

Amino acid substitutions, carried out to investigate the contribution of particular amino acids to peptide antifungal activity, will not only change parameters such as peptide charge and hydrophobicity, but can also affect the secondary structure of the peptide. Changes in secondary structure, in turn, may alter the amphipathic properties of a peptide (the distribution of hydrophilic and hydrophobic regions in the peptide in a particular secondary conformation). A commonly used technique to assess the secondary structure of antimicrobial peptides is circular dichroism since α-helical, β-sheet, and random coil peptide conformations display characteristic absorption profiles. Hydrophobic solvents, such as trifluorethanol (TFE), are frequently used to mimic the hydrophobic lipid bilayer environment. It should be noted that subtle differences between peptides in the inducibility of secondary conformations might be obscured in pure TFE. For example, histatin 5 and histatin variants with improved activity are all α-helical in pure TFE, but clearly demonstrate differences in the inducibility of helical structures when they are dissolved in solutions with gradually increasing solvent hydrophobicity (Helmerhorst et al., 2001b). The relationship between structure and antifungal activity remains difficult to interpret from studies that use a particular solvent as the lipid matrix. Histatin substitution analogs that are virtually inactive against *C. albicans* demonstrated comparable helical contents in pure TFE to that of native histatin (Tsai et al., 1996). Substitution of proline with alanine in tritrpticin changed the peptide conformation in the presence of 30-mM sodium dodecyl sulphate (SDS) micelles from β-sheet to α-helix without affecting its antibacterial or antifungal activity (Yang et al., 2002). Introduction of three helix-breaking proline residues in histatin 5 markedly reduced the helical portion in pure TFE without causing a significant decrease in activity (Situ, Balasubramanian, and Bobek, 2000). Likewise, the helical content of four substitution analogs of the murine cathelicidin peptide CRAMP-18 in 50%TFE or 30-mM SDS micelles did not parallel their antifungal

Table 10.2. *Amino acid sequences of some antifungal peptides*

		Sequence				
Peptide[1]	Source	1	10	20	30	40
AFP	Insect	Q H G H G G Q D Q	H G Y G H G Q Q A V	Y G K G H E G H G V	N N L G Q D G H G Q	H –
		G Y A H G H S D Q	H G H G G Q H G Q H	D G Y K N R G Y		
α-MSH	Mammals	S Y S M E H F R W	G K P V			
Bactenecin 5	Mammals	R F R P P I R R P	P I R P P F Y P P F	R P P I R P P I F P	P I R P P F R P P L	G P F P
Hisatin 3	Mammals	D S H A K R H H G	Y K R K F H E K H H	S H R G Y R S N Y L	Y D N	
Histatin 5	Mammals	D S H A K R H H G	Y K R K F H E K H H	S H R G Y		
hLF(1–11)	Mammals	G R R R R S V Q W	C A			
HNP1	Mammals	A C Y C R I P A C	I A G E R R Y G T C	I Y Q G R L W A F C	C	
NP1	Mammals	V V C A C R R A L	C L P R E R R A G F	C R I R G R I H P L	C C R R	
Protegrin-1	Mammals	R G G R L C Y C R	R R F C V C V G R			
Tenecin 3	Insect	D H H D G H L G G	H Q T G H Q G G Q Q	G G H L G G Q Q G G	H L G G H Q G G Q P	G –
		G H L G G H Q G G	I G G T G G Q Q H G	Q H G P G T G A G H	Q G G Y K T H G H	
Tritrypticin	Mammals	V R R F P W W W P	F L R R			

[1]AFP, antifungal protein; α-MSH, α-melanocyte-stimulating hormone; hLF, human lactoferrin; HNP, human neutrophil peptide

activities (Shin et al., 2000). Although the large variability in secondary structures of antifungal peptides listed in Table 10.1 and the data from the preceding structure–activity studies provide inconclusive evidence as to whether a particular secondary structure is related to antifungal activity, amphipathicity changes that occur on conformational changes are relevant for activity (see Section 10.3).

A substantial number of studies do indicate that cysteine bridges are important for activity in defensins and cathelidicins. In protegrins, four cysteine bridges typically form two intramolecular disulfide bridges. At least one intramolecular disulfide bond is required for retaining optimal candidacidal activity at physiological salt concentrations (Cho et al., 1998). An antifungal peptide from plants, Ib-AMP1, containing two disulfide bridges, was four times more active in its oxidized form than in its reduced form (Lee et al., 1999). A cyclic defensin peptide from *Rhesus macaque*, designated the θ-defensin, is active in high salt buffers, but only in its circularized (Cys–Cys (C–C) oxidized) conformation (Tang et al., 1999). Furthermore, engineered circularization of an antimicrobial peptide containing multiple cysteine residues rendered its antibacterial activity salt insensitive (Yu, Lehrer, and Tam, 2000). These studies strongly indicate that the formation of a secondary structure that involves Cys–Cys oxidation increases antimicrobial activity, although the mechanism by which this occurs remains to be investigated.

10.2.4. Cell–Wall–Related Interaction Characteristics

Fungi have characteristically thick cell walls, and constituents of the cell wall are the first entity of the cell to interact with any antifungal peptide. This interaction may be weak or strong, but represents the first step in any antifungal mechanism. The yeast cell wall from outside to inside is made up of a fibrillar layer, a mannoprotein layer, a β-glucan layer, a β-glucan/chitin layer, and an inner mannoprotein layer. A peptide can reach targets such as the cell membrane and intracellular targets only if it is able to diffuse freely through the layer of the cell wall. This requires, first, sufficient permeability of the cell wall and, second, that the peptide is not neutralized by binding firmly to its constituents (Fig. 10.1).

Binding of antifungal peptides to individual components of the yeast cell wall has not been thoroughly investigated, but interference of purified yeast cell-wall constituents, such as chitin, with peptide antifungal activity has been reported. Chitin components, chitobiose and chitotriose, inhibit antifungal activity of bovine α-defensin peptide NP-1 in a dose-dependent

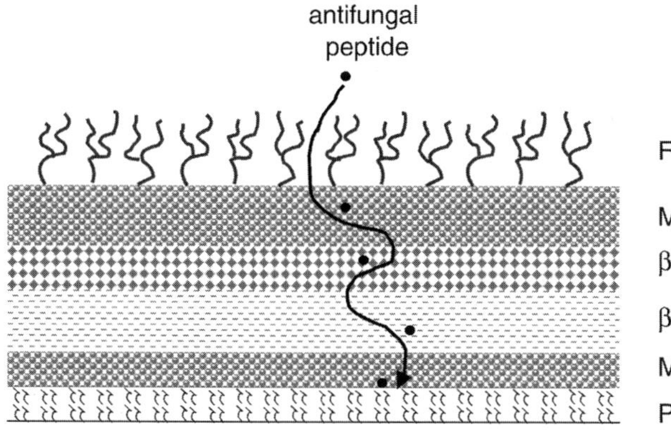

Figure 10.1. Route covered by an antifungal peptide that reaches the yeast cell membrane.

manner (Levitz et al., 1986). Tachystatin, an antifungal peptide from horseshoe crab with structural similarities to those of mammalian defensins binds chitin polymer at micromolar concentrations. In search for the chitin-binding domain in these peptides, it was established that both NP1 and tachystatin lack the common chitin-binding sequence, the "hevein domain" that is present in many chitin-binding proteins, such as chitinases (Beintema, 1994). In tachystatin, a phenylalanine residue contained in the planar surface of the β-turn in this peptide was postulated, based on structural modeling, to interact with the flat chitin surface. (Fujitani et al., 2002). It is feasible that phenylalanine residues, which are present in many other antifungal peptides, might play a crucial role in the primary association of antifungal peptides with the fungus and involve the chitin moiety.

A yeast cell-wall component known to interact with proteins is the outer mannoprotein layer. The mannoprotein moiety interacts with proteins such as fibronectin, laminin, fibrinogen, and complement factors expressed in mammalian tissues by recognizing the tripeptide sequence Arg-Gly-Asp (RGD) which is a common epitope in these proteins. Binding of yeast mannoproteins to the RGD sequence in these proteins generates a stable bond between yeast and mammalian cells (Fukazawa and Kagaya, 1997). By analogy, yeast mannoproteins could recognize RGD sequences if they were present in antimicrobial peptides. However, the RGD sequence is absent in peptides with antifungal activity (http://www.bbcm.univ.trieste.it/~tossi/pag1.htm), indicating that this recognition does not play a role in the fungus–antimicrobial peptide interaction.

Although binding partners for antifungal peptides on the fungal cell wall are poorly characterized, or may be absent, the human salivary antifungal peptide histatin 5 and the human defensin peptide HNP1 recognize a 67-kd protein in *C. albicans* that is present in whole cell lysate and crude membrane fractions (Edgerton et al., 1998). Despite large differences in primary and secondary structures between histatin 5 and HNP1 (Table 10.2), both peptides compete for binding, suggesting that rather nonspecific interactions are involved in the recognition of both histatin 5 and HNP1 (Edgerton et al., 2000). The nature of the histatin binding protein has recently been identified as the heat shock protein Ssa1/2p (Li, Reddy, Baer, et al., 2003). The potential recognition of both histatin 5 and HNP1 is of particular interest because these peptides share aspects of antifungal mechanisms as well (see Sections 10.4 and 10.5).

10.3. ANTIFUNGAL MECHANISTIC MODEL 1: DIRECT MEMBRANE PERMEABILIZATION

All peptides must bind to yeast cells to exert a biological effect. In some cases, binding leads directly to a biological effect. Several naturally occurring antifungal peptides with high amphipathicity from mammalian origin (e.g., protegrin) (Cho et al., 1998), amphibian origin (e.g., Peptide with N-terminal Glycine and C-terminal Leucine-amide (PGLa) and magainin 2) (Andreu et al., 1985; Zasloff, 1987), or insect origin (e.g., melittin) (Dufourcq and Faucon, 1977) cause instantaneous disruption of biological membranes. As model systems for membrane-permeabilizing activity, liposomes (Rex and Schwarz, 1998), oöcytes (Mangoni et al., 1996), and mitochondria (Westerhoff et al., 1989) have been used. Association of the peptides with these model systems causes disruption of membrane integrity, leading to leakage of encapsulated dyes (Ladokhin, Selsted, and White, 1997), dissipation of applied diffusion gradients (Gazit et al., 1995), or uncoupling of mitochondrial respiration that is due to dissipation of the transmembrane potential across the mitochondrial inner membrane (Helmerhorst et al., 2001b). Although it is questionable whether results obtained in model membrane systems can be extrapolated to activity toward living fungal cells, in some cases good correlations have been described (Edgerton et al., 1998; Helmerhorst et al., 2001b; Ruissen et al., 2001). Electron microscopy has enabled the direct visualization of the effect of a peptide on cellular membranes. Amphipathic model peptides were found to cause gross alterations in the plasma membrane of *C. albicans* and also to destroy intracellular, membrane-surrounded organelles (Ruissen et al., 2002).

Amino acid sequence analysis of membrane-disrupting α-helical proteins, such as mellittin, have indicated an unusually high amphipathicity, expressed as the hydrophobic moment (μ). The hydrophobic properties of an α-helix can be expressed in a hydrophobicity plot in which the vertical axis represents the mean hydrophobic moment (μ) and the horizontal axis represents the mean hydrophobicity (H) of a given peptide sequence (Eisenberg, 1984). Empirical data have allowed the assignment of domains in the hydrophobicity plot representing "surface-seeking," "globular," and "transmembrane" protein charactistics. It appears that the $[H, \mu]$ coordinates of a number of α-helical antifungal peptides that cause direct dye leakage from *C. albicans* cells and uncouple respiration of isolated mitochondria fall within the surface-seeking region. In contrast, antifungal peptides with much lower amphipathicity values (such as naturally occurring histatin 5) that do not plot in this region exhibit no pore-forming capacities (Helmerhorst et al., 2001b). It is assumed that the weak amphipathicity of histatins may preclude spontaneous insertion of this peptide into the membrane (Raj, Soni, and Levine, 1994). Thus peptide amphipathicity and hydrophobicity might determine the likelihood that a peptide will insert into the membrane and cause loss of cell integrity, whether by distinct pore formation or by formation of another membrane-destabilizing structure. Similarities in the $[H, \mu]$ coordinates rather than similarities in the primary or secondary peptide structure appear to determine this activity.

10.4. ANTIFUNGAL MECHANISTIC MODEL 2: PEPTIDE UPTAKE AND REACTIVE OXYGEN SPECIES FORMATION

10.4.1. Binding without Killing

Although a biological effect must imply that binding has occurred, binding is not always followed by membrane perturbation. In some cases, binding can occur without a fungicidal effect. For example, histatin 3 (Xu et al., 1999), histatin 5 (Gyurko et al., 2001), and NP1 (Lehrer et al., 1985) bind to *C. albicans* at 0 °C, but no killing occurs under these conditions. Likewise, in the presence of low concentrations of Ca^{2+} (0.5-mM) NP1 binds to *C. albicans* without causing a loss in cell viability (Lehrer et al., 1985; Selsted et al., 1985). Another discrepancy between peptide binding and killing is that one of the NP peptide family members, NP3a, binds most strongly to *C. albicans* cells, although it is only moderately fungicidal, whereas peptide NP1, which is the most fungicidal, binds as strongly as other moderately active peptide members (Lehrer et al., 1985). A similar discrepancy between binding and cell

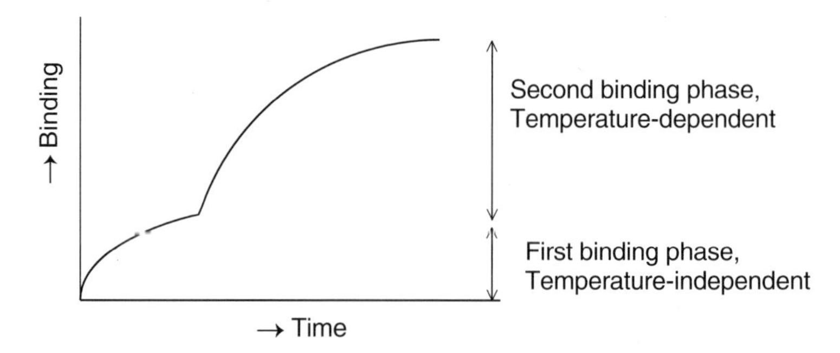

Figure 10.2. Two binding-phase characteristics, described for histatins and α-defensins.

killing was observed with lactoferrin-derived peptides (Lupetti et al., 2000). These results indicate that binding and activity appear to be two separate events, in contrast to results observed with the pore-forming peptides described in Section 10.3.

The binding of NP1 and histatins to the *C. albicans* cell surface is saturable (Lehrer et al., 1985; Edgerton et al., 1998; Xu et al., 1999) and can be competitively inhibited with structurally related peptide family members. Besides the fact that histatin 3 and NP1 do not necessarily kill on binding to yeast cells, another similarity between these peptides is that the binding seems to occur in two phases (Fig. 10.2). The first binding phase of NP1 occurs in minutes (Lehrer et al., 1985), whereas the first binding phase of histatin 3 lasts longer (Xu et al., 1999). In both cases, the first binding phase occurs independently of the incubation temperature whereas the second phase occurs only at 37 °C. Similar to NP and histatins, HNP1, an antimicrobial peptide from human neutrophils, binds to tumor cells in two phases, and the second phase does not occur at 0 °C (Lichtenstein et al., 1988). Because no killing of yeast cells or lysis of tumor cells occurs at 0 °C, binding and the initiation of cell death by these peptides are independent steps in the killing process. The second, temperature-sensitive binding phase appears to be necessary to initiate events leading to cell death. The two phase-binding characteristics observed with histatins and α-defensins interacting with yeast and mammalian cells may point toward shared aspects in their mechanisms of action.

10.4.2. Peptide Uptake

Because the second binding phase of histatins and α-defensins to *C. albicans* is temperature dependent, it is feasible that this phase represents an energy-dependent uptake process. Indeed, fluorescein isothio cyanate (FITC)-conjugated histatin 3 (Xu et al., 1999) and histatin 5 (Helmerhorst et al., 1999b; Gyurko et al., 2001) are internalized by *C. albicans* cells at 37 °C. Tenecin 3, an antifungal peptide from insects comprising 78 residues, also translocates across the yeast membrane in an energy- and temperature-dependent manner (Kim et al., 2001). These studies indicate that peptide binding is not the final stage of interaction and that a translocation event occurs that enables a peptide to reach intracellular targets. Interestingly, histatins as well as tenecin 3 are rich in histidine residues, but a specific role of these residues in peptide translocation has not yet been demonstrated.

There are several ways by which extracellular molecules can be internalized by fungi. These are simple diffusion, direct penetration across the membrane, facilitated diffusion, active transport, receptor-mediated endocytosis, and fluid-phase endocytosis. Internalization of a peptide could occur through diffusion, either directly or facilitated by a protein, or alternatively by endocytosis (Oehlke et al., 1997; Lindgren et al., 2000). Simple diffusion, penetration, and facilitated diffusion are considered to operate without cellular energy consumption, whereas endocytosis is an energy-driven process. The energy dependency of, for example, histatin translocation points toward (receptor-mediated) endocytosis or active uptake. Although the mechanism by which histatins and other antifungal peptides enter the yeast cell is as yet unknown, the activity of HNP1 against tumor cells, the binding of which also occurs in two phases, is inhibited by cytochalasin B, a blocker of receptor-mediated endocytosis (Lichtenstein et al., 1988). Therefore energy-dependent antifungal activity, observed with many mammalian antifungal peptides including histatins, defensins, and lactoferrin, might be related to adenosine triphosphate- (ATP-) dependent membrane invaginations that bud from the membrane to form vesicles, which represents the basis for endocytotic uptake mechanisms (Riezman et al., 1996; Prescianotto-Baschong and Riezman, 1998).

Alternatively, cationic peptides may not be internalized by endocytosis, but directly cross the fungal cell membrane and behave as cell membrane-permeant cations. Such cations are able to cross the membrane by virtue of a transmembrane potential because the intracellular environment is more negatively charged compared with the extracellular domain. In this case, cationic peptides are in fact electrophoretically translocated across the lipid bilayer (Furuya et al., 1991; Maduke and Roise, 1993). This mode of translocation is indirectly dependent on energy because the membrane potential is maintained by an expenditure of cellular energy. Evidence that such a mechanism of transportation might apply to the uptake of antifungal peptides is that the protonophore carbonyl cyanide *p*-chlorophenylhydrazone (CCCP), which rapidly dissipates proton membrane-potential gradients across the cytoplasmic membrane protects yeast cells against histatin 5 (Koshlukova et al., 1999), HNP1 (Lehrer et al., 1988), and also against yeast-killer toxin, an antifungal glycoprotein secreted by certain yeast strains (Skipper and Bussey, 1977). Although CCCP dissipates both the cytoplasmic and the mitochondrial transmembrane potential, Skipper and Bussey (1977) demonstrated that CCCP protects both respiring and fermenting cells against yeast-killer toxin, indicating that, at least for this peptide, the protective effect of CCCP is not at the mitochondrial level. Further support for the importance of a cytoplasmic transmembrane potential is that, besides protonophores, also chloride channel inhibitors such as diisothiocyanatostilbene-2,2′-disulfonic acid (DIDS), 2-(3-[trifluoromethyl]anilino) nicotinic acid (niflumic acid), and 5-nitro-2-(3-phenylpropylamino)benzoic acid (NPPB) are protective and inhibit, for example, histatin 5 activity (Baev et al., 2002). Although the membrane-potential dissipating effect of these agents was not demonstrated, these agents do interfere with ion pumps that are involved in the maintenance of the transmembrane potential. Therefore transmembrane-potential-driven translocation of antifungal peptides across the yeast membrane is an attractive alternative to endocytosis as a possible cellular peptide uptake mechanism.

10.4.3. Active Cell Participation: The Role of Cellular Respiration

Uptake of some antifungal peptides is temperature dependent and is required for killing. Therefore the cell seems to be actively involved in its own demise. Support for active cell participation in the killing mechanism triggered by a number antifungal peptides is provided by the observation that agents or conditions that inhibit cell metabolism strongly reduce cellular susceptibility. At 0 °C, *C. albicans* cells are insensitive to NP peptides (Selsted et al., 1985) and histatins (Gyurko et al., 2001). Furthermore, stationary phase cells are less sensitive than logarithmic phase cells to the killing effects of six NP peptides (Selsted et al., 1985), histatins (Helmerhorst, Troxler, and Oppenheim, 2001a) or yeast-killer toxin (Skipper and Bussey, 1977). This can be explained by the reduced permeability of the thicker cell wall of stationary phase cells which reduces the accessibility of internal targets, but also by lower cell metabolism. For example, the respiratory rate of stationary phase cells is approximately four times lower than that of logarithmic phase cells (Helmerhorst et al., 2001a). NP peptides are active against swollen spores of *A. fumigatus* and *Rhizopus oryzae*, but not against resting spores (Levitz et al.,

1986), another indication that cell metabolism is a determining factor in cell sensitivity.

A number of chemicals have been described to provide protection against the fungicidal action of some antifungal peptides. As described in Subsection 10.4.2, these agents include uncouplers such as CCCP and dinitrophenol (DNP) (Lehrer et al., 1988; Koshlukova et al., 1999; Skipper and Bussey, 1977), and also chloride channel inhibitors such as DIDS and nuflimic acid (Baev et al., 2002). Another class of agents that protect fungal cells against a number of antifungal peptides are respiratory chain inhibitors. These agents are in general not toxic to fungal cells, as most fungi are fully capable of fermentation. Many fungi, including *C. albicans*, possess a split respiratory pathway, with an alternative pathway branching off from the conventional respiratory pathway at the coenzyme Q level (Shepherd, Chin, and Sullivan, 1978; Helmerhorst et al., 2002). Millimolar amounts of sodium azide, a mitochondrial inhibitor of terminal oxidases in the conventional and the alternative respiratory pathway in *C. albicans* (Shepherd et al., 1978), protect *C. albicans* cells against the detrimental effects exerted by tenecin 3 (Kim et al., 2001), histatin 5 (Helmerhorst et al., 1999a; Gyurko et al., 2001), HNP1 (Lehrer et al., 1988), lactoferrin (Lupetti et al., 2000), SMAP-29 (Lee et al., 2002), and also against other, larger antifungal proteins such as protamine and bovine serum albumin (Olson, Hansing, and McClary, 1977). Sodium cyanide and antimycin A, which inhibit only the conventional respiratory pathway, partly protect *C. albicans* cells against histatin 5 and HNP1, respectively (Lehrer et al., 1988; Helmerhorst et al., 1999a), whereas antimycin A in combination with salicyl hydroxamic acid (SHAM), an inhibitor of the alternative respiratory pathway, provides full protection against HNP1 (Lehrer et al., 1988). These data indicate that chemical inhibition of respiration renders fungal cells insensitive to a rather large group of unrelated antifungal peptides.

Alternative approaches that show the relationship between respiration and antifungal sensitivity employ respiratory-deficient yeast mutants. Nonrespiring "petite" mutants of *C. albicans*, generated with acriflavine at elevated incubation temperatures, are much less sensitive to histatin 5 than are the respiring wild-type cells (Gyurko et al., 2000). Conclusive evidence that cellular respiration is of crucial importance for peptide activity is that aerobically grown *C. albicans* cells are largely protected from histatins or HNP1 activity when these cells are exposed to these peptides under anoxic conditions in an anaerobic chamber (Lehrer et al., 1988; Helmerhorst et al., 1999a). These experiments show that the changes in activity are due only to the absence of oxygen and respiration and do not represent artifacts induced by the addition of chemicals or genetic modification. In conclusion, the sensitivity of *C. albicans* cells to a large number of antifungal peptides, including histatins and HNP1, appears to be highly dependent on the mitochondrial activity of the target cell.

10.4.4. Evidence for Intracellular Targets: The Mitochondria

The use of fluorescent labels in conjunction with confocal fluorescence microscopy and fluorescence activated cell sorter (FACS) technology have enabled studies on the cellular target of antifungal peptides. Such studies have demonstrated that some antifungal peptides are internalized by yeast cells and have revealed their intracellular distribution. In *C. albicans* cells exposed to FITC-labeled tenecin 3, the peptide appears homogeneously distributed in the cytoplasm (Kim et al., 2001). On the other hand, FITC-histatin 3 (Xu et al., 1999) and FITC-histatin 5 (Helmerhorst et al., 1999a) show "patchy" and "granular" distribution patterns, respectively, suggesting the presence of specific target organelles in the yeast cytoplasm. Colocalization studies that use FITC histatin 5 and a marker that specifically stains the mitochondria, mitotracker orange, have indicated that the granular organelles targeted by histatin 5 represent the mitochondria (Fig. 10.3) (Helmerhorst et al., 1999a). Histatin 5 also exerts a direct biological effect on isolated mitochondria (Helmerhorst et al., 2001a, 2001b), providing further evidence that an interaction between histatin 5 and this organelle occurs, in vitro as well as in situ. Targeting of histatin 5 to the mitochondria is not due to conjugation of the peptide to FITC because FITC-labeled negative control peptides do not show such targeting. Furthermore, when *C. albicans* cells are exposed to unlabeled histatin 5 and subsequently fractionated, the peptide is contained in large amounts in the

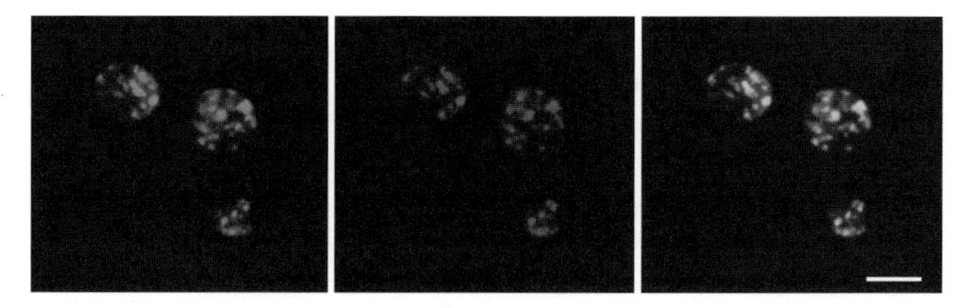

Figure 10.3. Mitochondrial targeting of histatin 5 in *C. albicans*. Left, FITC-histatin 5; middle, mitotracker orange; right, colocalization. Bar: 5 μm. [From: Helmerhorst et al. (1999a).] (See color section.)

cellular fraction enriched with mitochondria (Gyurko et al., 2001). These data indicate that the mammalian antifungal peptide histatin 5 translocates across the yeast plasma membrane and associates with mitochondria *in situ*.

Unequivocal proof that histatin 5 has an intracellular target and that association with this target leads to killing stems from studies with recombinantly engineered *C. albicans* cells that contain chromosomally encoded histatin 3 and 5 (Baev, Li, and Edgerton, 2001). In this system, the mechanism of histatin action could be investigated independently of binding and translocation events. Intracellular expression of either histatin 3 or histatin 5 induced cell killing, indicating that cell death could be initiated on exposure of only intracellular sites to histatins. Therefore extracellular binding and internalization of these peptides are preceding events, which is consistent with observations made with histatins and α-defensins that binding can occur without killing. There is indirect evidence that α-defensin peptides are also internalized, as their activity is energy dependent and shows many similarities with histatins, but the existence of an intracellular target for this peptide has yet to be elucidated. The strong increase in NP activity against liposomes containing cardiolipin (Hristova et al., 1997), a phospholipid which is specifically enriched in mitochondrial membranes (Hovius et al., 1990), is consistent with a potential association of this peptide with the mitochondrion as well.

10.4.5. Inhibition of Respiration

The biological consequence of the mitochondrial targeting of histatin 5 in situ has been investigated in functional assays by measurements of cellular respiration. Histatin 5 inhibits the respiration of logarithmic phase cells (Helmerhorst et al., 2001a), but not of slower respiring stationary phase cells (Koshlukova et al., 1999; Helmerhorst et al., 2001a). Inhibition of *C. albicans* respiration has also been reported for defensin peptides MCP-1 (NP1) and MCP-2 (NP2) in two separate studies (Patterson-Delafield et al., 1981; Selsted et al., 1985). Inhibition of respiration can be a direct result of the interference of an antifungal peptide with the mitochondrial respiratory chain. Alternatively, inhibition of respiration may occur as an indirect result of cell disintegration and therefore be a secondary effect to other detrimental effects of the peptide on the cell. Histatin 5 inhibits not only cellular respiration, but also inhibits respiration of isolated yeast mitochondria (Helmerhorst et al., 2001a, 2001b). This, together with the mitochondrial targeting of this peptide, strongly suggests that inhibition of cellular respiration is based on a direct interaction of histatin 5 with the mitochondria in situ.

10.4.6. Toxic Mediators: Reactive Oxygen Species

Inhibition of Respiration and Reactive Oxygen Species

It was not clear at first sight how blockage of the respiratory chain by histatin 5 could lead to yeast cell death, because *Candida*, like many other yeasts, is not dependent on respiration for its survival because of its ability to ferment substrates. Furthermore, in the case of histatins, mitochondrial respiration is targeted, while respiration is also required for cellular susceptibility to histatins. A solution for this apparent paradox was the discovery that histatins induce the formation of reactive oxygen species (ROS). The generation of ROS in *C. albicans* cells is strongly correlated with cell killing (Helmerhorst et al., 2001a), and killing and ROS formation are inhibited in the presence of oxygen radical scavengers. Histatin 5 also generates ROS in isolated mitochondria, suggesting that histatin generates ROS in cells at the mitochondrial level. In this respect it is interesting that histatin 5 is also a respiratory inhibitor. Although the process by which histatin 5 generates ROS in mitochondria and cells might be related to inhibition of respiration, the precise mechanism of ROS formation has yet to be elucidated. Certainly ROS formation requires the presence of oxygen and explains the protection of fungal cells against histatin 5 killing activity under anoxic conditions or in the presence of oxygen scavengers. Histatin 5 exerts neither fungicidal activity (Helmerhorst et al., 1999a; Gyurko et al., 2001) nor ROS-generating activity in the presence of the respiratory inhibitor sodium azide. Thus nonrespiring cells are protected from histatins by virtue of their inability to form ROS. In summary, histatin 5 interferes with active mitochondrial respiration, leading to ROS production that results ultimately in cell death. This mechanism explains the insensitivity of cells with conditionally, chemically, or genetically impaired mitochondrial respiration.

Inhibition of ROS-Neutralizing Enzymes

ROS formation is not unique to histatins: At least two other antifungal peptides form ROS in yeast cells. The plant antifungal peptide osmotin generates ROS in *Saccharomyces cerevisiae* cells (Narasimhan et al., 2001). In contrast to the situation with histatin 5, there is no evidence that the mitochondrion is involved in the osmotin-killing mechanism, as respiratory-deficient rho^0 mutants of *S. cerevisiae* are equally as sensitive as the wild-type strain. Another peptide that induces ROS in yeast cells is the N-terminal 11-residue fragment of lactoferrin, hLF(1–11). In *C. albicans* cells exposed to hLF(1–11), ROS are rapidly generated in a concentration-dependent fashion, and, as with histatins, ROS levels are highly correlated with cell killing. The ROS scavenger *N*-acetyl cysteine protects against hLF(1–11)-induced ROS generation and

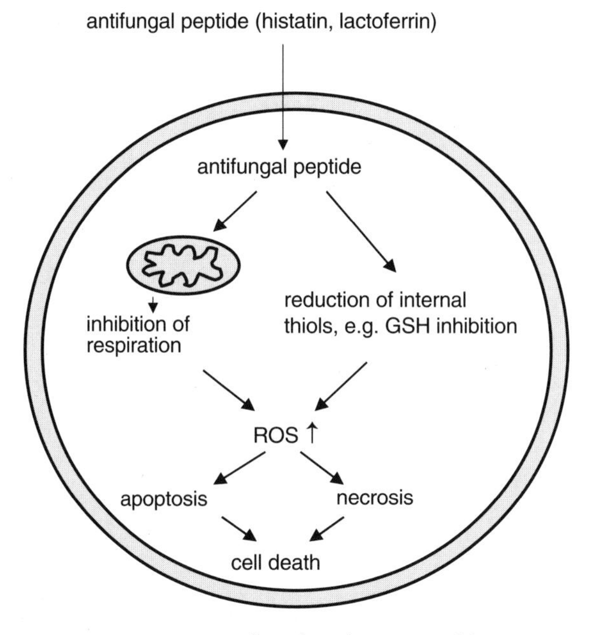

Figure 10.4. Antifungal mechanistic model 2.

cell killing. Furthermore, diamide, a chemical that inactivates intracellular ROS-scavenging thiols, such as α-glutamyl-cysteinylglycine (GSH), increases hLF(1–11)-induced cell death and ROS formation. This strongly indicates that intracellular GSH is a natural protector against hLF(1–11) and possibly also against other ROS-generating antifungal peptides. Because hLF(1–11) directly reduces internal thiol levels by 20%, it has been proposed that the antifungal mechanism of hLF(1–11) is based on ROS formation resulting from inhibition of naturally occurring ROS-neutralizing components rather than from a direct interference with electron transfer in the respiratory chain (Lupetti et al., 2002).

It is feasible that the histatin-derived model of mitochondrial ROS formation and the lactoferrin-derived model of ROS formation work in conjunction (Fig. 10.4). Antifungal peptides such as histatins, lactoferrin, and possibly HNP-1 enter the yeast cell through an unknown, but energy-dependent, mechanism. Once internalized, peptides are targeted to the respiratory apparatus. Interference with the respiratory machinery leads to out-of-sequence electron transfer to molecular oxygen and the generation of ROS. In addition, the peptide binds to internal thiols such as GSH, thereby reducing the oxygen-radical-scavenging capacity of the cell, which also leads to an increase in ROS levels.

10.5. ANTIFUNGAL MECHANISTIC MODEL 3: ADENOSINE TRIPHOSPHATE RELEASE PRECEEDES KILLING

Cells exposed to histatin 5, hLF(1–11), or HNP-1 virtually release all their intracellular ATP (Koshlukova et al., 1999, 2000; Edgerton et al., 2000; Lupetti et al., 2000), and ATP release and cell killing appear to occur virtually at the same time. It has been debated whether ATP release occurs as a consequence of cell death, for example, immediately following ROS-induced membrane-permeability changes, or whether ATP release preceedes cell death. Some studies suggest that ATP release might occur before the cell actually dies and that binding of ATP to purinergic $P2X_7$-like ATP receptors on the *C. albicans* membrane triggers cell killing (Koshlukova et al., 1999, 2000). Support for this is that suramin and pyridoxalphosphate-6-azophenyl-2′,4′-disulfonic acid (PPAD), which are ATP receptor antagonists, protect cells from histatin-induced killing (Koshlukova et al., 1999). Furthermore, periodate-oxidized ATP, which irreversibly blocks the interaction between extracellular ATP and cytoplasmic membrane ATP receptors in mycobacteria, significantly reduces hLF(1–11) killing activity against *C. albicans* (Lupetti et al., 2000). These data suggest that an interaction between extracellur ATP and its receptors is essential in histatin 5 and hLF(1–11)-provoked cell killing; however, it remains to be demonstrated that suramin, PPAD, or oxidized ATP has no effect on peptide-induced ATP release itself, which would unequivocally prove that ATP release preceedes cell death. Apyrase, which is a phosphatase, provides protection of cells against histatin 5 activity and ATP analogs BzATP or ATPγS themselves exhibit some fungicidal activity, although at concentrations (100–500 μM) exceeding those that can theoretically be released from cell numbers typically present in fungicidal assays. In conclusion, extracellular ATP could be a toxic mediator rather than just a sign of cell death. The fact that ATP release and cell killing are indistinguishable in time indicates that the binding of ATP to its putative $P2X_7$-like receptors is rapidly succeeded by a cell-death signal.

10.6. ANTIFUNGAL MECHANISTIC MODEL 4: PEPTIDE-GENERATED SIGNAL TRANSDUCTION

Recently it was discovered that α-melanocyte-stimulating hormone (α-MSH) exhibits potent antifungal activity, and this peptide appears to act by another distinct antifungal mechanism (Cutuli et al., 2000). α-MSH is a 13-residue melanocortin peptide that arises from the posttranslational processing of pro-opiomelanocortin (POMC) and shares the 1–13-residue sequence with adrenocorticotropic hormone (ACTH). α-MSH is produced by many cell types and exerts anti-inflammatory activity through inhibition of chemotaxis by neutrophils and nitric oxide production by macrophases. In addition to anti-inflammatory effects, it possesses potent killing activity against *C. albicans* and *Staphylococcus aureus.* The C-terminal tripeptide, Lys-Pro-Val, is as active as intact α-MSH in anti-inflammatory effects as well as in antimicrobial effects. The anti-inflammatory effects of α-MSH on neutrophils and macrophages are believed to be related to a rise in cyclic adenosine mono phosphate (AMP) (c-AMP). α-MSH also raises c-AMP in yeast cells, just like the adenylyl cyclase activator forskolin, which is also fungicidal. Evidence that yeast cell killing by α-MSH is related to a rise in c-AMP levels is that killing effects could be prevented by the adenylyl cyclase inhibitor ddAdo. It has been proposed that α-MSH interacts with the yeast cell through a G-protein-linked α-MSH receptor equivalent and that the intracellular effects such as a rise in c-AMP are exerted by signal transduction. Even though ACTH, which contains the Lys-Pro-Val sequence, is inactive against *C. albicans* (Cutuli et al., 2000), it is of interest that the Lys-Pro-Val sequence is present in many antimicrobial peptides, such as thenatin, SMAP-29, PR-29, LL-37, indolicidin, eCATH1–3, bactenecin 5, and bactenecin 7.5 (http://www.bbcm.univ.trieste.it/~tossi/pag1.htm). Therefore it is feasible that signal transduction and c-AMP effects may play a role in antibacterial and/or antifungal properties exerted by these peptides.

10.7. CELL-DEATH MECHANISMS: APOPTOSIS OR NECROSIS

After exposure of fungal cells to fungicidal peptides, exerting any of the four described mechanisms of action (Fig. 10.5), cell death is the ultimate outcome. The well-described types of cell death are apoptosis and necrosis. Apoptotic cell death involves the execution of a preprogrammed sequence of cellular events that are dependent on intracellular ATP, and this sequence usually takes several hours to fully develop. In contrast to apoptosis, cell necrosis is the consequence of membrane disruption that leads to ATP release and occurs within minutes (Lemasters et al., 1998).

Even though yeasts lack established apoptotic genes such as caspases and caspase-activated deoxyribunocleases, which fulfill crucial roles in metazoan apoptosis, an apoptotic phenotype has been observed in *S. cerevisiae* on oxidative stress (Madeo et al., 1999) and aging (Laun et al., 2001). During apoptosis, the yeast cytoplasmic membrane remains intact and impermeable to the necrosis marker propidium iodide (PI), a probe frequently used to

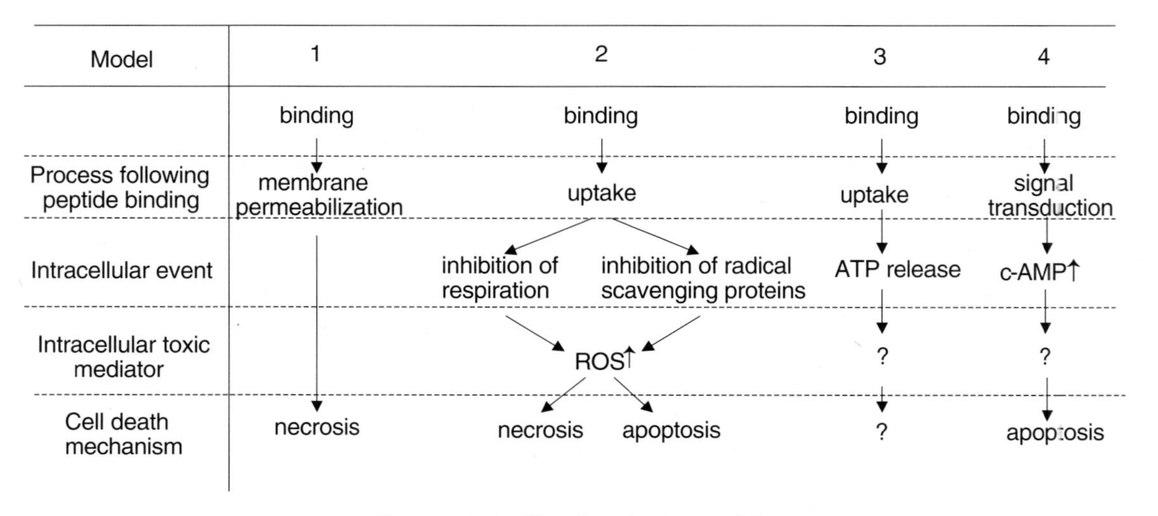

Figure 10.5. Antifungal mechanistic models.

discriminate between apoptotic and necrotic cell death (Darzynkiewicz et al., 1992).

In Fig. 10.5, the four antifungal mechanistic models are summarized. In mechanistic model 1, binding of the amphipathic antifungal peptide to the yeast cell instantly destroys cellular integrity because of perturbations in the cell membrane that lead to the nonselective efflux of intracellular constituents and influx of PI. The immediate cell death that follows on exposure of cells to peptide is representative of cell necrosis. In mechanistic model 2, ROS are formed either by interference of the peptide with mitochondrial respiration or by the inhibitory effect of the peptide on ROS-neutralizing enzymes. ROS have been identified as membrane-disrupting effector molecules in necrotic cell death and, at lower concentrations, as signaling molecules in the apoptotic cell-death cascade in mammalian cells (Reed, Jurgensmeier, and Matsuyama, 1998) or in *S. cerevisiae* cells (Madeo et al., 1999). Indeed, the antifungal peptide osmotin generates ROS and is able to initiate apoptosis as well as necrosis in *S. cerevisiae* cells, depending on peptide concentrations (Narasimhan et al., 2001). It is postulated that both types of cell death can also be induced by ROS in *C. albicans* cells. Histatin 5 and hLF(1–11) form ROS in *C. albicans* cells. Cell death occurs rapidly and is concomitant with the influx of PI (Helmerhorst et al., 1999a; Lupetti et al., 2000), pointing toward necrosis. Furthermore, histatin 5 and hLF(1–11) cause the efflux of more than 99% of all intracellular ATP (Koshlukova et al., 1999; Lupetti et al., 2000), which would be required intracellularly for operating an apoptosis program. Therefore histatin 5- and hLF(1–11)-provoked cell death following ROS formation is necrotic rather than apoptotic in nature, even though it cannot be excluded that apoptosis might occur at lower peptide concentrations. In mechanistic model 3, antifungal peptides induce the nonlytic release of ATP, and subsequent binding of ATP to extracellular purinergic receptors on *C. albicans* generates a cell-death signal. The nature of this signal and the intracellular cytotoxic mediators have not yet been established in *C. albicans*. Because virtually all intracellular ATP is released and cell death follows so quickly on ATP release, it is unlikely, however, that the cell-death mechanism is apoptotic in nature. In mechanistic model 4, peptide binding leads to a rise in c-AMP levels. In these cells, PI uptake occurs only after prolonged exposure of cells to the peptide, which points toward an apoptotic mechanism of cell killing.

10.8. CONCLUDING REMARKS

Based on the different events that occur when fungal cells are incubated with fungicidal peptides, four models for their modes of action can be distinguished (Fig. 10.5). The variety of these models clearly demonstrates that mere membrane permeabilization is just one of the potential antifungal mechanisms. The categorization into four models inevitably still is a simplification of reality, as one peptide may exhibit properties described in more than one model, and future investigations may uncover additional pathways of antifungal peptide mechanisms.

Antifungal mechanistic model 2 requires that cells be metabolically active. This requirement suggests that fungal cells escape the detrimental effect of peptides that act through this mechanism if cells are in a dormant, metabolically inactive form. The importance of cellular respiration for cell sensitivity suggests that fungal cells also escape peptide activity if they use organic compounds rather than oxygen as terminal electron acceptors (fermentative metabolism). Because *C. albicans* cells in various sites of infection, including deep-seated mycoses, are neither likely to grow rapidly in a log-phase-like fashion nor to respire at a considerable rate, cells under these conditions may well be insensitive to antifungal peptides that act strictly according to mechanistic model 2. The redundancy in peptides that display antifungal activity and the fact that there are at least four ways by which these peptides can exert their killing activity may prove to be an advantage to the host: Where one peptide might fail, another peptide could still be effective. An interesting discovery is that some peptides with different, known biological functions, for example, hormones and chemokines, appear to exhibit antifungal properties as well. Such multifunctionality of peptides would be a tremendeous expansion of the host's antifungal potential. In addition, synergism between peptides that act through different antifungal mechanisms may enhance the overall host's efficiency in defending against fungal infections.

Mechanistically, any of the four antifungal models becomes particularly interesting when peptides from various origins, differing in primary and secondary structure, exert their antifungal activity according to the same model. In such cases, the model may represent a very effective, non-peptide-selective killing mechanism toward which fungal evolution has not been able to develop an effective defense.

ACKNOWLEDGMENTS

The authors thank Robert F. Troxler for his helpful advice and critical review of the manuscript. Grant support from the National Institute of Health/the National Institute of Dental and Craniofacial Research (NIDCR) grants DE07652 and DE05672 is thankfully acknowledged.

REFERENCES

Andrä, J., Berninghausen, O., and Leippe, M. (2001). Cecropins, antibacterial peptides from insects and mammals, are potently fungicidal against *Candida albicans. Medical Microbiology and Immunology,* 189, 169–73.

Andrä, J. and Leippe, M. (1999). Candidacidal activity of shortened synthetic analogs of amoebapores and NK-lysin. *Medical Microbiology and Immunology,* 188, 117–24.

Andreu, D., Aschauer, H., Kreil, G., and Merrifield, R. B. (1985). Solid-phase synthesis of PYLa and isolation of its natural counterpart, PGLa [PYLa-(4–24)] from skin secretion of *Xenopus laevis. European Journal of Biochemistry,* 149, 531–35.

Andreu, D. and Rivas, L. (1998). Animal antimicrobial peptides: An overview. *Biopolymers,* 47, 415–33.

Baev, D., Li, X. S., Dong, J., Keng, P., and Edgerton, M. (2002). Human salivary histatin 5 causes disordered volume regulation and cell cycle arrest in *Candida albicans. Infection and Immunity,* 70, 4777–84.

Baev, D., Li, X., and Edgerton, M. (2001). Genetically engineered human salivary histatin genes are functional in *Candida albicans*: Development of a new system for studying histatin candidacidal activity. *Microbiology,* 147, 3323–34.

Beintema, J. J. (1994). Structural features of plant chitinases and chitin-binding proteins. *FEBS Letters,* 350, 159–63.

Bellamy, W., Wakabayashi, H., Takase, M., Kawase, K., Shimamura, S., and Tomita, M. (1993). Killing of *Candida albicans* by lactoferricin B, a potent antimicrobial peptide derived from the N-terminal region of bovine lactoferrin. *Medical Microbiology and Immunology (Berlin),* 182, 97–105.

Cho, Y., Turner, J. S., Dinh, N.-N., and Lehrer, R. I. (1998). Activity of protegrins against yeast-phase *Candida albicans. Infection and Immunity,* 66, 2486–93.

Cutuli, M., Cristiani, S., Lipton, J. M., and Catania, A. (2000). Antimicrobial effects of α-MSH peptides. *Journal of Leukocyte Biology,* 67, 233–9.

Darzynkiewicz, Z., Bruno, S., Del Bino, G., Gorczyca, W., Hotz, M. A., Lassota, P., and Traganos, F. (1992). Features of apoptotic cells measured by flow cytometry. *Cytometry,* 13, 795–808.

Diamond, G., Zasloff, M., Eck, H., Brasseur, M., Maloy, L., and Bevins, C. L. (1991). Tracheal antimicrobial peptide, a cysteine-rich peptide from mammalian tracheal mucosa: Peptide isolation and cloning of a cDNA. *Proceedings of the National Academy of Sciences USA,* 88, 3952–6.

Dufourcq, J. and Faucon, J. F. (1977). Intrinsic fluorescence study of lipid-protein interactions in membrane models. Binding of melittin, an amphipathic peptide, to phospholipid vesicles. *Biochimica et Biophysica Acta,* 467, 1–11.

Edgerton, M., Koshlukova, S. E., Araujo, M. W. B., Patel, R. C., Dong, J., and Bruenn, J. A. (2000). Salivary histatin 5 and human neutrophil defensin 1 kill *Candida albicans* via shared pathways. *Antimicrobial Agents and Chemotherapy*, 44, 3310–16.

Edgerton, M., Koshlukova, S. E., Lo, T. E., Chrzan, B. G., Straubinger, R. M., and Raj, P. A. (1998). Candidacidal activity of salivary histatins. *Journal of Biological Chemistry*, 273, 20438–47.

Eisenberg, D. (1984). Three dimensional structure of membrane and surface proteins. *Annual Reviews in Biochemistry*, 53, 595–623.

Fujitani, N., Kawabata, S., Osaki, T., Kumaki, Y., Demura, M., Nitta, K., and Kawano, K. (2002). Structure of the antimicrobial peptide tachystatin A. *Journal of Biological Chemistry*, 277, 23651–7.

Fukazawa, Y. and Kagaya, K. (1997). Molecular bases of adhesion of *Candida albicans*. *Journal of Medical and Veterinary Mycology*, 35, 87–99.

Furuya, S., Mihara, K., Aimoto, S., and Omura, T. (1991). Cytosolic and mitochondrial surface factor-independent import of a synthetic peptide into mitochondria. *European Molecular Biology Organization (EMBO) Journal*, 10, 1759–66.

Gazit, E., Boman, A., Boman, H. G., and Shai, Y. (1995). Interaction of the mammalian antibacterial peptide cecropin P1 with phospholipid vesicles. *Biochemistry*, 34, 11479–88.

Gururaja, T. L., Levine, J. H., Tran, D. T., Naganagowda, G. A., Ramalingam, K., Ramasubbu, N., and Levine, M. J. (1999). Candidacidal activity prompted by N-terminus histatin-like domain of human salivary mucin (MUC7). *Biochimica et Biophysica Acta*, 1431, 107–19.

Gyurko, C., Lendenmann, U., Helmerhorst, E. J., Troxler, R. F., and Oppenheim, F. G. (2001). Killing of *Candida albicans* by histatin 5: Cellular uptake and energy requirement. *Antonie van Leeuwenhoek*, 79, 297–309.

Gyurko, C., Lendenmann, U., Troxler, R. F., and Oppenheim, F. G. (2000). *Candida albicans* mutants deficient in respiration are resistant to the small cationic salivary antimicrobial peptide histatin 5. *Antimicrobial Agents and Chemotherapy*, 44, 348–54.

Harder, J., Bartels, J., Christophers, E., and Schröder, J.-M. (2001). Isolation and characterization of human β-defensin-3, a novel human inducible peptide antibiotic. *Journal of Biological Chemistry*, 276, 5707–13.

Helmerhorst, E. J., Van 't Hof, W., Simoons-Smit, A. M., Veerman, E. C. I., and Nieuw Amerongen, A. V. (1997). Synthetic histatin analogs with broad spectrum antimicrobial activity. *Biochemical Journal*, 326, 39–45.

Helmerhorst, E. J., Breeuwer, P., Van 't Hof, W., Walgreen-Weterings, E., Oomen, L. C. J. M., Veerman, E. C. I., Nieuw Amerongen, A. V., and Abee, T. (1999a).

The cellular target of histatin 5 on *Candida albicans* is the energized mitochondrion. *Journal of Biological Chemistry*, 274, 7286–91.

Helmerhorst, E. J., Reijnders, I. M., Van 't Hof, W., Simoons-Smit, A. M., Veerman, E. C. I., and Nieuw Amerongen, A. V. (1999b). Amphotericin B and fluconazole-resistant *Candida* spp., *Aspergillus fumigatus*, and other newly emerging pathogenic fungi are susceptible to basic antifungal peptides. *Antimicrobial Agents and Chemotherapy*, 43, 702–4.

Helmerhorst, E. J., Troxler, R. F., and Oppenheim, F. G. (2001a). The human salivary peptide histatin 5 exerts its antifungal activity through the formation of reactive oxygen species. *Proceedings of the National Academy of Sciences USA*, 98, 14637–42.

Helmerhorst, E. J., Van 't Hof, W., Breeuwer, P., Veerman, E. C. I., Abee, T., Troxler, R. F., Nieuw Amerongen, A. V., and Oppenheim, F. G. (2001b). Characterization of histatin 5 with respect to amphipathicity, hydrophobicity and effects on cell and mitochondrial membrane integrity excludes a candidacidal mechanism of pore formation. *Journal of Biological Chemistry*, 276, 5643–9.

Helmerhorst, E. J., Murphy, M. P., Troxler, R. F., and Oppenheim, F. G. (2002). Characterization of the mitochondrial respiratory pathways in *Candida albicans*. *Biochimica et Biophysica Acta*, 155, 73–80.

Hong, S. Y., Park, T. G., and Lee, K.-H. (2001). The effect of charge increase on the specificity and activity of a short antimicrobial peptide. *Peptides*, 22, 1669–74.

Hovius, R., Lambrechts, H., Nicolay, K., and De Kruijff, B. (1990). Improved methods to isolate and subfractionate rat liver mitochondria. Lipid composition of the inner and outer membrane. *Biochimica et Biophysica Acta*, 1021, 217–26.

Hristova, K., Selsted, M. E., and White, S. H. (1997). Critical role of lipid composition in membrane permeabilization by rabbit neutrophil defensins. *Journal of Biological Chemistry*, 272, 24224–33.

Iijima, R., Kurata, S., and Natori, S. (1993). Purification, characterization, and cDNA cloning of an antifungal protein from the hemolymph of *Sarcophaga peregrina* (flesh fly) larvae. *Journal of Biological Chemistry*, 268, 12055–61.

Kim, D.-H., Lee, D. G., Kim, K. L., and Lee, Y. (2001). Internalization of tenecin 3 by a fungal cellular process is essential for its fungicidal effect on *Candida albicans*. *European Journal of Biochemistry*, 268, 4449–58.

Kirkpatrick, C. H., Green, I., Rich, R. R., and Schade, A. L. (1971). Inhibition of growth of *Candida albicans* by iron-unsaturated lactoferrin: Relation to host-defense mechanisms in chronic mucocutaneous candidiasis. *Journal of Infectious Diseases*, 124, 539–44.

Klotz, S. A. (1994). The contibution of electrostatic forces to the process of adherence of *Candida albicans* yeast cells to substrates. *FEMS Microbiology Letters*, 120, 257–62.

Koshlukova, S. E., Araujo, M. W. B., Baev, D., and Edgerton, M. (2000). Released ATP is an extracellular cytotoxic mediator in salivary histatin-5 induced killing of *Candida albicans*. *Infection and Immunity*, 68, 6848–56.

Koshlukova, S. E., Lloyd, T. L., Araujo, M. W. B., and Edgerton, M. (1999). Salivary histatin 5 induces non-lytic release of ATP from *Candida albicans* leading to cell death. *Journal of Biological Chemistry*, 274, 18872–9.

Krause, A., Liepke, C., Meyer, M., Adermann, K., Forssmann, W. G., and Maronde, E. (2001). Human natriuretic peptides exhibit antimcirobial activity. *European Journal of Medical Research*, 6, 215–18.

Ladokhin, A. S., Selsted, M. E., and White, S. H. (1997). Sizing membrane pores in lipid vesicles by leakage of co-encapsulated markers: Pore formation by melittin. *Biophysical Journal*, 72, 1762–6.

Laun, P., Pichova, A., Madeo, F., Fuchs, J., Ellinger, A., Kohlwein, S., Dawes, I., Fröhlich, K.-U., and Breitenbach, M. (2001). Aged mother cells of *Saccharomyces cerevisiae* show markers of oxidative stress and apoptosis. *Molecular Microbiology*, 39, 1166–73.

Lawyer, C., Pai, S., Watabe, M., Borgia, P., Mashimo, T., Eagleton, L., and Watabe, K. (1996). Antimicrobial activity of a 13 amino acid tryptophan-rich peptide derived from a putative porcine precursor protein of a novel family of antibacterial peptides. *FEBS Letters*, 390, 95–8.

Lee, D. G., Kim, D.-H., Park, Y., Kim, H. K., Kim, H. N., Shin, Y. K., Choi, C. H., and Hahm, K.-S. (2001). Fungicidal effect of antimicrobial peptide, PMAP-23, isolated from porcine myeloid against *Candida albicans*. *Biochemical and Biophysical Research Communications*, 282, 570–4.

Lee, D. G., Kim, P. I., Park, Y., Park, S.-C., Woo, E.-R., and Hahm, K.-S. (2002). Antifungal mechanism of SMAP-29 (1–18) isolated from sheep myeloid mRNA against *Trichosporon beigelii*. *Biochemical and Biophysical Research Communications*, 295, 591–6.

Lee, D. G., Shin, S. Y., Kim, D.-H., Seo, M. Y., Kang, J. H., Lee, Y., Kim, K. L., and Hahm, K.-S. (1999). Antifungal mechanism of a cysteine-rich antimicrobial peptide, Ib-AMP1, from *Impatiens balsamina* against *Candida albicans*. *Biotechnology Letters*, 21, 1047–50.

Lehrer, R. I., Ganz, T., Szklarek, D., and Selsted, M. E. (1988). Modulation of the *in vitro* candidacidal activity of human neutrophil defensins by target cell metabolism and divalent cations. *Journal of Clinical Investigation*, 81, 1829–35.

Lehrer, R. I., Szklarek, D., Ganz, T., and Selsted, M. E. (1985). Correlation of binding of rabbit granulocyte peptides to *Candida albicans* with candidacidal activity. *Infection and Immunity*, 49, 207–11.

Lemasters, J. J., Nieminen, A.-L., Qian, T., Trost, L. C., Elmore, S. P., Nishimura, Y., Crowe, R. A., Cascio, W. E., Bradham, C. A., Brenner, D. A., and Herman, B. (1998). The mitochondrial permeability transition in cell death: A common mechanism in necrosis, apoptosis and autophagy. *Biochimica et Biophysica Acta*, 1366, 177–96.

Levitz, S. M., Selsted, M. E., Ganz, T., Lehrer, R. I., and Diamond, R. D. (1986). *In vitro* killing of spores and hyphae of *Aspergillus fumigatus* and *Rhizopus oryzae* by rabbit neutrophil cationic peptides and bronchoalveolar macrophages. *Journal of Infectious Diseases*, 154, 483–9.

Li, X. S., Reddy, M. S., Baev, D., and Edgerton, M. Candida albicans Ssa1/2p is the cell envelope binding protein for human salivary histatin 5. *Journal of Biological Chemistry*, 278, 28533–61.

Lichtenstein, A. K., Ganz, T., Nguyen, T.-M., Selsted, M. E., and Lehrer, R. I. (1988). Mechanism of target cytolysis by peptide defensins. *Journal of Immunology*, 140, 2686–94.

Lindgren, M., Hällbrink, M., Prochiantz, A., and Langel, Ü. (2000). Cell-penetrating peptides. *Trends in Pharmacological Sciences*, 21, 99–103.

Liu, B., Rayment, S. A., Gyurko, C., Oppenheim, F. G., Offner, G. D., and Troxler, R. F. (2000). The recombinant N-terminal region of human salivary mucin MG2 (MUC7) contains a binding domain for oral Streptococci and exhibits candidacidal activity. *Biochemical Journal*, 345, 557–64.

Lugardon, K., Chasserot-Golaz, S., Kieffer, A.-E., Maget-Dana, R., Nullans, G., Kieffer, B., Aunis, D., and Metz-Boutique, M.-H. (2001). Structural and biological characterization of chromofungin, the antifungal chromogranin A-(47–66)-derived peptide. *Journal of Biological Chemistry*, 276, 35875–82.

Lugardon, K., Raffner, R., Goumon, Y., Corti, A., Delmas, A., Bulet, P., Aunis, D., and Metz-Boutique, M.-H. (2000). Antibacterial and antifungal activities of vasostatin-1, the N-terminal fragment of chromogranin A. *Journal of Biological Chemistry*, 275, 10745–53.

Lupetti, A., Paulusma-Annema, A., Senesi, S., Campa, M., Van Dissel, J. T., and Nibbering, P. H. (2002). Internal thiols and reactive oxygen species in candidacidal activity exerted by an N-terminal peptide of human lactoferrin. *Antimicrobial Agents and Chemotherapy*, 46, 1634–9.

Lupetti, A., Paulusma-Annema, A., Welling, M. M., Senesi, S., Van Dissel, J. T., and Nibbering, P. H. (2000). Candidacidal activities of human lactoferrin peptides derived from the N-terminus. *Antimicrobial Agents and Chemotherapy*, 44, 3257–63.

Madeo, F., Fröhlich, E., Ligr, M., Grey, M., Sigrist, S. J., Wolf, D. H., and Fröhlich, K.-U. (1999). Oxygen stress: A regulator of apoptosis in yeast. *Journal of Cellular Biology*, 145, 757–67.

Maduke, M. and Roise, D. (1993). Import of mitochondrial presequence into protein-free phospholipid vesicles. *Science*, 260, 364–7.

Mangoni, M. E., Aumelas, A., Charnet, P., Roumestand, C., Chiche, L., Despaux, E., Grassy, G., Calas, B., and Chavanieu, A. (1996). Change in membrane permeability induced by protegrin 1: Implication of disulphide bridges for pore formation. *FEBS Letters*, 383, 93–8.

Mahoney, M. M., Lee, A. Y., Brezinski-Caliguri, D. J., and Huttner, K. M. (1995). Molecular analysis of the sheep cathelin family reveals a novel antimicrobial peptide. *FEBS Letters*, 377, 519–22.

McCafferty, D. G., Cudic, P., Yu, M. K., Behenna, D. C., and Kruger, R. (1999). Synergy and duality in peptide antibiotic mechanisms. *Current Opinion in Chemical Biology*, 3, 672–80.

Nagaoka, I., Hirota, S., Yomogida, S., Ohwada, A., and Hirata, M. (2000). Synergistic actions of antibacterial neutrophil defensins and cathelicidins. *Inflammation Research*, 49, 73–9.

Narasimhan, M. L., Damsz, B., Coca, M. A., Ibeas, J. I., Yun, D.-J., Pardo, J. M., Hasegawa, P. M., and Bressan, R. A. (2001). A plant defense response effector induces microbial apoptosis. *Molecular Cell*, 8, 921–30.

Newman, S. L., Gootee, L., Gabay, J. E., and Selsted, M. E. (2000). Identification of constituents of human neutrophil azurophil granules that medidate fungistasis against *Histoplasma capsulatum*. *Infection and Immunity*, 68, 5668–72.

Oehlke, J., Beyermann, M., Wiesner, B., Melzig, M., Berger, H., Krause, E., and Bienert, M. (1997). Evidence for extensive and non-specific translocation of oligopeptides across plasma membranes of mammalian cells. *Biochimica et Biophysica Acta*, 1330, 50–60.

Olson, V. L., Hansing, R. L., and McClary, D. O. (1977). The role of metabolic energy in the lethal action of basic proteins on *Candida albicans*. *Canadian Journal of Microbiology*, 23, 166–74.

Oppenheim, F. G., Xu, T., McMillian, F. M., Levitz, S. M., Diamond, R. D., Offner, G. D., and Troxler, R. F. (1988). Histatins, a novel family of histidine-rich proteins in human parotid secretion. *Journal of Biological Chemistry*, 263, 7472–7.

Ösapay, K., Tran, D., Ladokhin, A. S., White, S. H., Henschen, A. H., and Selsted, M. E. (2000). Formation and characterization of a single Trp-Trp cross-link in indolicidin that confers protease stability without altering antimicrobial activity. *Journal of Biological Chemistry*, 275, 12017–22.

Patterson-Delafield, J., Szklarek, D., Martinez, R. J., and Lehrer, R. I. (1981). Microbicidal cationic proteins of rabbit alveolar macrophages: Amino acid composition and functional attributes. *Infection and Immunity*, 31, 723–31.

Pollock, J. J., Denepitiya, L., MacKay, B. J., and Iacono, V. J. (1984). Fungistatic and fungicidal activity of human parotid salivary histidine-rich polypeptides on *Candida albicans*. *Infection and Immunity*, 44, 702–7.

Prescianotto-Baschong, C. and Riezman, H. (1998). Morphology of the yeast endocytic pathway. *Molecular Biology of the Cell*, 9, 173–89.

Raj, P. A. and Edgerton, M. (1995). Functional domain and poly-L-proline II conformation for candidacidal activity of bactenecin 5. *FEBS Letters*, 368, 526–30.

Raj, P. A., Soni, S.-D., and Levine, M. J. (1994). Membrane-induced helical conformation of an active candidacidal fragment of salivary histatins. *Journal of Biological Chemistry*, 269, 9610–19.

Reed, J. C., Jurgensmeier, J. M., and Matsuyama, S. (1998). Bcl-2 family proteins and mitochondria. *Biochimica et Biophysica Acta*, 1366, 127–37.

Rex, S., and Schwarz, G. (1998). Quantitative studies on the melittin-induced leakage mechanism of lipid vesicles. *Biochemistry*, 37, 2336–45.

Riezman, H., Munn, A., Geli, M. I., and Hicke, L. (1996). Actin-, myosin- and ubiquitin-dependent endocytosis. *Experientia*, 52, 1033–41.

Ruissen, A. L. A., Groenink, J., Helmerhorst, E. J., Walgreen-Weterings, E., Van 't Hof, W., Veerman, E. C. I., and Nieuw Amerongen, A. V. (2001). Effects of histatin 5 and derived peptides on *Candida albicans*. *Biochemical Journal*, 356, 361–8.

Ruissen, A. L. A., Groenink, J., Van 't Hof, W., Walgreen-Weterings, E., Van Marle, J., Van Veen, H. A., Voorhout, W. F., Veerman, E. C. I., and Nieuw Amerongen, A. V. (2002). Histatin 5 and derivatives: Their localization and effects on the ultrastructural level. *Peptides*, 23, 1391–9.

Scheit, K. H. and Bhargava, P. M. (1985). Effect of seminalplasmin, an antimcirobial protein from bovine semen, on growth and macromolecular synthesis in *Candida albicans*. *Indian Journal of Biochemistry and Biophysics*, 22, 1–7.

Schonwetter, B. S., Stolzenberg, E. D., and Zasloff, M. A. (1995). Epithelial antibiotics induced at sites of inflammation. *Science*, 267, 1645–8.

Schröder, J.-M. and Harder, J. (1999). Molecules in focus. Human beta-defensin-2. *The international Journal of Biochemistry and Cell Biology*, 31, 645–51.

Selsted, M. E., Szklarek, D., Ganz, T., and Lehrer, R. I. (1985). Activity of rabbit leukocyte peptides against *Candida albicans*. *Infection and Immunity*, 49, 202–6.

Shepherd, M. G., Chin, C. M., and Sullivan, P. A. (1978). The alternative respiratory pathway of *Candida albicans*. *Archives of Microbiology*, 116, 61–7.

Shi, J., Zhang, G., Wu, H., Ross, C., Blecha, F., and Ganz, T. (1999). Porcine epithelial α-defensin 1 is expressed in the dorsal tongue at antimicrobial concentrations. *Infection and Immunity*, 67, 3121–7.

Shin, S. Y., Kang, S.-W., Lee, D. G., Eom, S. H., Song, W. K., and Kim, J. I. (2000). CRAMP analogues having potent antibiotic activity against bacterial, fungal, and tumor cells without hemolytic activity. *Biochemical and Biophysical Research Communications*, 275, 904–9.

Situ, H., Balasubramanian, S. V., and Bobek, L. A. (2000). Role of α-helical conformation of histatin 5 in candidacidal cativity examined by proline variants. *Biochimica et Biophysica Acta*, 1475, 377–82.

Skerlavaj, B., Scocchi, M., Gennaro, R., Risso, A., and Zanetti, M. (2001). Structural and functional analysis of horse cathelicidin peptides. *Antimicrobial Agents and Chemotherapy*, 45, 715–22.

Skipper, N. and Bussey, H. (1977). Mode of action of yeast toxins: Energy requirement for *Saccharomyces* killer toxin. *Journal of Bacteriology*, 129, 668–77.

Tang, Y.-Q., Yuan, J., Ösapay, G., Ösapay, K., Tran, D., Miller, C. J., Ouellette, A J., and Selsted, M. E. (1999). A cyclic antimicrobial peptide produced in primate leukocytes by the ligation of two truncated β-defensins. *Science*, 286, 498–502.

Tsai, H., Raj, P. A., and Bobek, L. A. (1996). Candidacidal activity of recombinant human salivary histatin 5 and variants. *Infection and Immunity*, 64, 5000–7.

Van 't Hof, W., Veerman, E. C. I., Helmerhorst, E. J., and Nieuw Amerongen, A. V. (2001). Antimicrobial peptides: Properties and applicability. *Biological Chemistry*, 382, 597–619.

Westerhoff, H. V., Hendler, R. W., Zasloff, M., and Juretić, D. (1989). Interactions between a new class of eukaryotic antimicrobial agents and isolated rat liver mitochondria. *Biochimica et Biophysica Acta*, 975, 361–9.

Xu, Y., Ambudkar, I., Yamagishi, H., Swaim, W., Walsh, T. J., and O'Connell, B. C. (1999). Histatin 3-mediated killing of *Candida albicans*: Effect of extracellular salt concentration on binding and internalization. *Antimicrobial Agents and Chemotherapy*, 43, 2256–62.

Xu, T., Levitz, S. M., Diamond, R. D., and Oppenheim, F. G. (1991). Anticandidal activity of major human salivary histatins. *Infection and Immunity*, 59, 2549–54.

Xu, T., Telser, E., Trolxer, R. F., and Oppenheim, F. G. (1990). Primary structure and anticandidal activity of the major histatin from parotid secretion of the subhuman primate, *Macaca fascicularis*. *Journal of Dental Research*, 69, 1717–23.

Yan, H. and Hancock, R. E. W. (2001). Synergistic interactions between mammalian antimicrobial defense peptides. *Antimicrobial Agents and Chemotherapy*, 45, 1558–60.

Yang, S.-T., Shin, S. Y., Kim, Y.-C., Kim, Y., Hahm, K.-S., and Kim, J. I. (2002). Conformation-dependent antibiotic activity of tritrpticin, a cathelicidin-derived antimicrobial peptide. *Biochemical and Biophysical Research Communications*, 296, 1044–50.

Yu, Q., Lehrer, R. I., and Tam, J. P. (2000). Engineered salt-insensitive β-defensins with end-to-end circularized structures. *Journal of Biological Chemistry*, 275, 3943–9.

Zasloff, M. (1987). Magainins, a class of antimicrobial peptides from Xenopus skin: Isolation, characterization of two active forms, and partial cDNA sequence of a precursor. *Proceedings of the National Academy of Sciences USA*, 84, 5449–53.

CHAPTER 11

Antimicrobial peptides from platelets in defense against cardiovascular infections

Michael R. Yeaman

11.1. INTRODUCTION

The access of potential pathogens to the bloodstream emphasizes the requisite for rapid and potent antimicrobial defense mechanisms to defend against such challenges. In this regard, platelets are becoming increasingly recognized for their likely multiple roles in antimicrobial host defense. Platelets share many structural and functional archetypes with leukocytes, and, once activated, platelets respond in specific ways that emphasize their antimicrobial functions, including accumulation at sites of tissue injury or infection, direct interaction with microbial pathogens, and liberation of antimicrobial peptides. Platelet microbicidal proteins (PMPs) exert potent and direct antimicrobial effects against organisms that commonly gain access to the bloodstream. In experimental models in vitro and in vivo, antimicrobial peptides from platelets appear to play important roles in limiting or preventing infection that is due to bacteria and fungi. Furthermore, certain antimicrobial proteins from platelets are chemokines that recruit professional phagocytes to sites of trauma or infection and potentiate the antimicrobial mechanisms of these cells. Thus, although their physiologic relevance has yet to be fully defined, PMPs likely play significant roles in antimicrobial host defense through direct and indirect mechanisms. Importantly, synthetic peptides designed in part based on the functional determinants of PMPs exert potent and durable antimicrobial activity in complex biomatrices such as whole blood, plasma, and serum ex vivo. From these perspectives, recognition of the structures and functions of PMPs advances our understanding of their likely roles in antimicrobial host defense, and provides a relevant basis for development of novel anti-infective agents and strategies.

11.2. MAMMALIAN PLATELETS ARE STRATEGIC HOST DEFENSE CELLS

The trophism of potentially pathogenic microorganisms for the bloodstream by invasion or trauma corresponds to the need for two key host responses: (1) rapid and potent antimicrobial systems that function without necessarily evoking inflammation, and (2) minimization of blood loss and initiation of tissue repair. Invertebrate hemocytes are dual-function cells that perform both of these roles (Nachum et al., 1980; Yeaman, 1997). In mammals, thrombocytes and leukocytes have historically been viewed to mediate hemostasis or inflammation, respectively (Tocantins, 1938; Weksler, 1971). Although a principal function is the maintenance of hemostasis, mammalian platelets have retained important features of cell-mediated effector cells that contribute multiple functions in antimicrobial host defense (Yeaman, 1997; Yeaman and Bayer, 1999, 2000).

Mammalian platelets are small (2–4 μm), ephemeral cells that derive from megakaryocyte lineage (Heyssel, 1961; Murphy, Robinson, and Roswell, 1967; Marcus, 1969; Harker and Finch, 1969; White, 1972). Because platelets are devoid of nuclei, they are capable of only limited translation by stable megakaryocyte mRNA templates (Booyse and Rafelson, 1967a, 1967b; Warshaw, Laster, and Shulman, 1967). Thus the majority of bioactive molecules carried by platelets are stored in premature or mature forms within granules. Platelets contain three distinct cytoplasmic granule types. Dense (δ) granules store mediators of vascular tone, including serotonin, adenosine diphosphate (ADP), eicosanoids, thromboxane A_2 (TXA_2), calcium, and phosphate (Day, Ang, and Holmsen, 1972; MacFarlane and Mills, 1975; Colman, 1991). By comparison, alpha (α) granules contain proteins important to hemostatis, including adhesion (e.g., fibrinogen, thrombospondin, vitronectin, and von Willebrand factor), coagulation (e.g., plasminogen and α_2-plasmin inhibitor), and endothelial cell repair [e.g., platelet-derived growth factor (PDGF), permeability factor, transforming growth factors α and β (TGF-α and TGF-β)]. Lysosomal (λ) granules store constituents that mediate thrombus dissolution. Degranulation is subject to agonist specificity and potency. For example, low levels of thrombin or ADP induce δ and α degranulation, whereas λ granules secrete only when these agonists are present at considerably higher levels (MacFarlane et al., 1975; Day and Rao, 1986; Davies et al., 1993). As detailed in the next section, platelets are now known to contain an arsenal of antimicrobial peptides. From these perspectives, platelets are the earliest and predominant cells in the bloodstream that respond to microorganisms or tissue infection and are capable of liberating stored antimicrobial constitutents rapidly in these settings.

Platelets have structural archetypes of antimicrobial effector cells, supporting the concept that platelets play multiple roles to prevent or limit infection. Furthermore, platelets perform functions that reflect key contributions to immunity. The structures and functions of platelets as they relate to antimicrobial host defense have been the focus of recent reviews (Yeaman, 1997; Yeaman and Bayer, 1999; Klinger and Jelkmann, 2002). Mammalian platelets have likely retained these antimicrobial characteristics over the course of their evolutionary history. Illustrating this concept, many invertebrates and insects possess hemocytes that maintain hemostasis, and simultaneously act as professional phagocytes. The following discussion focuses on the antimicrobial peptides that have been isolated and characterized over the past decade and the emerging view that these peptides are integral to antimicrobial host defense through direct and indirect mechanisms.

11.3. PLATELETS CONTAIN AN ARRAY OF ANTIMICROBIAL PEPTIDES

The relationship between platelet structure and function provides a compelling basis for their likely integral roles in antimicrobial host defense. Yet, until recently, the specific molecules contributing to these activities were not known. Over the past decade, contemporary techniques in protein chemistry and molecular biology have been applied to resolve the identification and characteristics of these antimicrobial constituents. Thus it is now clear that platelets contain a diverse and complementary group of peptides that exert explicit antimicrobial activities.

11.3.1. Early Studies Implicated Antimicrobial Molecules in Platelets

Specific interactions with microbial pathogens substantiate a thematic role for platelets in antimicrobial host defense. Early evidence pointed to platelet antimicrobial effects as being mediated by release of antimicrobial constituents. As early as the nineteenth century, Fodor reported a heat-stable bactericidal action of serum, termed β-lysin, distinguishable from heat-labile α-lysin complement proteins (Fodor, 1887). Later, Gengou (1901) verified that serum β-lysin activity was derived from cells associated with blood coagulation and functioned independently of complement. Tocantins (1938) published a comprehensive review of information relating to the immune functions of platelets. Subsequently several investigators demonstrated that platelets, not leukocytes, reconstituted the bactericidal activity of rabbit serum and exert bacteriostatic or bactericidal effects in vitro against a variety of pathogens, including *Bacillus* spp., *Staphylococcus* spp., *Listeria* spp. and *Salmonella* spp., (see reviews in Yeaman, 1997; Yeaman and Bayer, 2000).

In attempts to isolate the constituents responsible for platelet antimicrobial activities, Myrvik reported that two platelet-derived agents in serum were associated with killing of *Bacillus subtilis* (Myrvik, 1956). Likewise, Myrvik and Leake (1960) and Jago and Jacox (1961) isolated two heat-stable proteins from platelets that were associated with bactericidal action of serum. Extending these results, Johnson and Donaldson (1968) and Weksler and Nachman (1971) resolved two cationic proteins from rabbit platelets with in vitro bactericidal activity against *B. subtilis* or *Staphylococcus aureus*. Tew and co-workers (Tew, Roberts, and Donaldson, 1974; Donaldson and Tew, 1977) later reported that at least one β-lysin was released from rabbit platelets stimulated with thrombin in vitro. Overall, these studies pointed to antimicrobial effector molecules from platelets as being small and cationic, with masses ranging from 6 to 40 kd. Carroll and Martinez (1981a, 1981b, and 1981c) described the isolation, microcompositional analysis, microbicidal spectrum, and mechanisms of action of an antibacterial peptide (PC-III) from rabbit serum. More recently, Darveau et al. (1992) described peptides related to human platelet factor 4 (PF-4) that exert antimicrobial activity in combination with conventional antibiotics against Gram-negative bacteria.

11.3.2. Identification and Characterization of Antimicrobial Peptides from Platelets

Studies performed over the past decade have specifically identified and characterized the antimicrobial peptides from mammalian platelets. Recently, a group of antimicrobial peptides, termed platelet microbicidal proteins (PMPs), has been isolated from supernatants of rabbit platelets with or without thrombin stimulation (Yeaman et al., 1992, 1997). Reversed-phase high-performance liquid chromatography (RP-HPLC) has revealed that rabbit platelets contain a family of PMPs liberated upon thrombin induction, termed thrombin-induced PMPs (tPMPs). Mass spectroscopy showed that PMPs and tPMPs range in size from 6.0 to 9.0 kd (Yeaman, 1997; Yeaman et al., 1997). The fact that thrombin is generated at sites of endovascular infection, and is a potent stimulant of platelet degranulation, is relevant to PMP release in such settings (Drake, Rodgers, and Sande, 1984; Drake and Pang, 1988; Bancsi, Thompson, and Bertina, 1994). Amino acid compositional analyses show that these polypeptides are rich in the basic residues, lysine, arginine, and histidine (total content ~25%), corresponding to cationicity. The antimicrobial activities of PMPs and tPMPs are stable upon heating to 56 °C and mitigated by anionic adsorption (Yeaman, Puentes et al., 1992). Dankert et al. (1995) confirmed that platelets release bactericidal substances upon thrombin stimulation. Sequence, molecular mass, lysine content, and a distinct disulfide array distinguish PMPs or tPMPs from the well-characterized neutrophil defensins (Yeaman, 1997; Yeaman et al., 1997; Yeaman and Bayer, 2000). Moreover, PMPs and tPMPs are distinguishable from platelet lysozyme in mass, amino acid composition, and antimicrobial activity. PMPs and tPMPs exert strong antimicrobial activities in vitro against bacteria and fungi (Yeaman et al., 1993; Yeaman et al., 1997). As in the case of human platelet antimicrobial peptides (HPAPs) (see subsequent discussion), it is possible that tPMPs represent either native or processed forms of PMPs present in unstimulated rabbit platelets.

Recently, Yount et al. (2003) characterized the predominant PMPs contained within rabbit platelets. Matrix assisted laser desorption ionization time of flight (MALDI-TOF) mass spectroscopy, N-terminal amino acid sequencing, and cloning methods showed that two forms of PMP-1 are recoverable from rabbit platelets in the presence or absence of thrombin stimulation. Amino acid sequencing and MALDI-TOF mass spectroscopy data demonstrated native PMP (nPMP-1) and tPMP-1 species to be identical proteins (PMP-1) that display amino-terminal polymorphism. N-terminal amino acid sequence data (NH_2-[S]$D^1DPKE^5SEGDL^{10}HCVCV^{15}KTTSL^{20}$) enabled cloning of PMP-1 from rabbit bone marrow, and characterization of its full-length cDNA. Results indicate that PMP-1 is synthesized as a 106-amino-acid precursor, processed to yield the 72-residue mature peptide. The calculated mass for mature PMP-1 (7957 d) was within standard confidence intervals of its experimentally determined value (7951 d). BLAST searches of the Swiss Protein database and multiple-sequence alignments of retrieved proteins revealed significant homology of PMP-1 to PF-4 proteins. Based on phylogenetic relatedness, congruent-sequence motifs, and predicted three-dimensional structures, PMP-1 exhibited greatest homology to human PF-4 (hPF-4). tPMP-1 and hPF-4 also have similar in vitro antimicrobial activities, corroborating the functional significance of their homologous physicochemical and phylogenetic properties. Based on these data, nPMP-1 and tPMP-1 were identified as rabbit immunologs of hPF-4, a peptide with known microbicidal and chemokine roles (Tang et al., 2002). Integrating their equivalent structural and antimicrobial properties, PMP-1 was identified as the rabbit immunolog of hPF-4, a peptide with recognized microbicidal and chemokine activities. These results establish the identity of PMP-1 and provide a structural basis for the multifunctional roles of mammalian PF-4 homologs. Moreover, these findings support the broader concept that stimuli present in the setting of infection prompt platelets to release microbicidal peptides that contribute to coordinated antimicrobial host defense.

Although PMPs were initially isolated from rabbit platelets, analogous peptides have now been identified in human platelets. Tang et al. (2002) isolated and characterized HPAPs, released from human platelets following thrombin stimulation. These peptides are structural and functional analogs of rabbit PMPs. Characterized HPAPs include PF-4, platelet basic protein (PBP) and derivatives, connective-tissue-activating peptide 3 (CTAP-3), a neutrophil-activating peptide 2 (NAP-2), RANTES ("released upon activation, normal T cell expressed and secreted"), thymosin-β-4 (Tβ-4), and fibrinopeptides A and B (FP-A and FP-B). These novel findings demonstrated for the first time the direct antimicrobial functions of this group of human platelet proteins, even though these proteins had been structurally characterized previously. Krijgsveld et al. (2000) found that carboxy-terminal diamino acid truncated versions of NAP-2 or CTAP-3 exerted antimicrobial activity in vitro. These peptides have been termed thrombocidins 1 and 2, respectively. It is interesting to note that FP-A and FP-B were not detectable in total protein extracts of platelets. This finding suggests these HPAPs are generated by thrombin stimulation of platelets. Conceivably, such peptides may be generated by thrombin-cleavable Arg-Gly sites in platelet fibrinogen (Turner et al., 2002). Thus thrombin (a serine protease), platelet-derived proteases, proteases generated by tissue injury, phagocytes (e.g., cathepsin G) or inflammation, or even microbial proteases, may process precursor proteins and contribute to generation of multiple antimicrobial peptides from platelets. These concepts are likely relevant to the multifunctional roles of platelets in antimicrobial host defense (Yeaman, 1997; Yeaman and Bayer, 2000; Klinger and Jelkmann, 2002).

For consistency, the term platelet microbicidal proteins (PMPs) is used in this review to encompass PMPs, tPMPs, HPAPs, thrombocidins, or other antimicrobial peptides from platelets. Comparisons of the structures and functions of these antimicrobial polypeptides from platelets with those from other sources are reviewed elsewhere (Yeaman, 1997; Yeaman and Bayer, 2000; Tang et al., 2002). Phylogenetically, there are three predominant lineages of microbicidal chemokine PMPs: (1) PF-4 and variants, (2) PBP and the derivatives CTAP-3 and NAP-2, and (3) RANTES (Yeaman, 1997; Yeaman and Bayer, 2000). PF-4, PBP, CTAP-3, and NAP-2 are α- or cysteine-X-cysteine CXC-chemokines, whereas RANTES is a β-, or cysteine-cysteine CC-chemokine. It should be emphasized that, in addition to their now-recognized potent and direct antimicrobial effects, such chemokines recruit and potentiate the antimicrobial mechanisms of leukocytes and lymphocytes (see reviews addressing these topics: Yeaman, 1997; Yeaman and Bayer, 2000). Thus PMPs likely contribute direct and indirect mechanisms of antimicrobial host defense, as outlined in the next section.

11.4. PMPs EXERT DIRECT AND INDIRECT ANTIMICROBIAL ACTIVITIES IN VITRO

Identification of the diverse array of antimicrobial peptides in platelets prompted intensive investigation regarding their antimicrobial properties and mechanisms of action. As a result, these peptides have been shown to affect rapid and potent antimicrobial activities against a spectrum of organisms relevant to the cardiovascular compartment.

11.4.1. PMPs Exhibit Potent Microbicidal Activities Against Vasculotrophic Organisms

PMPs exert potent and even synergistic in vitro activity against organisms that frequently gain access to the bloodstream. For example, potent microbicidal activities against *S. aureus, Staphylococcus epidermidis, B. subtilis, Escherichia coli, Candida albicans,* and *Cryptococcus neoformans* have been documented for PMPs from rabbits or humans (Yeaman, 1997; Yeaman and Bayer, 2000; Tang et al., 2002). PF-4 appears to exert the broadest spectrum of activity against pathogens studied to date (Yeaman, 1997; Tang et al., 2002; Yount et al., 2003). Its antimicrobial spectrum includes Gram-positive cocci and bacilli, Gram-negative bacilli (e.g., family Enterobacteriaceae and *Pseudomonas* spp.), and fungal pathogens *C. albicans* and *C. neoformans.* Likewise, RANTES is highly active against bacteria and fungi, but is considerably less abundant than PF-4 in human platelets (Tang et al., 2002). In contrast, CTAP-3 is the quantitatively predominant PMP in human platelets, but exhibits antimicrobial potency substantially less than that of PF-4 or RANTES. The microbicidal effects of PMPs are achieved at nanomolar to micromolar concentrations (1–5 μg/ml) in vitro. In addition, these peptides are active at physiological ranges of pH (5.5–8.0) and act synergistically against microbial pathogens in vitro (see next subsection). Thus the antimicrobial activities of PMPs observed in vitro are relevant to conditions that exist in vivo. These findings also demonstrate that antimicrobial peptides are conserved among mammalian platelets. Thus the structural and functional congruence of rabbit and human PMPs also enable opportunities to examine the roles of PMPs in antimicrobial host defense in relevant models.

11.4.2. Mechanisms of PMP Action are Distinguishable From These of Other Antimicrobial Peptides

The effects of PMPs on intact microbial cells, protoplasts, and lipid bilayers in vitro have been studied by use of transmission and scanning electron microscopy (TEM and SEM, respectively) and biophysical techniques (Koo, Bayer et al., 1996; Yeaman, Bayer et al., 1998; Xiong, Yeaman, and Bayer, 1999). Generally, these data indicate that PMPs initially target and perturb microbial cell membranes. For example, in *S. aureus*, rapid cytoplasmic membrane permeabilization occurs, but membrane depolarization may or may not ensue. Subsequently the cytoplasmic membrane appears to condense, corresponding to hypertrophy of the cell wall of *S. aureus* following exposure to PMPs for 60–90 min. Typically, perturbations in cell ultrastructure precede detectable bactericidal and bacteriolytic effects of PMPs. Fungal pathogens are affected in a similar manner by tPMP-1 in vitro (Yeaman et al., 1993). Koo, Bayer et al. (1996, 1999) reported that tPMP-1 achieves antistaphylococcal effects by means of a mechanism that involves voltage-dependent membrane permeabilization. Protoplasts derived from *S. aureus* whole cells exhibit tPMP-1 susceptibility or resistance characteristics corresponding to those of the phenotype from which they were prepared, suggesting that antimicrobial effects are independent of the cell wall. Studies of isogenic strain pairs suggested the potential involvement of autolytic enzymes in *S. aureus* killing by PMPs; however, death was not attributable solely to autolytic enzyme activation (Xiong et al., 1999). Activation of abnormal autolysin function has been suggested as a mechanism of cationic antimicrobial peptides (Ginsburg, 1988), but remains to be proven.

The bactericidal mechanisms of PMP action against isogenic tPMP-1-susceptible (tPMP-1^S) and tPMP-1-resistant (tPMP-1^R) *S. aureus* are distinct from those of human defensin NP1 (HNP1) or other cationic antimicrobial agents in key parameters, including role of transmembrane potential ($\Delta\psi$), permeabilization, and bactericidal activity. For example, rabbit PMP-2 rapidly permeabilized and depolarized a tPMP-1^S strain, with extent of permeabilization inversely related to pH (Yeaman et al., 1997). However, tPMP-1 did not significantly depolarize the tPMP-1^S strain, but permeabilized this strain in a manner directly correlated with pH. Depolarization, permeabilization, and killing of the tPMP-1^R strain by PMP-2 and tPMP-1 was significantly reduced compared with that of the tPMP-1^S counterpart. Moreover, culture in menadione reconstituted the tPMP-1^R $\Delta\psi$ to a level equivalent to the tPMP-1^S strain, increased depolarization that was due to PMP-2 (but not tPMP-1), and restored permeabilization and killing of the tPMP-1^R strain.

Thus mechanisms of distinct PMPs differ, involving pH-dependent membrane permeabilization with or without membrane depolarization. These mechanisms were distinguishable from hNP-1 or the cationic antibacterial agents protamine or gentamicin.

11.4.3. PMPs Appear to Interfere with Microbial Targets beyond the Cytoplasmic Membrane

The ultrastructural effects described in the preceding subsection suggest that an initial target of action for PMPs is the microbial cytoplasmic membrane. However, the delay ($\sim$ 2 h) between initial exposure and microbicidal action implicate other, likely intracellular, targets of action for these peptides. Xiong et al. (1996, 1999) demonstrated that preexposure of tPMP-1^S *S. aureus* with tetracycline, a 30S ribosomal subunit inhibitor, significantly decreased the ensuing staphylocidal effect of tPMP-1 over a concentration range of 0.16–1.25 μg/ml. In these studies, preexposure to novobiocin (aninhibitor of bacterial DNA gyrase subunit B), azithromycin, quinupristin, or dalfopristin (inhibitors of 50S ribosomal subunits) mitigated the staphylocidal effect of tPMP-1 over a concentration range of 0.31–1.25 μg/ml. These data suggested that tPMP-1 exerts anti-*S. aureus* activities, in part, through mechanisms involving inhibition of macromolecular synthesis.

Further investigations by this group supported this concept (Xiong, Bayer, and Yeaman, 2002). In tPMP-1^S *S. aureus* strains, purified tPMP-1 caused a significant reduction in DNA and RNA synthesis that temporally corresponded to the extent of staphylocidal activity. In contrast, tPMP-1 exerted substantially less inhibition of macromolecular synthesis in an isogenic tPMP-1^R strain, paralleling reduced staphylocidal effects. However, tPMP-1 caused equivalent degrees of protein synthesis inhibition in these strains. Collectively, these data suggest that at least some PMPs inhibit specific macromolecular synthesis pathways as part of their overall staphylocidal mechanism(s).

11.5. PLATELETS FUNCTION AS ANTIMICROBIAL HOST DEFENSE CELLS

The natural history of endovascular infection, particularly when involving highly pathogenic organisms such as *S. aureus*, has conventionally been viewed as the following series of events (Yeaman and Bayer, 2000): (1) access of microorganisms to the bloodstream, (2) adhesion of bloodborne pathogens to normal or abnormal vascular endothelium (e.g., endocardium associated with rheumatic heart disease or prosthetic cardiac valves), (3) tissue expression of soluble stimuli and ligands that promote platelet activation and deposition, (4) subsequent deposition of circulating platelets in response to secondary agonists generated by initial platelet adhesion and activation [e.g., ADP, platelet activating factor (PAF)], or ligands (e.g., GPIIb-IIIa), and (5) further platelet deposition, degranulation, and liberation of PMPs and other granule contents. Thus platelets represent arguably the earliest opportunity for host defenses to intervene in the establishment and progression of endovascular infection. The outcome of this initial interaction between microorganisms and platelets may largely determine whether infection is initiated within the cardiovascular compartment. The precise mechanisms of PMPs generation at sites of infection remain to be fully delineated. Likewise, the specific contributions of PMPs to the prevention or attenuation of infections have not yet been established. Yet a compelling body of evidence exists supporting the view that platelet elaboration of PMPs provides a physiologically relevant mechanism of antimicrobial host defense.

11.5.1. Early Studies Demonstrated Platelet Response to Settings of Infection

Platelets represent the most rapid and abundant inflammatory cells that respond to endothelial cell damage or microbial colonization. Thus platelets may be viewed as the earliest of opportunities for inflammatory responses to intercede in microbial pathogenesis and effect antimicrobial host defense against induction and evolution of endovascular infection. However, the contemporary view that platelets are integral host defense cells has not always been the case.

It is clear that platelets are present at sites of infection involving the vascular endothelium. Early interpretations of their role in this setting hypothesized that platelets promote the establishment and evolution of endovascular infection. For example, several investigators suggested that platelets facilitate microbial adhesion to fibrin matrices or endothelial cells in vitro (Durack, Beeson, and Petersdorf, 1973; Calderone, Rotondo, and Sande, 1978; Scheld, Valone, and Sande, 1978; Klotz, Harrison, and Misra, 1989). Herzberg et al. (1990) have proposed that increased streptococcal binding to and aggregation of platelets is correlated with increased virulence of these strains in experimental animal models of endocarditis. Platelet aggregation has also been suggested to be detrimental to the host, as massive endovascular vegetations may be associated with clinical severity such as emboli and infarcts (Clawson, 1977; Nicolau et al., 1993). Others have suggested that platelet aggregation and internalization of microorganisms may protect pathogens from exposure to antibiotics or clearance by neutrophils or other leukocytes (Clawson and White, 1971).

The preceding observations demonstrate that platelets recognize and react to microorganisms upon entry into the vascular compartment. Undoubtedly, platelets can interact with microorganisms through specific interactions (Yeaman, 1997; Bayer et al., 1995), suggesting that platelets monitor the bloodstream for the presence of potential pathogens. Thus platelets aggressively respond in host defense in the presence of potential pathogens. Moreover, there are no data substantiating a concept that platelets inherently facilitate microbial pathogenesis (e.g., enhanced pathogen survival, endothelial cell penetration, or dissemination into tissue parenchyma) or negatively influence the immune response. In this respect, host cells that promote detrimental infection would be highly disadvantageous from an evolutionary perspective. On the contrary, a substantial body of evidence now emphasizes that platelets are important participants in antimicrobial host defense against infection.

11.5.2. Platelets Recognize and Target Settings of Infection

Microbial colonization induces rapid thromboplastin (tissue factor) expression by vascular endothelial cells and monocytes. Reviews of these responses as they may influence antimicrobial host defenses are available elsewhere (Smith, 1993; Yeaman, 1997; Yeaman and Bayer, 2000; Klinger and Jelkmann, 2002; Yeaman and Bayer, 1999). In brief, soluble tissue factor prompts an intrinsic pathway proteolytic cascade, ultimately resulting in thrombin generation. Thrombin is a potent platelet agonist, activating platelets to undergo a rapid morphologic change from discoid to amoeboid and to display microtubule and granule organization in preparation for degranulation. In turn, inducible platelet receptors, such as the fibrinogen (GPIIb-IIIa) and P-selectin receptors, are expressed by activated platelets. Degranulation follows, releasing ADP, TXA_2 and PAF through activation of membrane phospholipase A_2. Such agonists stimulate ensuing waves of platelet deposition, activation, and degranulation at sites of endothelial infection. Beyond the scope of this review, phospholipase A2 may exert antimicrobial effects independent of, but complementary to, other nonoxidative mechanisms (Weiss, Bayer, and Yeaman, 2000).

Platelets also navigate toward soluble signals generated in the setting of vascular infection or complement fixation. Such endovascular infections include infective endocarditis, suppurative thrombophlebitis, mycotic aneurysm, septic endarteritis, catheter and dialysis access site infections, and infections of vascular prostheses and stents (Yeaman and Bayer, 2000). Tissue factor and ensuing thrombin stimulation increase platelet adhesion to infected vascular endothelial cells, promoting platelet accumulation at these sites (Carney, 1992). In this context, adherent platelets become activated to liberate antimicrobial peptides that may contribute to host defense against microbial colonization and deeper tissue invasion. The rapid and numerically significant presence of platelets at these sites has been well established. Osler (1886) made the earliest observations of platelets accumulating upon filaments introduced into animal veins. Similarly, Cheung and Fischetti (1990) and others (Roberts and Buchbinder, 1972; Piguet et al., 1993) showed that platelets are the first cells to adhere to indwelling vascular catheters. Furthermore, platelets rapidly deposit upon cardiac valve prostheses and endovascular stents and are the earliest and most abundant cells in endocarditis vegetations in rabbits and humans. Thus platelets rapidly target surfaces, as well as sites of injury to vascular endothelium, that are vulnerable to infection by organisms that may gain access to the bloodstream.

Endothelial cell ligands assist in targeting platelets to sites of infection. Platelets exhibit specific receptors that sense agonists and bind to ligands characteristic of injured endothelial cells or exposed subendothelial stroma resulting from infection. Ligands recognized by platelet membrane glycoprotein (GP) receptors include (see Yeaman, 1997) collagen (GPIa-IIa, or VLA-2), fibronectin (GPIc-IIa, or VLA-5), von Willebrand factor (GPIb-IX-V), laminin (GPIc-IIa, or VLA-6), vitronectin [$\alpha_V(\beta_3$ integrin], and thrombin.

11.5.3. Platelets Liberate PMPs in Relevant Settings of Infection

To date, the focus of antimicrobial peptide discovery and characterization has been on peptides evolved to be secreted onto mucosal surfaces or contained within professional phagocytes. Many of these peptides exert potent antimicrobial activity under defined test conditions in vitro, but cause significant toxicity to mammalian tissues in vitro or in vivo. Such toxicity has been viewed as an impediment to development of these antimicrobial peptides as novel anti-infective agents.

Selective toxicity among antimicrobial peptides involves complex and specific interactions between peptide and target pathogen (Yeaman and Yount, 2003). However, it is also likely that these peptides may be rendered less harmful to the host simply through strategic localization or expression that minimizes their interaction with potentially vulnerable host tissues. Three paradigms illustrate this conceptual model. Among vertebrates, many antimicrobial peptides are secreted onto relatively inert epithelial surfaces, such as the tracheal, lingual, or intestinal mucosa of mammals, or the skin of amphibians. This localization – along with rapidly inducible expression – places such peptides in key positions to respond to potential pathogens present on mucosal barriers, yet protects more sensitive tissues from host cytotoxicity. A similar, albeit perhaps more complex, mechanism likely contributes to selective toxicity of antimicrobial peptides found in granules of phagocytic leukocytes. The key antimicrobial functions of professional phagocytes include internalization (phagocytosis) of pathogens, subjecting them to the harsh microenvironment of the phagolysosome. Neutrophils, monocytes, and macrophages contain an array of antimicrobial peptides, including defensins. However, defensins may also exhibit poor selective toxicity, exerting membrane permeabilizing and other harmful effects on microorganisms and mammalian cells alike. To protect the host against autocidal effects, phagocytes normally internalize and expose pathogens to lethal concentrations of these peptides in the maturing phagolysosome, rather than degranulating these potentially injurious components into the extracellular milieu. Within these restricted confines of the phagolysome, defensins and other antimicrobial peptides are present in very high relative concentrations, in which they may act harshly and synergistically with one another, along with oxidative killing mechanisms. In this way, antimicrobial peptides may be constrained to granules of mammalian phagocytes to minimize their potential for host cytotoxicity. Moreover, Yeaman (1997) suggested that antimicrobial activities of PMPs are potentiated in mildly acidic conditions, such as those found in the maturing phagolysosome (see next section).

Antimicrobial peptides from platelets represent a distinct paradigm likely optimized for function in the vascular compartment, without concomitant host cytotoxicity. PMPs exert potent microbicidal activities against pathogens that commonly enter the bloodstream. Numerous studies in humans have shown that levels of PMPs such as PF-4 increase markedly (4-fold–6-fold; up to 5 μg/ml) during acute phase infections with viruses, bacteria, fungi, or protozoa (Essien and Ebhota, 1983; Lorenz and Brauer, 1988; Srichaikul et al., 1989; Mezzano et al., 1992; Yamamoto, Klein, and Friedman, 1997). Thus PF-4 plasma levels increase during acute phases of cytomegaloviremia, bacterial septicemia, streptococcal nephritis, candidiasis, and malaria. Likewise, PMP levels increase in rabbit plasma as a result of staphylococcal challenge in vivo and in models of experimental infective endocarditis in vitro (Mercier et al., 2000). These findings correspond to data demonstrating that staphylococcal cells or α-toxin prompt PMP release from rabbit platelets in vitro (Azizi et al., 1996; Bayer et al., 1997). Shahan and co-workers (1998) showed that expression of other members of the PF-4 family of chemokines is also increased following stimulation with pathogenic and nonpathogenic fungi. In this regard, transcription of the CXC or α-chemokines is preferentially upregulated over CC β-chemokines temporally and in a concentration-dependent manner. These findings suggest that chemokines, including certain PMPs, rapidly respond to microbial pathogens. Similarly, Wilson et al. (2001) have recently shown that endotoxin prompts striking increases in circulating levels of soluble P-selectin, an indicator of platelet degranulation. Interestingly, unlike most other markers of inflammation studied, plasma concentrations of soluble P-selectin progressively increased in these settings for up to 8 h following endotoxin injection in human volunteers. These findings support the concept that PMPs are released from platelets in response to hallmark signals of infection and inflammation or upon interaction with microorganisms themselves.

Recent investigations have shown that microorganisms prompt the release of antimicrobial peptides from platelets in vitro and in vivo through specific response pathways. For example, Trier, Bayer, and Yeaman (2000) demonstrated that *S. aureus* triggers release of platelet antimicrobial responses likely through interaction with the platelet ADP P2X receptor. Inhibition of this receptor by pyridoxal-5′-phosphate derivative (P5PD; a specific P2X inhibitor) or nonspecific inhibitors (e.g., suramin), inhibited platelet-associated antimicrobial activity in response to *S. aureus* exposure. Thus P5PD prevented the anti-*S. aureus* effects attributable to *S. aureus* stimulation of platelets. In contrast, 3′-phosphoadenosine-5′-phosphosulfonate (P5PS; a platelet ADP / P2Y receptor inhibitor) failed to inhibit *S. aureus*-stimulated antimicrobial responses of platelets. Therefore, in the setting of P5PS, platelets generated an uninhibited response that resulted in killing of the stimulating *S. aureus* organism. These findings support the concept that PMPs are released in physiologically relevant contexts from platelets in response to infection.

11.6. THE PHYSIOLOGIC RELEVANCE OF PMPs IN ANTIMICROBIAL HOST DEFENSE

The hypothesis that PMPs exert optimal antimicrobial activity in context-specific settings of infection is supported by extensive data in vitro and in vivo. As described in the preceding section, platelet deposition and degranulation at sites of infection and tissue injury likely result in locally high concentrations of these antimicrobial peptides. Also, many PMPs are cationic peptides and likely accumulate upon electronegative pathogen surfaces. Thus measurements of their free concentration in blood, plasma, or sera probably underestimate the antimicrobial activities of PMPs at foci of infection (Yeaman, 1997; Yeaman and Yount, 2003). Furthermore, human PF-4 is ultimately concentrated within the spleen and liver (Rucinski et al., 1990), likely related to sequestration of pathogens during reticuloendothelial clearance. Supporting this concept, studies with tPMP-1^S and tPMP-1^R *S. aureus* or *C. albicans* pairs indicate that the tPMP-1^S phenotype negatively influences proliferative ability within splenic versus renal tissues (Yeaman et al., 1996; Dhawan, Bayer, and Yeaman, 1998; Bates et al., 2003). These data suggest that PMP antimicrobial activities are influenced by specific microenvironmental milieu at local sites of infection. The discovery that the microbicidal activity of certain PMPs is diminished in hypernatremic conditions simulating those in the kidney (Koo et al., 1996) may well explain these differences. PMPs may also exit the cardiovascular compartment to function in antimicrobial host defense. For example, Frohm et al. (1996) found that the human PMP Tβ-4 and other antimicrobial peptides are detectable within human wound and blister fluid. These findings imply that PMPs may function to prevent or limit infection beyond the vascular compartment. Therefore the varying abundance, antimicrobial potencies, and spectra, along with conditional optima for antimicrobial activity, suggest complementary roles for PMPs in antimicrobial host defense.

11.6.1. PMPs Exhibit Conditional Optima Relevant to Host Defense

PMPs exert rapid, direct, and potent antimicrobial activities in vitro against bacterial and fungal pathogens that often gain access to the bloodstream. Thus these peptides likely optimize microbicidal activity within this physiologic context, without concomitant host cytotoxicity. The conditional optima of PMPs in vitro suggest they provide rapid, potent, and complementary host defense functions in relevant contexts in vivo. For example, human PF-4, PBP, and Tβ-4 exert substantially greater in vitro activity against *E. coli* than against *S. aureus* (Tang et al., 2002). Moreover, these peptides exert greater activity against bacteria than against fungi under identical test conditions in vitro. By comparison, these peptides exhibited substantially greater antimicrobial activities than those of CTAP-3 under the same conditions. Except for Tβ-4, all human PMPs exert moderately to substantially greater antimicrobial activity at mildly acidic pH than at pH 7.5. In comparison, at pH 7.5, Tβ-4 caused significant and rapid killing of *S. aureus* or *E. coli*. Yet this peptide exerted no significant bactericidal or fungicidal activity under this assay condition in vitro. It is notable that Tβ-4 exhibits an overall net anionic charge at neutral pH, whereas PF-4, CTAP-3, and RANTES are cationic. However, FP-A and FP-B, also anionic, exerted greater antimicrobial activities and spectra under slightly acidic conditions. Thus the structure–activity relationships of individual PMPs may be optimal in specific microbiological or biochemical contexts.

Other potentially relevant antimicrobial properties have also been detected for certain PMPs in vitro. For example, when purified, most PMPs are generally more potent in solution than agar diffusion assays. Thus PMPs may have greater antimicrobial potency in the fluid phase of the bloodstream (e.g., whole blood or at the blood/endothelial cell interface), rather than deeper within solid lesions such as vegetations of infective endocarditis. This concept has been supported by recent evidence from an in vitro model of infective endocarditis (see next subsection; Mercier et al., 2000). In addition, PF-4 has been shown to exhibit a bimodal antifungal effect against *C. albicans* (Tang et al., 2002). In vitro, PF-4 concentrations of less than 5 nmol/ml caused significant antifungal activity. However, at concentrations >5 nmol/ml, PF-4 failed to exert significant anticandidal activity. This dose-response relationship was highly reproducible and specific to PF-4 versus *C. albicans*. The biophysical characteristics of PF-4 may provide insights into this result. For example, Mayo and Chen (1989) found that hPF-4 exists in a monomer–dimer–trimer equilibrium in solution. Decreases in peptide concentration, pH, or ionic strength favor the monomeric form of PF-4, whereas the opposite conditions favor tetramer formation. Hypothetically, PF-4 may exert its greatest microbicidal effect at relatively low local concentrations, with diminishing direct antimicrobial potency, as the concentration increases to favor multimers. This concentration-dependent antimicrobial activity could be inversely related to a gradient of concentration necessary for PF-4 or other chemokine PMPs to perform optimal functions as leukocyte chemoattractants (Yeaman, 1997).

Perhaps even more importantly, the PMPs PF-4 and CTAP-3 exert in vitro synergistic killing of *E. coli* (Tang et al., 2002). Synergy was demonstrated when these peptides were combined at individually sublethal concentrations, as increasing concentrations of CTAP-3 were supplemented with 0.1 nmol of PF-4. These striking results argue that the release of multiple PMPs in the local setting of infection, even at sublethal concentrations of individual peptides, may yield a potent antimicrobial milieu that significantly contributes to host defense. Collectively, the observations previously described emphasize the likelihood that PMPs exert context-specific activities that optimize and coordinate antimicrobial host defenses.

11.6.2. PMPs Modulate Pathogen Interactions with Platelets and Endothelial Cells

The fact that PMPs interact with microbial surfaces is consistent with the concept that these peptides may alter microorganism interactions with host cells. For example, *S. aureus* clinical isolates exhibit heterogeneity regarding platelet adhesion, aggregation, and susceptibility to tPMP-1 (Yeaman et al., 1992b). A significant, positive correlation is observed between platelet adherence and aggregation among such strains. For example, exposure of tPMP-1^S or tPMP-1^R *S. aureus* strains to sublethal concentrations of tPMP-1 reduces the velocity and magnitude of platelet aggregation by *S. aureus* (Bayer et al., 1995). Whether tPMP-1 achieves this effect by altering bacterium-to-platelet binding, or other mechanisms, has not been established as yet. However, no correlation is observed between either platelet adherence or aggregation and in vitro susceptibility to tPMP-1. These data imply that susceptibility to tPMP-1 is likely independent of other platelet–microbe interactions.

Generally, bacterial and fungal adhesion to platelets in vitro is through rapid, saturable, and reversible interactions, suggesting receptor–ligand mechanisms. Modified Scatchard analyses indicate that the number of binding sites per platelet vary somewhat for distinct *S. aureus* strains (Yeaman et al., 1992). In these studies, binding of individual *S. aureus* cells to platelets was influenced more by the number of binding sites on platelets than on platelet binding affinities of bacterial cells. These findings suggest organism-specific platelet interactions. Protease K treatment did not significantly alter adherence of *S. aureus* to rabbit platelets in vitro. However, exposure to periodate or to tPMP-1 significantly reduced staphylococcal adherence to rabbit platelets. These findings suggest that *S. aureus* adhesion to platelets may be multimodal, involving tPMP-1-sensitive and carbohydrate surface ligands; it is possible that such ligands may be identical. Parallel investigations demonstrate that in vitro exposure to tPMP-1, alone or combined with classical antibiotics, significantly reduces adhesion of *S. aureus* to platelets in vitro, irrespective of tPMP-1^S or tPMP-1^R phenotype (Yeaman et al., 1994).

Complementary evidence indicates that platelets protect against endovascular infections by interfering with pathogen interactions with host cells. For example, superimposed on their direct antimicrobial effects, PMPs appear to interfere with microbe–endothelial cell interactions (Yeaman et al., 1995). In recent studies, Filler et al. (1999) demonstrated that platelets in vitro protect human umbilical vein endothelial cells (HUVECs) from injury that was due to *C. albicans*. In these experiments, ^{51}chromium release from HUVECs that was due to a tPMP-1^S *C. albicans* strain was reduced by nearly 50% in the presence of a platelet-to-fungus ratio of 20:1. Furthermore, HUVEC protection by platelets was associated with a 37% reduction in germ tube length in *C. albicans* after a 2-h exposure. In contrast, HUVEC damage by an isogenic tPMP-1^R strain was uninhibited by platelets.

11.7. PMPs LIKELY PLAY MULTIPLE ROLES IN ANTIMICROBIAL HOST DEFENSE IN VIVO

There is abundant evidence that platelets target and interact directly with microbial pathogens of every class (viruses, bacteria, fungi, and protozoa). Reviews focusing on these interactions are available elsewhere (Yeaman et al., 1997; Klinger and Jelkmann, 2002; Yeaman and Bayer, 2000). Moreover, platelets specifically respond to organisms and stimuli generated in the setting of infection by releasing an array of PMPs. Chemonavigation of platelets to local settings of infection, localized PMP release in these settings, and affinity of these peptides for pathogens likely promote conditions consistent with relevant antimicrobial activities of PMPs in vivo (Yeaman et al., 1997; Yeaman and Yount, 2003): (1) local concentrations sufficient for microbicidal potency and (2) the establishment of gradients promoting leukocyte chemotactic navigation. These functions are analogous to other chemokines, such as interleukin-8 (IL-8). For example, IL-8 is generated and introduced into plasma during infection, but exerts localized effects targeting sites of infection. The fact that PMPs share structural similarities to such cytokines underscores the likelihood that they function in a similar manner, in addition to having direct microbicidal properties.

A compelling body of evidence now points to platelets as being integral to multiple functions relating to direct and indirect antimicrobial host defense. PMPs are felt to represent a significant component of these functions. The fact that platelets contain an array of antimicrobial peptides, along with other antimicrobial effector functions, presents challenges to demonstrating the specific antimicrobial roles for PMPs individually or in combination in vivo. Thus the physiologic conditions under which PMPs alone or synergistically contribute to microbiostatic or microbiocidal host defense mechanisms have yet to be defined. Nonetheless, numerous studies supporting multiple roles for platelets and PMPs in antimicrobial host defense have now been performed. The following discussion focuses on evidence that points to key functions for PMPs in limiting the establishment and progression of endovascular infection in vivo.

11.7.1. Deficiencies in Platelet Quantity and Quality Increase Risk of Infection

It has long been recognized that appropriate platelet quantity and quality are important to homeostasis and host defense. For example, inherited conditions such as Wiscott–Aldrich Syndrome, May–Hegglin Anomaly, Gray–Platelet Syndrome, and related platelet disorders strongly correlate with increased morbidity and mortalilty that are due to infection (Yeaman, 1997). However, these conditions may represent a convergence of multiple disorders in cell-mediated immunity, making the definition of platelet contributions to host defense more difficult to independently define.

More specific roles of platelets and PMPs in antimicrobial host defense in vivo have been investigated by use of complementary approaches. For example, Sullam et al. (1993) employed an experimental animal model to examine the role of platelets in defense against infective endocarditis in vivo. In this study, a tPMP-1^S viridans Strep. strain was used to induce endocarditis in animals either with normal platelet counts or those with selective immune thrombocytopenia. Of note, there were no differences in these groups of animals regarding other leukocyte quantity or quality or complement activity. Animals with thrombocytopenia exhibited significantly higher streptococcal densities in vegetations compared with their counterparts with normal platelet counts. Reports of Dankert et al. (2001) likewise support the concept that platelets are active in host defense against infective endocarditis. These data substantiate the concept that platelets and PMPs are integral to host defense mechanisms that limit establishment and/or evolution of endovascular infections.

In humans, thrombocytopenia is emerging as a significant, independent indicator of worsened morbidity and mortality that is due to infection. In patients undergoing cytotoxic cancer chemotherapy as well as nonneoplastic conditions, thrombocytopenia is believed to put patients at increased risk of morbidity and mortality that are due to bacterial or fungal infection (Feldman et al., 1991; Kirkpatrick, Reeves, and MacGowan, 1994; Viscoli et al., 1994; Chang et al., 2000). For example, in the absence of neutropenia, thrombocytopenia has been positively correlated with increased incidence and severity of lobar bacterial pneumonia in elderly individuals (Kirkpatrick et al., 1994). Thus neutropenia in the setting of a normal platelet count does not appear to diminish host defense against endovascular infection in vivo. Collectively, the majority of data suggest that the platelet/microbe interaction is beneficial in attenuating infection in vivo.

Data pertaining to other human diseases have also implicated platelets as being crucial to antimicrobial host defense against infection. For example, Chang et al. (2000) examined the impact of thrombocytopenia on morbidity and mortality that were due to infection in liver transplant recipients. By several measures of outcome, thrombocytopenia was determined to be a significant and independent predictor associated with increased infection and related morbidity and mortality. For example, nadir platelet counts were significantly lower in nonsurvivors compared with survivors. Nearly 50% of patients with nadir platelet counts of $\leq 30 \times 10^9$/L presented with a major infection within 30 days of transplantation, compared with only 17% of patients with nadir platelet counts exceeding this threshold. Similarly, fungal infections were observed in 14% of patients exhibiting nadir platelet counts below this breakpoint versus 0% in those with nadir platelet counts above it. It is notable that every fungal infection occurred within patients with nadir platelet counts below this threshold before the presentation of fungal infection. In each case, nadir platelet counts preceded the first major infection by a median of 7 days. These findings strongly implicate thrombocytopenia in the worsened morbidity and mortality that occur due to infections in such patient populations.

Recently thrombocytopenia has been proposed as a predictor of invasive bacterial infection risk among children with cancer, fever, and neutropenia (Santolaya et al., 2002). Among these children, five potential risk factors were analyzed for correlation with invasive bacterial infection: C-reactive protein serum level, hypotension, leukemia relapse, thrombocytopenia (platelet count $\leq$50,000/mm^3 blood), and chemotherapy. Trombocytopenia was identified as a sole risk of invasive bacterial infection in 12% of children in this cohort study (sensitivity, specificity, positive and negative predictive values of the model were 92%, 76%, 82%, and 90%, respectively).

The preceding findings are consistent with the importance of platelet antimicrobial responses in the setting of human infection. Mavrommatis et al. (2000) demonstrated two distinguishable levels of platelet and coagulation cascade responses to Gram-negative infection in humans. In the initial state, uncomplicated sepsis is associated with increases in blood levels of FBP-A and PF-4–two known PMPs from human platelets. Increases in the release of these antimicrobial peptides were temporally associated with a reduction in platelet count, suggesting that activated platelets are cleared following degranulation. However, in severe sepsis, and particularly in the context of septic shock, platelet number and responsivity are reduced, and coagulation factors are depleted. These consequences indicate that, in cases of profound sepsis, platelet responses to microbial challenge may be suppressed or overwhelmed.

Global inhibitors of platelet functions have been shown to increase levels of bacteremia and mortality in some experimental animal models (Korzweniowski et al., 1979; Yeaman, 1997). However, antiplatelet therapies that reduce the adhesive functions of platelets, without interfering with their antimicrobial responses, may enhance host defense against endovascular infection. Nicolau et al. (1993) found that aspirin administered prophylactically reduced the induction phase and extent of *S. aureus* infective endocarditis in the rabbit model. However, as aspirin is a global inhibitor of cyclooxygenase function in endothelial cells as well as in platelets, these findings remain difficult to interpret. The recent studies of Kupferwasser et al. (1999) shed light on this situation. These investigators demonstrated that acetylsalicylic acid (ASA) exerted direct antimicrobial activities against *S. aureus* in an experimental animal model. In these studies, ASA at 8 (mg/kg)/day [but not 4 or 12 (mg/kg)/day] was responsible for significant decreases in vegetation weight, echocardiographic-verified vegetation size, vegetation and renal *S. aureus* densities, and renal embolic lesions, compared with untreated controls. Reduced aggregation was observed when platelets were preexposed to ASA, or when *S. aureus* was preexposed to salicylate. Moreover, *S. aureus* adhesion to sterile vegetations, platelets in suspension, fibrin matrices, or fibrin–platelet matrices was also significantly reduced following bacterial exposure to salicylate. Thus it is highly likely that aspirin or salicylates suppress the ability of *S. aureus to* adhere to platelets, but do not impede the antimicrobial responses of platelets, such as their ability to release PMPs.

11.7.2. Evidence for Direct Antimicrobial Effects of PMPs In Vivo

Evidence from a variety of investigators broadly supports the view that PMPs are integral to host defense against infection in vivo. This evidence integrates both in vitro and in vivo studies, providing a complementary perspective of platelets in this role. Dankert and colleagues (1995) indirectly studied the influence of putative bactericidal factors released from thrombin-stimulated platelets on the natural history of infective endocarditis that is due to viridans group streptococci. Platelet lysate-susceptible streptococci adhered to endocarditis vegetations as well or better than the releasate-resistant test strains 5 min postchallenge. However, between 5 min and 48 h after inoculation, lysate-susceptible strains exhibited a dramatically reduced ability to proliferate in this setting compared with their lysate-resistant counterparts ($P < .001$). Moreover, studies in vitro indicated that clearance of the lysate-susceptible strains were not caused by complement bactericidal activity or surface phagocytosis by neutrophils.

Abundant evidence exists substantiating the concept that PMPs specifically contribute to the antimicrobial mechanisms of platelets in vivo. For example, Yeaman et al. (1993) demonstrated that susceptibility to tPMP-1 negatively affects the establishment and proliferation of *C. albicans* infection. In this rabbit model, tPMP-1^S *C. albicans* exhibits significantly less proliferation in cardiac vegetations compared with a tPMP-1^R strain in the same genetic background. Moreover, the tPMP-1^S *C. albicans* strain exhibited a reduced incidence of metastatic dissemination to the spleen compared with the tPMP-1^R counterpart. Likewise, Dhawan et al. (1998) found that an in vitro tPMP-1 susceptibility phenotype correlated with reduced virulence in experimental endocarditis due to *S. aureus*. After coinoculation of animals with isogenic tPMP-1^R and tPMP-1^R strains, significantly lower densities of the tPMP-1^S strain were present in vegetations, kidneys, and spleens, compared with those for the tPMP-1^R strain. However, there were no differences in the ability of the strains to adhere to platelet–fibrin matrices or in their clearance from the bloodstream. These data suggest that tPMP-1 limits postadhesion survival, proliferation, and hematogenous dissemination of *S. aureus* in this model of infection. Alternatively, a tPMP-1^R phenotype may confer a selective advantage associated with the enhanced progression of this infection or serve as a surrogate for *S. aureus* virulence factors that mediate this effect. These results support the concept that PMPs significantly contribute to the antimicrobial functions of platelets in the context of microbial colonization of the vascular endothelium.

11.7.3. Evidence for Immunopotentiating Effects of PMPs In Vivo

Although it is beyond the scope of this review, some PMPs are known to perform other important functions that contribute significantly to antimicrobial host defense, including recruitment of leukocytes to sites of infection and potentiating their antimicrobial activites (Yeaman, 1997; Yeaman and Bayer, 1999; Cole et al., 2001; Tang et al., 2002). This nomenclature reflects the fact that these proteins and peptides exert direct microbicidal activities and mediate leukocyte navigation to sites of infection.

Many PMPs identified are members of the intercrine family of chemokines (Oppenheim et al., 1991). Therefore, PF-4, CTAP-3, NAP-2, and

RANTES represent what can now be termed microbicidal chemokines. These peptides are known to chemoattract leukocytes to sites of infection and potentiate the antimicrobial mechanisms of these cells. It follows that the inflammatory milieu may be ideal for the antimicrobial functions of certain PMPs in vivo. For example, the pH of the leukocyte phagolysosomes can be reduced to as low as 4.5 after phagocytosis of microorganisms (Spitznagel, 1984). Similarly, abscess exudate, serum, and interstitial fluid containing leukocytes may achieve acidic pH (Spitznagel, 1984; Shafer, Martin, and Spitznagel, 1986). Recently a model has been proposed integrating the direct antimicrobial, chemokine, and leukocyte potentiating functions of PMPs in relevant contexts and conditional optima (Yeaman, 1997). PMPs active under mildly acidic conditions may enhance the ability of leukocytes to kill microbial pathogens through nonoxidative mechanisms, particularly as the maturing phagolysosome becomes acidified. Walz et al. (1989) previously found, consistent with this model, that chemokine PMPs amplify potential antimicrobial responses in leukocytes. Moreover, Cocchi et al. (1995) showed that the β-chemokine RANTES suppresses human immunodeficiency virus proliferation or pathogenesis through direct antiviral effects and/or modulation of T-cell function. Likewise, Cole et al. (2001) found that some CXC chemokines have direct antibacterial activity. Thus PMPs likely play two key roles in antimicrobial host defense: (1) direct inhibition or killing of pathogens, as platelets target PMPs to accumulate at sites of infection and (2) recruiting and amplifying antimicrobial mechanisms of leukocytes before or following phagocytosis of organisms exposed previously to these peptides.

Interactions with leukocytes provide an additional mechanism by which platelets and PMPs contribute to antimicrobial host defense. For example, in addition to chemokine PMPs, an array of bioactive molecules released from activated platelets serve as chemoattractants for monocytes and neutrophils. These stimuli include PF-4, PAF, PDGF, and eicosanoids such as 12-HETE (Tzeng et al., 1985). Subcutaneous injection of PF-4 prompts a rapid neutrophil infiltration in experimental animal models (Deuel et al., 1981). Thus, in consort with recruiting monocytes and neutrophils to sites of infection, platelet activation and degranulation are associated with potentiation of leukocyte antimicrobial functions. Mandell and Hook (1969) showed that activated platelets facilitate phagocytosis of *Salmonella* spp. by mouse peritoneal macrophages. Others have shown that PF-4 amplifies neutrophil fungicidal activities in vitro (Walz et al., 1989). Christin et al. (1996) demonstrated that platelets and neutrophils act synergistically in vitro to damage and kill *Aspergillus* spp. In turn, molecules produced by activated monocytes or neutrophils may activate platelets. For example, monocytes exposed to bacterial components generate tissue factor, which elicits thrombin production and subsequent platelet activation. Oxygen metabolites, myeloperoxidase, and halides generated by leukocytes may prompt rapid platelet degranulation (Clark and Klebanoff, 1979). Likewise, neutrophil- or monocyte-derived PAF is a potent agonist of platelet activation, triggering shape change, receptor expression (e.g., activated GPIIb-IIIa), and platelet degranulation. Neutrophil leukotrienes C4, D4, or E4 may also amplify platelet aggregation and degranulation, alone or in combination with epinephrine or thrombin (Mehta, Mehta, and Lawson, 1986). Taken together, these findings underscore the likelihood that the interplay among platelets, PMPs, and leukocytes are important for optimal host defense against infection.

11.7.4. Efficacy of PMPs in Host Defense is Substantiated by Epidemiological Evidence

Under normal conditions, bacteria and other potential microbial pathogens enter the human bloodstream multiple times per day. Yet the incidence of cardiovascular infection is vanishingly low in comparison with the frequency of microorganism access to the bloodstream (Bayer and Scheld, 2000). This fact illustrates the reality that mechanisms of host defense countering the establishment and proliferation of intravascular infection are generally extremely effective. It is believed that PMPs significantly contribute to these mechanisms. Because of the likely multiple antimicrobial functions of PMPs, the potential for organisms with reduced PMP susceptibility to exhibit survival advantages in settings of intravascular infection have recently been investigated.

Wu, Yeaman, and Bayer (1994) also observed a correlation between infective endocarditis (IE) source and diminished susceptibility to tPMP-1 in vitro. These data suggest that tPMP-1^S organisms are less frequently associated with endovascular infection in humans compared with their tPMP-1^R counterparts. Similar observations have correlated *Salmonella* resistance to defensins, antimicrobial peptides present in neutrophils, with increased virulence (Fields, Groisman, and Heffron, 1989; Groisman et al., 1992). More recently, Fowler et al. (2000) examined the relationship between *S. aureus* IE and in vitro resistance to tPMP-1. These investigators evaluated the in vitro tPMP-1 susceptibility phenotype of *S. aureus* isolates from 58 prospectively identified patients with definite IE. On multivariate analyses, *S. aureus* IE complicating an infected intravascular device was significantly more likely to be caused by a tPMP-1^R strain ($P = 0.02$). Among the *S. aureus* strains studied, no correlations were detected between in vitro tPMP-1^R phenotype and the severity of IE. Collectively, these findings were interpreted to indicate that direct PMP activity is an integral component of host defense against endovascular infections such as endocarditis. This theme was recently confirmed in studies focusing on viridans group streptococci (Dankert et al., 2001).

Despite the overall profound degree of success in antimicrobial host defense, occasionally pathogens succeed in causing intravascular infections. Thus it is possible that certain organisms are capable of exploiting the platelet as an adhesive surface if they circumvent the antimicrobial functions of platelets. Conceivably, virulence properties or strategies of microbial pathogens may circumvent the antimicrobial functions of platelets. If so, pathogens may gain an advantage in pathogenesis of endovascular infections. Thus pathogens capable of resisting the antimicrobial functions of platelets may exploit exhausted platelets in the evolution of IE or other vascular infections. For example, Dhawan et al. (1998, 1999, 2000) and Fowler et al. (2000) have implicated this likelihood in pathogenesis and therapy of experimental and human staphylococcal IE, respectively (also see subsequent discussion). However, in this context, PMPs appear to be involved in effective host defense against *S. aureus* strains that exhibit a PMP-susceptible phenotype in vitro.

Another potential strategy of platelet exploitation by microbial pathogens may be exemplified in viridans group streptococci. Through molecular mimicry of structural domains of collagen important in hemostasis, *Streptococcus sanguis* triggers platelets to aggregate in vitro (Meyer, Gong, and Herzberg, 1998). Increased blood pressure, intermittent electrocardiographic abnormalities, and changes in blood catecholamine concentration rapidly followed inoculation of an aggregation-positive strain of *S. sanguis* into experimental animals. These effects were associated with acute thrombocytopenia and accumulation of ^{111}indium-labeled platelets in the lungs. Moreover, platelet aggregation-rendered thrombi that were interpreted to be responsible for the observed hemodynamic changes, acute pulmonary hypertension, and cardiac abnormalities. In contrast, a *S. sanguis* strain incapable of inducing platelet aggregation failed to yield such effects. Thus it would be predicted that the aggregation-positive *S. sanguis* strain may have reduced susceptibility to the antimicrobial properties of platelets, exploiting these cells as adhesive surfaces in IE pathogenesis.

The preceding observations suggest correlations between reduced susceptibility to tPMPs and increased propensity for involvement in endovascular infections. However, it should be understood that the concept of resistance to tPMPs is distinct from classical antibiotic resistance and represents low-level and perhaps artificial resistance to an isolated peptide under relatively austere conditions in vitro. For example, unlike the logarithmic resistance frequencies often observed when pathogens are exposed to high levels of conventional antibiotics, the in vitro tPMP^R phenotype represents a modest and arithmetic increase in survival of ~40% when tPMP-1 is tested at a concentration of 1–2 μg/ml. Even with very small increases in tPMP-1 concentration (e.g., 4 μg/ml), the in vitro difference in tPMP^S-versus tPMP^R phenotype is abolished, as either phenotype is completely killed (Xiong, Bayer, and Yeaman, 2002). Thus the in vitro tPMP-1^R phenotype represents a laboratory breakpoint for which relevance to microbial pathogenesis and host defense has not been determined.

The preceding facts emphasize several important points regarding likely functions of PMPs in antimicrobial host defense: (1) PMPs are cationic and likely accumulate upon microbial membranes or envelopes that are electronegative; (2) platelets target sites of endovascular injury through vectored chemotaxis along with recognition and adhesion to injured tissues and pathogens themselves, further concentrating the liberation of PMPs to the setting of infection. Thus the concentrations of these peptides within infective loci are likely to be significantly greater than in blood or plasma (i.e., underestimated in whole blood or plasma) and exceed the low levels used to differentiate susceptibility and resistance in vitro. Recent studies by Trier, Bayer, and Yeaman (2000) are consistent with these concepts, as isogenic tPMP-1^S and tPMP-1^R *S. aureus* strains were equivalent in susceptibility to killing in whole plasma ex vivo; (3) platelet quantity and/or quality with respect to PMP content or delivery to sites of infection appears to be an important variable influencing the natural history of endovascular infection. From these perspectives, the observed correlations between tPMP-1^R phenotype and propensity to be involved in intravascular infection may rest with equal weight on either platelet content, liberation, and targeting of PMPs to sites of infection, any putative inherent or adaptive capacity for pathogens to evade susceptibility to PMPs, or a combination of these parameters; and (4) interaction of PMPs with other host defense mechanisms (e.g., other PMPs, antimicrobial peptides, complement, or phagocytic responses) is not accounted for in the in vitro tPMP susceptibility assay. For example, the in vitro bioassay used to differentiate tPMP-1^S and tPMP-1^R isolates is conducted with tPMP-1 alone at relatively low levels. Recent studies have demonstrated that human PMPs act synergistically at sublethal concentrations to effect killing (Tang, Yeaman, and Selsted, 2002). Thus correlates relating to reduced susceptibility to PMPs should be considered in the context of low-level resistance to tPMP-1 in the absence of other antimicrobial host defenses (also see next section).

11.8. ADVANCES AND POTENTIAL APPLICATIONS OF ANTIMICROBIAL PEPTIDES FROM PLATELETS

The burst of new information in recent years regarding the antimicrobial roles of platelets and their antimicrobial peptides has provided insights into antimicrobial host defense. In addition, the fact that platelets liberate their antimicrobial peptide arsenal into the bloodstream suggests that the structure–activity relationships inherent to these molecules are directly relevant to immunity in this context. These perspectives have identified new initiatives that may capitalize on the evolutionary optimization of such molecules to exert multiple antimicrobial functions.

11.8.1. New Models Will Facilitate Studies of PMPs and Related Antimicrobial Peptides

In comparison with leukocytes or epithelial mucosa, platelets respond to microorganisms or soluble mediators of inflammation by liberating PMPs directly into the bloodstream. The potential interactions between an antimicrobial peptide and components of the cardiovascular compartment are multifactorial. For example, whole blood or derived matrices may contain binding or blocking proteins that inactivate such peptides. Similarly, blood may contain peptidases capable of degrading a peptide over time. Alternatively, blood or blood fractions may contain constituents that amplify the actions of PMPs against target pathogens or are enhanced by PMPs, resulting in potentiated microbicidal effects. Understanding such factors is directly relevant to the roles of PMPs in host defense against infection and potential therapeutic applications of antimicrobial peptides.

Development of an ex vivo assay has provided an important new screen in which to test the activities of antimicrobial peptides relevant to the cardiovascular environment (Yeaman et al., 2002). This assay was designed to provide a rigorous challenge to the extent and duration of antimicrobial peptide efficacy in complex biomatrices including whole blood and plasma. In addition, the assay can be performed in normal and heat-inactivated pooled normal human serum to assess the influence of heat-labile components on antimicrobial peptide efficacy. Plasma differs from whole blood in that plasma is essentially devoid of leukocytes and erythrocytes. Moreover, serum is distinct from whole blood or plasma as it is a cell-free biomatrix generated by the coagulation of whole blood and plasma components, with associated activation of proteases and other factors. The assay design includes simultaneous introduction of antimicrobial agent and organism into biomatrices or media or 2-h preincubation of the antimicrobial peptide in test matrices or media before inoculation of organisms. This approach avoids difficulties in differentiating microbiostatic versus microbicidal effects, as some organisms (e.g., Gram-negative bacteria) often exhibit a "regrowth" phenomenon after exposure to cationic agents (e.g., aminoglycoside antibiotics) in conventional antimicrobial assays (Xiong et al., 1996). Thus the biomatrix assay integrates quantitative culture over a 24-h period to verify microbicidal effects. This assay system differs from other methods that focus on whole blood (Wallis et al., 2001) and do not include the preincubation step integral to the preceding assay.

Similarly, an in vitro model of infective endocarditis has been established by Mercier et al. (2000) to examine the relevance of PMPs in host defense. This model introduces platelet–fibrin clots that simulate endocarditis vegetations into a circulating medium simulating the bloodstream. At any time point, organisms, platelets, platelet agonists, or anti-infective agents may be introduced. Likewise, the chamber fluid or simulated vegetations may be removed at any point and assessed histologically or microbiologically for the quantitative or qualitative parameters of infection. This model has been used to examine the ability of platelets to limit the colonization and proliferation of *S. aureus* strains that differ in their in vitro susceptibility to tPMP-1. In time–kill studies, early platelet activation (thrombin stimulation 30 min before bacterial inoculation) correlated with a significant bactericidal effect against tPMP-1^S ($r^2 > .90$, $P < .02$) but not against tPMP-1^R strains. There were no detectable differences among the test strains in initial colonization of simulated vegetations. These data underscore the temporal relationship between thrombin-induced platelet activation and *S. aureus* killing likely relevant to PMPs roles in antimicrobial host defense. Moreover, as they mirror outcomes in the experimental rabbit model, these results validate this model as an important new tool in evaluating platelet and PMP antimicrobial functions in a complex milieu containing multiple constituents.

11.8.2. PMPs Appear to Potentiate the Actions of Conventional Antibiotics In Vivo

PMPs potentiate the action of many classes of conventional antibiotics in vitro. For example, against inocula representing *S. aureus* densities present within early, developing, and established endocarditis vegetations in vitro, tPMP-1 exerted synergistic bactericidal effects in combination with anti-staphylococcal antibiotics oxacillin or vancomycin (Yeaman et al., 1992a). This synergistic effect was observed regardless of whether the *S. aureus* strain exhibited an intrinsic tPMP-1^S or tPMP-1^R phenotype in vitro. Additionally, the staphylocidal effects of tPMP-1 were augmented by pretreatment of *S. aureus* with either penicillin or vancomycin (Xiong, Yeaman, and Bayer, 1999). Platelet β-lysin has also been shown to act synergistically with ampicillin in vitro against *Listeria* spp. (Asensi and Fierer, 1991). Moreover, exposure to sublethal levels of tPMP-1 prolongs the growth inhibitory period against *S. aureus*, similar to classic "postantibiotic" effects characteristic of antibiotics such as vancomycin and oxacillin (Yeaman et al., 1992a).

These in vitro results have also been demonstrated in vivo. For example, Dhawan et al. compared the efficacies of vancomycin prophylaxis and treatment outcomes in the rabbit model of experimental infective endocarditis (Dhawan et al., 1999, 2000). In infections that are due to isogenic tPMP-1^S or tPMP-1^R strain pairs of *S. aureus*, vancomycin therapy (selected for its relatively slow bactericidal activity) reduced tPMP-1^S, but not tPMP-1^R densities in vegetations as compared with untreated controls ($P < .01$). However, prophylactic administration of vancomycin produced no difference in infectivity or severity of infection by the two *S. aureus* strains. In contrast, treatment with oxacillin (more rapidly staphylocidal than vancomycin) amplified the clearance of tPMP-1^S or tPMP-1^R organisms from vegetations; the extent of clearance was greater for tPMP-1^S cells. A similar theme was observed when tPMP-1^S or tPMP-1^R *C. albicans* strains were studied in experimental endocarditis treated with fluconazole (Yeaman et al., 1998). Collectively, these data suggest that susceptibility to tPMP-1 corresponds with greater efficacy of antimicrobial therapy, but not prophylaxis, in the setting of experimental IE. These results may be explained, in part, by the requirement for microbicidal activity in the treatment of established infective endocarditis, whereas prophylactic efficacy would be expected to rely more on microbiostatic and antiadhesion effects. However, in either case, the preceding observations point to the potential for strategic use of antibiotics that function synergistically or in complement with PMPs or synthetic derivatives thereof.

11.8.3. Synthetic Peptides Modeled from PMP Templates

Peptides that exert antimicrobial activity in artificial media may lack activity within blood or other complex biological matrices. For this reason, Yeaman et al. (2002) developed the ex vivo assay previously described and used it as a rigorous test for peptides designed to exert antimicrobial activity in blood and blood fractions. Novel antimicrobial peptides (RP-1 and RP-11) were designed based in part on the structure–activity relationships present in PMPs. RP-1, RP-11, or gentamicin were introduced into biomatrices either coincident with, or 2 h before, inoculation with a serum-resistant strain of *E. coli*. The antimicrobial activities of these peptides were assessed by quantitative culture 2 h after bacterial inoculation and compared with peptide-free and gentamicin controls. In whole blood and homologous plasma or serum, coincident introduction of RP-1 or RP-11 with *E. coli* caused a significant reduction in viable count versus peptide-free controls. Furthermore, significant antimicrobial efficacy remained when RP-1 or RP-11 were placed into whole blood or plasma 2 h before, organism inoculation. These results suggest the peptides were not rapidly inactivated within these biomatrices. In some test conditions, RP-1 exerted anti-*E. coli* effects that were equivalent to or exceeded that of gentamicin. The anti-*E. coli* efficacies of these peptides were negatively affected by preincubation in serum or in heat-inactivated serum. However, the peptides were consistently effective at lower concentrations in biomatrices than in artificial media, indicating favorable antimicrobial interactions with components of blood or blood fractions. Interestingly, RP-11 exerted little or no anti-*E. coli* activity in artificial media at either pH 5.5 or 7.2. Nonetheless, as intended from functional design (i.e., to be active in the cardiovascular milieu), RP-11 exhibited striking antimicrobial activity in whole blood, plasma, and serum that endured in whole blood and plasma following preincubation for 2 h before organism inoculation. Thus the peptides exerted potent and remarkably durable antimicrobial activity in whole blood and homologous plasma when coincubated with serum and as compared with conventional artificial media. Collectively, these findings support the concept that synthetic peptides can be designed to exert potent antimicrobial activities in relevant biological contexts, including whole blood, plasma, and serum.

It should be noted that blood components may interact with the peptides or the target organism to potentiate killing. Such interactions have been observed with other conventional antibiotics (Darveau et al., 1992; Pruul and McDonald, 1992; Miglioli, Schoffel, and Gianfranceschi, 1996; Hostacka, 1998) or tPMPs in vitro (Koo et al., 1996). Likewise, Yan and Hancock (2001) identified synergistic interactions of antimicrobial peptides and lysozyme in vitro. Varra et al. (1984) suggested that an outer-membrane disorganizing peptide [polymyxin B nonapeptide (PMBN)] sensitizes *E. coli* to bactericidal activity of serum in vitro, hypothetically by making the organism vulnerable to the lytic effects of complement fixation. Subsequently, this same group demonstrated that a PMBN acted synergistically with human serum in bactericidal activity against Gram-negative bacteria in vitro (Viljanen et al., 1986). However, from the previously discussed studies, RP-1 does not require complement to effect antimicrobial activity, as evidenced by activity in defined artificial media. In contrast, neither RP-1 nor RP-11 demonstrated antimicrobial activity in heat-inactivated serum. Oh, Hong, and Lee (1999) found that, consistent with these results, short peptides consisting of L-amino acids lost antimicrobial activity in heat-inactivated serum. Additionally, low-density lipoprotein has been implicated as an inhibitor of small α-helical peptide activity in serum (Peck-Miller, Darveau, and Fell, 1993). Heat inactivation may also be hypothesized to inactivate potentiating factors in serum, but the loss of RP-1 antimicrobial activity in this matrix implies influences beyond this effect.

The preceding considerations emphasize and confirm the importance of assays that assess peptide antimicrobial activities under conditions potentially relevant to their original source context or to their intended functions as therapeutic agents in vivo. However, the differences between such assays and conventional methods should be understood. For example, in preincubation assays, the antimicrobial agent and target organism are incubated for 2 h before quantitative culture. A significant amount of inactivation of the peptide may occur over this period, leading to false-negative results (e.g., apparent lack of antimicrobial activity). The potential for this effect can be minimized if shortened antimicrobial agent–organism exposure or incubation periods are sufficient to detect efficacy. Additionally, the quantitative expression of antimicrobial activity of the agent of interest is more difficult than assays in artificial media, as the biological matrices themselves (e.g., whole blood and serum) may exhibit differential antimicrobial effects. Alternatively, potencies of distinct peptides may vary in biomatrices under differing conditions such as pH, reflecting structure–activity relationships of antimicrobial peptides. These influences may be significant in specific contexts in which peptides may exert optimal antimicrobial activities in vivo.

11.8.4. PMPs and Related Peptides May Be Useful in Detecting Loci of Infection

Along with their localized release in the setting of infection, antimicrobial peptides from platelets appear to concentrate on, or be in close proximity to, pathogenic organisms themselves. These characteristics not only underscore their relevance to host defense, but provide the basis for using such peptides or derivatives thereof as diagnostic or imaging reagents in detecting infection.

Moyer et al. (1996) developed a ^{99}Technetium-labeled peptide from the heparin-binding domain of human PF-4 as a potential imaging agent to label neutrophils in vitro and localize infection in vivo. In vitro, the labeled peptide associated with high specificity to neutrophils and other leukocytes. In an *E. coli* model of infection in rabbits, this reagent identified focal infections *in vivo* within 4 h versus conventional imaging agents.

More recently, Welling et al. (2001) substantiated the hypothesis that cationic antimicrobial peptides may discriminate between microbial cells and host tissues in vivo. Studies evaluated whether such peptides specifically accumulate in sites of infection, as compared with sterile inflammatory lesions, because of preferential avidity for microorganisms. Peptide affinity and specificity for pathogens in vivo was assessed by intravenous injection of ^{99}mTc-labeled synthetic derivatives of human ubiquicidin or lactoferrin into animals experimentally infected with *S. aureus, Klebsiella pneumoniae*, or *C. albicans*. As controls, sterile inflammatory sites were induced by the introduction of heat-killed microorganisms or purified lipopolysaccharide into thigh muscle. Labeled human defensin, human polyclonal IgG, and ciprofloxacin were examined as comparative agents. The ^{99}Technetium-labeled peptides and defensin accumulated at a significantly higher rate and to a greater extent in bacterial- and *C. albicans*- infected lesions in mice and rabbits compared with noninfected but inflamed tissues. These data were interpreted to indicate that the peptides distinguish between microorganisms and host tissues and, in doing so, accumulate at sites of infection in vivo.

11.9. SUMMARY AND PROSPECTS

Mammalian platelets are unique and multipurpose inflammatory cells that exhibit archetypal features indicative of their roles in antimicrobial host defense. When stimulated in the context of infection, platelets release an array of polypeptides, collectively termed PMPs that exert potent and broad-spectrum antimicrobial activities. PMPs include certain chemokines that also potentiate the antimicrobial functions of leukocytes. Thrombocytopenia, defects in platelet quality, and inhibitors of platelet degranulation increase susceptibility to and severity of certain cardiovascular infections. Likewise, pathogens susceptible to PMPs produce less severe infections in vivo compared with counterpart strains that exhibit reduced susceptibility to PMPs. Thus platelets play key roles in antimicrobial host defense. The multiple antimicrobial properties of PMPs appear to function in relevant contexts of infection within the vascular compartment, without concomitant host cytotoxic effects. Substantiating this concept are synthetic peptides modeled on structure activity relationships identified in PMPs that exert potent antimicrobial activities in whole blood, plasma, and serum. From these perspectives, platelets and PMPs represent an intriguing and multifaceted component of antimicrobial host defense. In addition to their biological significance, PMPs could lead to new diagnostic and therapeutic interventions in human infections. For example, PMPs amplify the action of conventional antimicrobial agents, and provide templates for the development of novel anti-infective agents that act against pathogens exhibiting multiple antibiotic resistance.

ACKNOWLEDGMENTS

Several individuals have influenced this review through their insights and efforts, including Arnold Bayer, Eric Brass, Paul Sullam, Steve Projan, Nannette Yount, L. Iri Kupferwasser, Yan-Qiong Xiong, Kimberly Gank, Yi-Quan Tang, Michael Selsted, Tomas Ganz, and Jack Edwards. The efforts of Robert M. Delzell are recognized with gratitude. The author was supported by grants AI39001 and AI48031 from the National Institutes of Health.

REFERENCES

Asensi, V. and Fierer, J. (1991). Synergistic effect of human lysozyme plus ampicillin or β-lysin on the killing of *Listeria monocytogenes*. *Journal of Infectious Diseases*, 163, 574–8.

Azizi, N., Li, C., Shen, A. J., Bayer, A. S., and Yeaman, M. R. (1996). *Staphylococcus aureus* elicits release of platelet microbicidal proteins *in vitro*. Abstract 866. In abstracts of Thirty-Sixth Interscience Conference on Antimicrobial Agents and Chemotherapy. American Society for Microbiology, New Orleans, LA.

Bancsi, M. J. L. F., Thompson, J., and Bertina, R. M. (1994). Stimulation of monocyte tissue factor expression in an *in vitro* model of bacterial endocarditis. *Infection and Immunity*, 62, 5669–72.

Bates, D. M., von Eiff, C., McNamara, P. J., Peters, G., Yeaman, M. R., Bayer, A. S., and Proctor, R. A. (2003). *Staphylococcus aureus* mutants are as infective as the parent strains but the menadione biosynthetic mutant persists within the kidney. *Journal of Infectious Diseases*, 187, 1654–61.

Bayer, A. S. and Scheld, W. M. (2000). Endocarditis and intravascular infections. In *Principles and Practice of Infectious Diseases*, 5th ed., ed. G. L. Mandell, J. E. Bennet, and R. Dolin, pp. 857–902. New York: Churchill Livingstone.

Bayer, A. S., Ramos, M. D., Menzies, B. E., Yeaman, M. R., Shen, A. J., and Cheung, A. L. (1997). Hyperproduction of α-toxin by *Staphylococcus aureus* results in paradoxically-reduced virulence in experimental endocarditis: A host defense role for platelet microbicidal proteins. *Infection and Immunity*, 65, 4652–60.

Bayer, A. S., Sullam, P. M., Ramos, M., Li, C., Cheung, A. L., and Yeaman, M. R. (1995). *Staphylococus aureus* induces platelet aggregation via a fibrinogen-dependent mechanism which is independent of principal platelet GPIIb/IIIa fibrinogen-binding domains. *Infection and Immunity*, 63, 3634–41.

Booyse, F. and Rafelson, M. E. (1967a). Stable messenger RNA in the synthesis of contractile protein in human platelets. *Biochemica et Biophysica Acta*, 145, 188–92.

Booyse, F. and Rafelson, M. E. (1967b). *In vitro* incorporation of amino acids into the contractile protein of human blood platelets. *Nature (London)*, 215, 283–85.

Calderone, R. A., Rotondo, M. F., and Sande, M. A. (1978). *Candida albicans* endocarditis: Ultrastructural studies of vegetation formation. *Infection and Immunity*, 20, 279–89.

Carney, D. H. (1992). Postclotting cellular effects of thrombin mediated by interaction with high-affinity thrombin receptors. In *Thrombin: Structure and Function*, ed. L. J. Berliner, pp. 351–70. New York: Plenum.

Carroll, S. F. and Martinez, R. J. (1981a). Antibacterial peptide from normal rabbit serum. I. Isolation from whole serum, activity, and microbiologic spectrum. *Biochemistry*, 20, 5973–81.

Carroll, S. F. and Martinez, R. J. (1981b). Antibacterial peptide from normal rabbit serum. II. Compositional microanalysis. *Biochemistry*, 20, 5981–7.

Carroll, S. F. and Martinez, R. J. (1981c). Antibacterial peptide from normal rabbit serum. III. Inhibition of microbial electron transport. *Biochemistry*, 20, 5988–94.

Chang, F. Y., Singh, N., Gayowski, T., Wagener, M. M., Mietzner, S. M., Stout, J. E., and Marino, I. R. (2000). Thrombocytopenia in liver transplant recipients: Predictors, impact on fungal infections, and role of endogenous thrombopoietin. *Transplantation*, 69, 70–5.

Cheung, A. L. and Fischetti, V. A. (1990). The role of fibrinogen in staphylococcal adherence to catheters *in vitro*. *Journal of Infectious Diseases*, 161, 1177–86.

Christin, L, Wyson, D. R., Meshulam, T., Hastey, R., Simons, E. R., and Diamond, R. D. (1996). Mechanisms and target sites of damage in killing of *Candida albicans* hyphae by human polymorphonuclear neutrophils. *Journal of Infectious Diseases*, 176, 1567–78.

Clark, R. A. and Klebanoff, S. J. (1979). Myeloperoxidase-mediated platelet release reaction. *Journal of Clinical Investigation*, 63, 177–83.

Clawson, C. C. 1977. Role of platelets in the pathogenesis of endocarditis. In *Infectious Endocarditis*. American Heart Association Monograph, 52, 24–7.

Clawson, C. C. and White, J. G. (1971). Platelet interaction with bacteria. II. Fate of bacteria. *American Journal of Pathology*, 65, 381–98.

Cocchi, F., DeVico, A. L., Garzino-Demo, A., Arya, S. K., Gallo, R. C., and Lusso, P. (1995). Identification of RANTES, MIP-1α, and MIP-1β as the major HIV-suppressive factors produced by $CD8^+$ T cells. *Science,* 270, 1811–15.

Cole, A. M., Ganz, T., Liese, A. M., Burdick, M. D., Liu, L., and Strieter, R. M. (2001). IFN-inducible ELR-CXC chemokines display defensin-like antimicrobial activity. *Journal of Immunology,* 167, 623–7.

Colman, R. W. (1991). Receptors that activate platelets. *Proceedings of the Society for Experimental Biology and Medicine,* 197, 242–8.

Dankert, J., Krijgsveld, J., van Der Werff, J., Joldersma, W., and Zaat, S. A. (2001). Platelet microbicidal activity is an important defense factor against viridans streptococcal endocarditis. *Journal of Infectious Diseases,* 184, 597–605.

Dankert, J., van der Werff, J., Zaat, S. A. J., Joldersma, W., Klein, D., and Hess, J. (1995). Involvement of bactericidal factors from thrombin-stimulated platelets in clearance of adherent viridans streptococci in experimental infective endocarditis. *Infection and Immunity,* 63, 663–71.

Darveau, R. P., Blake, J., Seachord, C. L., Cosand, W. L., Cunninigham, M. D., Cassiano-Cough, L., and Maloney, G. (1992). Peptide related to the carboxy-terminus of human platelet factor IV with antibacterial activity. *Journal of Clinical Investigation,* 90, 447–55.

Davies, T. A., Fine, R. E., Johnson, R. J., Levesque, C. A., Rathbun, W. H., Seetoo, K. F., Smithe, S. J., Strohmeier, G., Volicer, L., Delva, L., and Simons, E. R. (1993). Non-age related differences in thrombin responses by platelets from male patients with advanced Alzheimer's disease. *Biochemical and Biophysical Research Communications,* 194, 537–43.

Day, H. J. and Rao, A. K. (1986). Evaluation of platelet function. *Seminars in Hematology* 23, 89–101.

Day, H. J., Ang, G. A. T., and Holmsen, H. (1972). Platelet release reaction during clotting of native human platelet rich plasma. *Proceedings of the Society for Experimental Biology and Medicine,* 139, 717–21.

Deuel, T. F., Senior, R. M., Chang, D., Griffith, G. L., Heinrikson, R. L., and Kaiser, E. T. (1981). Platelet factor-4 is chemotactic for neutrophils and monocytes. *Proceedings of the National Acadademy of Sciences USA,* 78, 4548–87.

Dhawan, V. K., Bayer, A. S., and Yeaman, M. R. (1998). *In vitro* resistance to thrombin-induced platelet microbicidal protein is associated with enhanced progression and hematogenous dissemination in experimental *Staphylococcus aureus* infective endocarditis. *Infection and Immunity,* 66, 3476–9.

Dhawan, V. K., Yeaman, M. R., and Bayer, A. S. (1999). Influence of *in vitro* susceptibility phenotype against thrombin-induced platelet microbicidal protein on treatment and prophylaxis outcomes of experimental *Staphylococcus aureus* endocarditis. *Journal of Infectious Diseases,* 180, 1561–8.

Dhawan, V. K., Bayer, A. S., and Yeaman, M. R. (2000). Thrombin-induced platelet microbicidal protein susceptibility phenotype influences the outcome of oxacillin prophylaxis and therapy of experimental *Staphylococcus aureus* endocarditis. *Antimicrobial Agents and Chemotherapy*, 44, 3206–9.

Donaldson, D. M. and Tew, J. G. (1977). β-lysin of platelet origin. *Bacteriological Reviews*, 41, 501–12.

Drake, T. A. and Pang, M. (1988). *Staphylococcus aureus* induces tissue factor expression in cultured human cardiac valve endothelium. *Journal of Infectious Diseases*, 157, 749–56.

Drake, T. A., Rodgers, G. M., and Sande, M. A. (1984). Tissue factor is a major stimulus for vegetation formation in enterococcal endocarditis in rabbits. *Journal of Clinical Investigation*, 73, 1750–53.

Durack, D. T., Beeson, P. B., and Petersdorf, R. G. (1973). Experimental bacterial endocarditis. III. Production of progress of the disease in rabbits. *British Journal of Experimental Pathology*, 54, 142–51.

Essien, E. M. and Ebhota, M. I. (1983). Platelet secretory activities in acute malaria (*Plasmodium falciparum*) infection. *Acta Haematologica*, 70, 183–8.

Feldman, C., Kallenbach, J. M., Levy, H., Thorburn, J. R., Hurwitz, M. D., and Koornhof, H. J. (1991). Comparision of bacteraemic community-acquired lobar pneumonia due to *Streptococcus pneumoniae* and *Klebsiella pneumoniae* in an intensive care unit. *Respiration*, 58, 265–70.

Fields, P. L., Groisman, E. A., and Heffron, F. (1989). A *Salmonella* locus that controls resistance to microbicidal proteins from phagocytic cells. *Science*, 243, 1059–62.

Filler, S. G., Joshi, M., Phan, Q. T., Diamond, R. D., Edwards Jr., J. E. E., and Yeaman, M. R. (1999). Platelets protect vascular endothelial cells from injury due to *Candida albicans*. Abstract 2163, In abstracts of Thirty-Ninth Conference on Antimicrobial Agents and Chemotherapy. American Society for Microbiology, San Francisco, CA.

Fodor, J. (1887). Die fahigkeit des blutes bakterien zu vernichten. *Deutsches Medizinische Wochenschrift*, 13, 745–7.

Fowler, V. G., McIntyre, L. M., Yeaman, M. R., Peterson, G. E., Reller, L. B., Corey, G. R., Wray, D., and Bayer, A. S. (2000). *In vitro* resistance to thrombin-induced platelet microbicidal protein in isolates of *Staphylococcus aureus* from endocarditis patients correlates with an intravascular device source. *Journal of Infectious Diseases*, 182, 1251–4.

Frohm, M., Gunne, H., Bergman, A.-C., Agerberth, B., Bergman, T., Boman, A., Lidén, S., Jörnvall, H., and Boman, H. G. (1996). Biochemical and antibacterial analysis of human wound and blister fluid. *European Journal of Biochemistry*, 237, 86–92.

Gengou, O. (1901). De l'origine de l'axenine de serums normaux. *Annales de L'Institut Pasteur (Paris)*, 15, 232–45.

Ginsburg, I. 1988. The biochemistry of bacteriolysis: Facts, paradoxes, and myths. *Microbiological Science*, 5, 137–42.

Groisman, E. A., Parra-Lopez, C., Salcedo, M., Lipps, C. J., and Heffron, F. (1992). Resistance to host antimicrobial peptides in necessary for *Salmonella* virulence. *Proceedings of the National Academy of Science USA*, 89, 11939–43.

Harker, L. A. and Finch, C. A. (1969). Thrombokinetics in man. *Journal of Clinical Investigation*, 48, 963–9.

Herzberg, M. C., Gong, K., MacFarlane, G. D., Erickson, P. R., Soberay, A. H., Krebsbach, P. H., Gopalraj, M., Schilling, K., and Bowen, W. H. (1990). Phenotypic characterization of *Streptococcus sanguis* virulence factors associated with bacterial endocarditis. *Infection and Immunity*, 58, 515–22.

Heyssel, R. M. (1961). Determination of human platelet survival utilizing ^{14}C-labelled serotonin. *Journal of Clinical Investigation*, 40, 2134–8.

Hostacka, A. (1998). Serum sensitivity and cell surface hydrophobicity of *Klebsiella pneumoniae* treated with gentamicin, tobramycin, and amikacin. *Journal of Basic Microbiology*, 38, 383–8.

Jago, R. and Jacox, R. F. (1961). Cellular source and character of a heat-stable bactericidal property associated with rabbit and rat platelets. *Journal of Experimental Medicine*, 113, 701–9.

Johnson, F. B. and Donaldson, D. M. (1968). Purification of staphylocidal β-lysin from rabbit serum. *Journal of Bacteriology*, 96, 589–95.

Kirkpatrick, B., Reeves, D. S., and MacGowan, A. P. (1994). A review of the clinical presentation, laboratory features, antimicrobial therapy and outcome of 77 episodes of pneumococcal meningitis occurring in children and adults. *Journal of Infection*, 29, 171–82.

Klinger, M. H. F. and Jelkmann, W. (2002). Role of blood platelets in infection and inflammation. *Journal of Interferon and Cytokine Research*, 22, 913–22.

Klotz, S. A., Harrison, J. L., and Misra, R. P. (1989). Aggregated platelets enhance adherence of *Candida* yeasts to endothelium. *Journal of Infectious Diseases*, 160, 669–77.

Koo, S. P., Bayer, A. S., Kagan, B. L., and Yeaman, M. R. (1999). Membrane permeabilization by thrombin-induced PMP-1 is modulated by transmembrane voltage polarity and magnitude. *Infection and Immunity*, 67, 2475–81.

Koo, S. P., Bayer, A. S., Sahl, H. G., Proctor, R. A., and Yeaman, M. R. (1996). Staphylocidal action of thrombin-induced platelet microbicidal protein is not solely dependent on transmembrane potential. *Infection and Immunity*, 64, 1070–4.

Koo, S. P., Yeaman, M. R., and Bayer, A. S. (1996). Staphylocidal action of thrombin-induced platelet microbicidal protein is influenced by microenvironment and target cell growth phase. *Infection and Immunity*, 64, 3758–64.

Korzweniowski, O. M., Scheld, W. M., Bithell, T. C., Croft, B. H., and Sande, M. A. (1979). The effect of aspirin on the production of experimental *Staphylococcus aureus* endocarditis [abstract]. In *Program Abstracts of the Nineteenth Interscience Conference on Antimicrobial Agents and Chemotherapy (Boston)*. Washington, D.C.: American Society for Microbiology.

Krijgsveld, J., Zaat, S. A., Meeldijk, J., van Veelen, P. A., Fang, G., Poolman, B., Brandt, E., Ehlert, J. E., Kuijpers, A. J., Engbers, G. H., Feijen, J., and Dankert, J. (2000). Thrombocidins, microbicidal proteins from human blood platelets, are C-terminal deletion products of CXC chemokines. *Journal of Biological Chemistry*, 275, 20374–81.

Kupferwasser, L. I., Yeaman, M. R., Shapiro, S. M., Nast, C. C., Sullam, P. M., Filler, S. G., and Bayer, A. S. (1999). Acetylsalicylic acid reduces vegetation bacterial density, hematogenous bacterial dissemination, and frequency of embolic events in experimental *Staphylococcus aureus* endocarditis through antiplatelet and antibacterial effects. *Circulation*, 99, 2791–97.

Lorenz, R. and Brauer, M. (1988). Platelet factor 4 (PF-4) in septicaemia. *Infection*, 16, 273–6.

MacFarlane, D. E. and Mills, D. C. B. (1975). The effects of ATP on platelets: Evidence against the central role of released ADP in primary aggregation. *Blood*, 46, 309–14.

MacFarlane, D. E., Walsh, P. N., Mills, D. C. B., Holmsen, H., and Day, H. J. (1975). The role of thrombin in ADP-induced platelet aggregation and release: A critical evaluation. *British Journal of Haematology*, 30, 457–64.

Mandell, G. L. and Hook, E. W. (1969). The interaction of platelets, *Salmonella*, and mouse peritoneal macrophages. *Proceedings of the Society for Experimental Biology and Medicine*, 132, 757–9.

Marcus, A. J. (1969). Platelet function. *New England Journal of Medicine*, 280, 1213, 1278, 1330.

Mavrommatis, A. C., Theodoridis, T., Orfanidou, A., Roussos, C., Christopoulou-Kokkinou, V., and Zakynthinos, S. (2000). Coagulation system and platelets are fully activated in uncomplicated sepsis. *Critical Care Medicine*, 28, 451–7.

Mayo, K. H. and Chen, M. J. (1989). Human platelet factor 4 monomer-dimer-tetramer equilibria investigated by ^{1}H NMR spectroscopy. *Biochemistry*, 28, 9469–78.

Mehta, P., Mehta, J., and Lawson, D. (1986). Leukotrienes potentiate the effects of epinephrine and thrombin on human platelet-aggregation. *Thrombosis Research*, 41, 731–8.

Mercier, R. C., Rybak, M. J., Bayer, A. S., and Yeaman, M. R. (2000). Influence of platelets and platelet microbicidal protein susceptibility on the fate of *Staphylococcus aureus* in an *in vitro* model of infective endocarditis. *Infection and Immunity*, 68, 4699–705.

Meyer, M. W., Gong, K., and Herzberg, M. C. (1998). *Streptococcus sanguis*-induced platelet clotting in rabbits and hemodynamic and cardiopulmonary consequences. *Infection and Immunity*, 66, 5906–14.

Mezzano, S., Burgos, M. E., Ardiles, L., Olavarria, F., Concha, M., Caorsi, I., Aranda, E., and Mezzano, D. (1992). Glomerular localization of platelet factor 4 in streptococcal nephritis. *Nephrology*, 61, 58–63.

Miglioli, P. A., Schoffel, U., and Gianfranceschi, L. (1996). The *in vitro* synergistic inhibitory effect of human amniotic fluid and gentamicin on growth of *Escherichia coli*. *Chemotherapy* 42, 206–9.

Moyer, B. R., Vallabhajosula, S., Lister-James, J., Bush, L. R., Cyr, J. E., Snow, D. A., Bastidas, D., Lipszyc, H., and Dean, R. T. (1996). Technetium-99m-white blood cell-specific imaging agent developed from platelet factor 4 to detect infection. *Journal of Nuclear Medicine*, 37, 673–9.

Murphy, E. A., Robinson, G. A., and Rowsell, A. (1967). The pattern of platelet disappearance. *Blood*, 30, 26–31.

Myrvik, Q. N. (1956). Serum bactericidins active against Gram-positive bacteria. *Annals of the New York Academy of Science*, 66, 391–400.

Myrvik, Q. N. and Leake, E. S. (1960). Studies on antibacterial factors in mammalian tissues and fluids. IV. Demonstration of two non-dialyzable components in the serum bactericidin system for *Bacillus subtilis*. *Journal of Immunology*, 84, 247–50.

Nachum, R., Watson, S. W., Sullivan Jr., J. D., and Seigel, S. E. (1980). Antimicrobial defense mechanisms in the horseshoe crab, *Limulus polyphemus*: Preliminary observations with heat-derived extracts of *Limulus* amoebocyte lysate. *Journal of Invertebrate Pathology*, 32, 51–8.

Nicolau, D. P., Freeman, C. D., Nightingale, C. H., Quintiliani, R., Coe, C. J., Maderazo, E. G., and Cooper, B. W. (1993). Reduction of bacterial titers by low-dose aspirin in experimental aortic valve endocarditis. *Infection and Immunity*, 61, 1593–5.

Oh, J. E., Hong, S. Y., and Lee, K. H. (1999). Structure-activity relationship study: Short antimicrobial peptides. *Journal of Peptide Research*, 53, 41–6.

Oppenheim, J. J., Zachariae, C. O. C., Mukaida, N., and Matsushima, K. (1991). Properties of the novel proinflammatory supergene "intercrine" cytokine family. *Annual Review of Immunology*, 9, 617–48.

Osler, W. (1886). On certain problems in the physiology of the blood corpuscles. *Medical News* 48, 365–70, 393–9, 421–5.

Peck-Miller, K. A., Darveau, R. P., and Fell, H. P. (1993). Identification of serum components that inhibit the tumorcidal activities of amphiphilic alpha-helical peptides. *Cancer Chemotherapy and Pharmacology*, 32, 109–15.

Piguet, P. F., Vesin, C., Ryser, J. E., Senaldi, G., Grau, G. F., and Tachini-Cottier, F. (1993). An effector role for platelets in systemic and local lipopolysaccharide-induced toxicity in mice, mediated by a CD11a- and CD54a-dependent interaction with endothelium. *Infection and Immunity*, 61, 82–7.

Pruul, H. and McDonald, P. J. (1992). Potentiation of antibacterial activity of azithromycin and other macrolides by normal human serum. *Antimicrobial Agents and Chemotherapy*, 36, 10–16.

Roberts, W. C. and Buchbinder, N. A. (1972). Right-sided valvular infective endocarditis: a clinicopathologic study of 12 necropsy patients. *American Journal of Medicine*, 53, 7–19.

Rucinski, B., Niewiarowski, S., Strzyzewski, M., Holt, J. C., and Mayo, K. H. (1990). Human platelet factor 4 and its C-terminal peptides: Heparin binding and clearance from the circulation. *Thrombosis and Haemostasis*, 63, 493–8.

Santolaya, M. E., Alvarez, A. M., Aviles, C. L., Becker, A., Cofre, J., Enriquez, N., O'Ryan, M., Paya, E., Salgado, C., Silva, P., Tordecilla, J., Varas, M., Villarroel, M., Viviani, T., and Zubieta, M. (2002). Prospective evaluation of a model of prediction of invasive bacterial infection risk among children with cancer, fever, and neutropenia. *Clinical Infectious Diseases*, 35, 678–83.

Scheld, W. M., Valone, J. A., and Sande, M. A. (1978). Bacterial adherence in the pathogenesis of infective endocarditis: interaction of dextran, platelets, and fibrin. *Journal of Clinical Investigation*, 61, 1394–1404.

Shafer, W. M., Martin, L. E., and Spitznagel, J. K. (1986). Late intraphagosomal hydrogen ion concentration favors the *in vitro* antimicrobial capacity of a 37-kilodalton cationic granule protein of human neutrophil granulocytes. *Infection and Immunity*, 53, 651–5.

Shahan, T. A., Sorenson, W. G., Paulauskis, J. D., Morey, R., and Lewis, D. M. (1998). Concentration- and time-dependent upregulation and release of the cytokines MIP-2, KC, TNF, and MIP-1-α in rat alveolar macrophages by fungal spores implicated in airway inflammation. *American Journal of Respiratory Cell and Molecular Biology*, 18, 435–40.

Smith, C. W. (1993). Leukocyte-endothelial cell interaction. *Seminars in Hematology*, 30, 45–55.

Spitznagel, J. K. (1984). Non-oxidative antimicrobial reactions of leukocytes. *Contemporary Topics in Immunobiology*, 14, 283–343.

Srichaikul, T., Nimmannitya, S., Sripaisarn, T., Kamolsilpa, M., and Pulgate, C. (1989). Platelet function during the acute phase of dengue hemorrhagic fever. *Southeast Asian Journal of Tropical Medicine and Public Health*, 20, 19–25.

Sullam, P. M., Frank, U., Yeaman, M. R., Tauber, M. G., Bayer, A. S., and Chambers, H. F. (1993). Effect of thrombocytopenia on the early course of streptococcal endocarditis. *Journal of Infectious Diseases*, 168, 910–14.

Tang, Y. Q., Yeaman, M. R., and Selsted, M. E. (2002). Antimicrobial peptides from human platelets. *Infection and Immunity*, 70, 6524–33.

Tew, J. G., Roberts, R. R., and Donaldson, D. M. (1974). Release of β-lysin from platelets by thrombin and by a factor produced in heparinized blood. *Infection and Immunity*, 9, 179–86.

Tocantins, L. M. (1938). The mammalian blood platelet in health and disease. *Medicine*, 17, 155–257.

Trier, D. A., Bayer, A. S., and Yeaman, M. R. (2000). *Staphylococcus aureus* elicits antimicrobial responses from platelets via an ADP-dependent pathway. Abstract 1010. In Program Abstracts of the Fortieth Interscience Conference on Antimicrobial Agents and Chemotherapy. Washington, D.C.: American Society for Microbiology.

Turner, R. B., Liu, L., Sazonova, I. Y., and Reed, G. L. (2002). Structural elements that govern the substrate specificity of the clot-dissolving enzyme plasmin. *Journal of Biological Chemistry*, 277, 33068–74.

Tzeng, D. Y., Deuel, T. F., Huang, J. S., and Baehner, R. L. (1985). Platelet-derived growth factor promotes polymorphonuclear leukocyte activation. *Blood*, 4, 1123–28.

Vaara, M., Viljanen, P., Vaara, T., and Makela, P. H. (1984). An outer membrane-disorganizing peptide PMBN sensitizes *Escherichia coli* to serum bactericidal action. *Journal of Immunology*, 132, 2582–9.

Viljanen, P., Kayhty, H., Vaara, M., and Vaara, T. (1986). Susceptibility of Gram-negative bacteria to the synergistic bactericidal action of serum and polymyxin B nonapeptide. *Canadian Journal of Microbiology*, 32, 66–9.

Viscoli, C, Bruzzi, P., Castagnola, E., Boni, L., Calandra, T., Gaya, H., Meuneir, F., Feld, R., Zinner, S., Klastersky J., et al. (1994). Factors associated with bacteraemia in febrile, granulocytopenic patients. The International Antimicrobial Therapy Cooperative Group (IATCG) of the European Organization for Research and Treatment of Cancer (EORTC). *European Journal of Cancer*, 30, 430–7.

Wallis, R. S., Palaci, M., Vinhas, S., Hise, A., Ribeiro, F. C., Landen, K., Cheon, S. H., Song, H. Y., Phillips, M., Dieteze, R., and Ellner, J. J. (2001). A whole blood bactericidal assay for tuberculosis. *Journal of Infectious Diseases*, 183, 1300–3.

Walz, A., Dewald, B., von Tscharner, V., and Baggiolini, M. (1989). Effects of neutrophil-activating peptide NAP-2, platelet basic protein, connective tissue-activating peptide III, and platelet factor 4 on human neutrophils. *Journal of Experimental Medicine*, 170, 1745–50.

Warshaw, A. L., Laster, L., and Shulman, N. R. (1967). Protein synthesis by human platelets. *Journal of Biological Chemistry*, 242, 2094–100.

Weiss, J., Bayer, A. S., and Yeaman, M. R. (2000). Antimicrobial peptides in host defense against *Staphylococcus aureus* and other Gram-positive pathogens. In *Gram-Positive Pathogens*, ed. V. Fischetti, R. Novick, J. Ferretti, D. Portnoy, and J. Rood, pp. 431–41. Washington, D.C.: American Society for Microbiology Press.

Weksler, B. B. and Nachman, R. L. (1971). Rabbit platelet bactericidal protein. *Journal of Experimental Medicine*, 134, 1114–30.

Welling, M. M., Lupetti, A., Balter, H. S., Lanzzeri, S., Souto, B., Rey, A. M., Savio, E. O., Paulusma-Annema, A., Pauwels, E. K., and Nibbering, P. H. (2001). 99mTc-labeled antimicrobial peptides for detection of bacterial and *Candida albicans* infections. *Journal of Nuclear Medicine*, 42, 788–94.

White, J. G. (1972). Platelet morphology and function. In *Hematology*, ed. W. J. Williams, E. Beutler, A. J. Erslev, and R. W. Rundles, pp. 1023–39. New York: McGraw-Hill.

Wilson, M., Blum, R., Dandona, P., and Mousa, S. (2001). Effects in humans of intravenously administered endotoxin on soluble cell-adhesion molecule and inflammatory markers: a model of human diseases. *Clinical and Experimental Pharmacology and Physiology*, 28, 376–80.

Wu, T., Yeaman, M. R., and Bayer, A. S. (1994). *In vitro* resistance to platelet microbicidal protein correlates with endocarditis source among staphylococcal isolates. *Antimicrobial Agents and Chemotherapy*, 38, 729–32.

Xiong, Y. Q., Bayer, A. S., and Yeaman, M. R. (2002). Inhibition of *Staphylococcus aureus* intracellular macromolecular synthesis by thrombin-induced platelet microbicidal proteins. *Journal of Infectious Diseases*, 185, 348–56.

Xiong, Y. Q., Caillon, J., Drugeon, H., Potel, G., and Baron, D. (1996). Influence of pH on adaptive resistance of *Pseudomonas aeruginosa* to aminoglycosides and their post-antibiotic effects. *Antimicrobial Agents and Chemotherapy*, 40, 35–39.

Xiong, Y. Q., Yeaman, M. R., and Bayer, A. S. (1999). *In vitro* antibacterial activities of platelet microbicidal protein and neutrophil defensin against *Staphylococcus aureus* are influenced by antibiotics differing in mechanism of action. *Antimicrobial Agents and Chemotherapy*, 43, 1111–17.

Yamamoto, Y., Klein, T. W., and Friedman, H. (1997). Involvement of mannose receptor in cytokine interleukin-1beta (IL-1beta), IL-6, and granulocyte-macrophage colony-stimulating factor responses, but not in chemokine macrophage inflammatory protein 1beta (MIP-1beta), MIP-2, and KC responses, caused by attachment of *Candida albicans* to macrophages. *Infection and Immunity*, 65, 1077–82.

Yan, H. and Hancock, R. E. W. (2001). Synergistic interactions between mammalian antimicrobial defense peptides. *Antimicrobial Agents and Chemotherapy*, 45, 1558–60.

Yeaman, M. R. 1997. The role of platelets in antimicrobial host defense. *Clinical Infectious Diseases*, 25, 951–70.

Yeaman, M. R. and Bayer, A. S. (1999). Antimicrobial peptides from platelets. *Drug Resistance Updates*, 2, 116–26.

Yeaman, M. R. and Bayer, A. S. (2000). *Staphylococcus aureus*, platelets, and the heart. *Current Infectious Disease Reports*, 2, 281–98.

Yeaman, M. R., Bayer, A. S., Koo, S. P., Foss, W., and Sullam, P. M. (1998). Platelet microbicidal proteins and neutrophil defensin disrupt the *Staphylococcus aureus* cytoplasmic membrane by distinct mechanisms of action. *Journal of Clinical Investigation*, 101, 178–87.

Yeaman, M. R., Cheng, D., Desai, B., Kupferwasser, L. I., Xiong, Y. Q., Edwards Jr., J. E., and Bayer, A. S. (1998). Influence of fluconazole therapy on infective endocarditis due to *Candida albicans* exhibiting *in vitro* susceptibility or resistance to platelet microbicidal protein. Abstract No. (V-83). In abstracts of the ninety-eighth General Meeting of the American Society for Microbiology, Atlanta, GA.

Yeaman, M. R., Ibrahim, A., Filler, S. G., Bayer, A. S., Edwards Jr., J. E., Bayer, A. S., and Ghannoum, M. A. (1993). Thrombin-induced rabbit platelet microbicidal protein is fungicidal *in vitro*. *Antimicrobial Agents and Chemotherapy*, 37, 546–53.

Yeaman, M. R., Ibrahim, A. S., Ritchie, J. A., Filler, S. G., Bayer, A. S., Edwards Jr., J. E., and Ghannoum, M. A. (1995). Sublethal concentrations of platelet microbicidal protein reduce *Candida albicans* interactions with vascular endothelial cells *in vitro*. Abstract 491. 1995 Infectious Diseases Society of American Annual Meeting, San Francisco, CA.

Yeaman, M. R., Norman, D. C., and Bayer, A. S. (1992a). Platelet microbicidal protein enhances antibiotic-induced killing of and post-antibiotic effect in *Staphylococcus aureus*. *Antimicrobial Agents and Chemotherapy*, 36, 1665–70.

Yeaman, M. R., Norman, D. C., and Bayer, A. S. (1992b). *Staphylococcus aureus* susceptibility to thrombin-induced platelet microbicidal protein is independent of platelet adherence or aggregation *in vitro*. *Infection and Immunity*, 60, 2368–74.

Yeaman, M. R., Puentes, S. M., Norman, D. C., and Bayer, A. S. (1992). Partial purification and staphylocidal activity of thrombin-induced platelet microbicidal protein. *Infection and Immunity*, 60, 1202–9.

Yeaman, M. R., Soldan, S. S., Ghannoum, M. A., Edwards Jr., J. E., Filler, S. G., and Bayer, A. S. (1996). Resistance to platelet microbicidal protein results in increased severity of experimental *Candida albicans* endocarditis. *Infection and Immunity*, 64, 1379–84.

Yeaman, M. R., Sullam, P. M., Dazin, P. F., and Bayer, A. S. (1994). Platelet microbicidal protein alone or in combination with antibiotics reduces *Staphylococcus aureus* adherence to platelets *in vitro*. *Infection and Immunity*, 62, 3416–23.

Yeaman, M. R., Sullam, P. M., Dazin, P. F., Norman, D. C., and Bayer, A. S. (1992). Characterization of *Staphylococcus aureus*-platelet interaction by quantitative flow cytometry. *Journal of Infectious Diseases*, 166, 65–73.

Yeaman, M. R., Tang, Y.-Q., Shen, A. J., Bayer, A. S., and Selsted, M. E. (1997). Purification and *in vitro* activities of rabbit platelet microbicidal proteins. *Infection and Immunity*, 65, 1023–31.

Yeaman, M. R., Gank, K. D., Bayer, A. S., and Brass, E. P. (2002). Synthetic peptides that exert antimicrobial activities in whole blood and blood-derived matrices. *Antimicrobial Agents and Chemotherapy*, 46: 3883–91.

Yeaman, M. R. and Yount, N. Y. (2003). Mechanisms of antimicrobial peptide action and resistance. *Pharmacologogical Reviews*, 54, 27–55.

Yount, N. Y., Gank, K. D., Xiong, Y. Q., Bayer, A. S., Welch, W. H., and Yeaman, M. R. (2003). Rabbit tPMP-1 is a structural and functional immunoanalogue of human platelet factor-4. Abstract 1991. In abstracts of one-hundred third General Meeting of the American Society for Microbiology, Washington, D.C.

CHAPTER 12

Mechanisms of bacterial resistance to antimicrobial peptides

R. Tamayo, A. C. Portillo, and J. S. Gunn

12.1. INTRODUCTION

Antimicrobial peptides, as a part of the human innate immune system, are a central component in the battle against invading pathogens. Antimicrobial peptides are present at skin and mucosal surfaces (including the upper and lower airways, gastrointestinal tract, and genitourinary tract), which are key locations of microbial invasion. As discussed in earlier chapters, antimicrobial peptides present at these sites participate in microbial killing and can be expressed constitutively or induced upon insult or injury to host tissues. If invading organisms are not eliminated by the innate immune system at the skin or mucosal surfaces, they are faced with another barrage of antimicrobial factors (including antimicrobial peptides) within phagocytic cells. Over the course of time, because of the constant interaction of human pathogens with antimicrobial peptides of the host, many microbes have evolved mechanisms of resistance to the action of these killer peptides (see Fig. 12.1 for an overview). Much like the response of the host to these invading organisms, the microbial antimicrobial peptide resistance mechanisms can be both constitutive and inducible, and the expression of these resistance phenotypes likely aids in the development and/or maintenance of disease (Table 12.1). In this chapter, we present the mechanisms of resistance to antimicrobial peptides that have been uncovered for various bacterial pathogens. Although the pathogens discussed will have preferences for different anatomic sites of invasion or colonization, various modes of pathogenesis, and different membrane structure, surprising commonalities will be revealed in their mechanisms of antimicrobial peptide resistance.

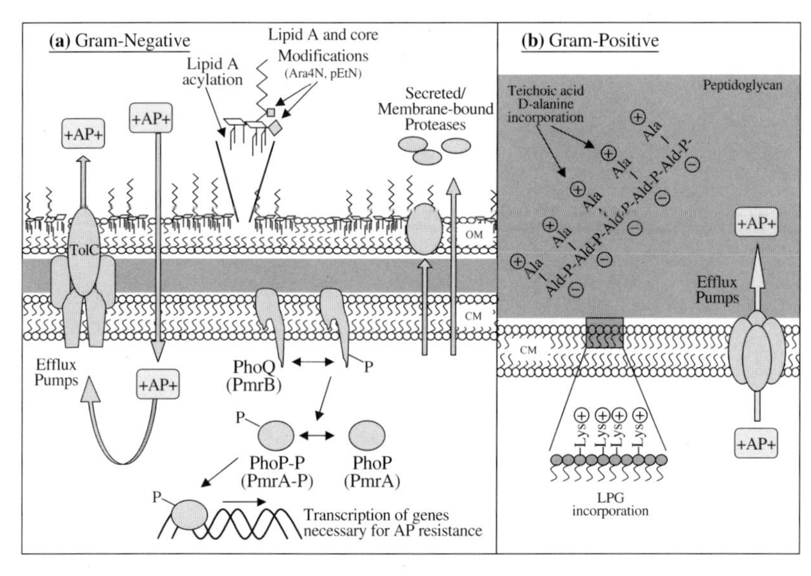

Figure 12.1. Known antimicrobial peptide resistance mechanisms in Gram-positive and Gram-negative bacteria. (a) Gram-negative bacteria express antimicrobial peptide (AP) resistance mechanisms involving proteolysis (e.g., *Salmonella enterica* serovar Typhimurium PgtE, *Escherichia coli* OmpT), efflux (e.g., *Neisseria gonorrhoeae* Mtr, *S. enterica* serovar Typhimurium Sap transporters), and regulated modifications to lipopolysaccharide (LPS) that alter initial AP–bacterium interactions. In *S. enterica* serovar Typhimurium, activation of the PhoPQ two-component regulatory system leads to acylation of lipid A with palmitate by PagP. Induction of the PmrAB two-component system, which functions similarly to PhoPQ, leads to transcription of genes involved in aminoarabinose (Ara4N) addition to lipid A (*pmrE, pmrHFIJKLM*) and phosphoethanolamine to lipid A and core (*pmr?*). Other Gram-negative bacteria also constitutively add Ara4N to the LPS. (b) Gram-positive bacteria such as *S. aureus* continuously produce substitutions to teichoic acids and lipoteichoic acids (*dltABCD*-mediated D-Ala addition) and phospholipids (*mprF*-mediated lysine addition) that are essential for AP resistance. Efflux pumps may also play a role in antimicrobial peptide resistance in Gram-positive bacteria (*S. aureus* QacA).

12.2. SALMONELLAE

The Salmonellae are acquired from contaminated food or water and interact initially with the mucosa of the gastrointestinal tract. Certain serotypes of *Salmonella* spp., which are facultative intracellular pathogens, are capable of causing systemic infections by penetrating into tissues and replicating within phagocytic cells, particularly in the liver and spleen. Intracellular pathogens must withstand the antimicrobial properties of a variety of cationic peptides

Table 12.1. *Summary of antimicrobial peptide resistance mechanisms*

Mechanism of antimicrobial peptide resistance	Pathogens utilizing mechanism[1]	Locus (loci) involved
Inducible LPS modification	*Salmonella*	*pagP*- palmitoylation of LPS
- PhoPQ mediated	*Escherichia coli*	*phoPQ*- 2-CRS[2]
	Shigella	
	Pseudomonas	
	Yersinia	
- PmrAB mediated[3]	*Salmonella*	*pmrAB*[4]- 2-CRS
	E. coli	*pmrE (pagA/ugd)*- Ara4N addition
		pmrHFIJKLM (*arnBCA-T–/pbgP1-4E1-3*)- Ara4N addition
Constitutive LPS modifications[5]	*Burkholderia*	*waaF*- LPS synthesis
		pmrE- Ara4N addition to LPS
		lytB- phospholipid/ peptidoglycan synthesis
	Proteus	Ara4N addition to LPS
		sap locus[6]
		A subunit of F^1F^0 ATPase[6]
		O-acetyltransferase[6]
Efflux pumps	*Staphylococcus*	*qacA*- resistance to tPMP1
	Yersinia	*rosAB*- activated at 37°C or by AP
	Neisseria	*mtrAB*- resistance to LL-37 and protegrin, bile salts, and fatty acids
	Salmonella	*sapABCDF*- amino acid similarity to other transporters
WTA and LTA modifications	*Staphylococcus*	*dltABCD*- D-alanine addition to teichoic acids
Phospholipid modifications	*Staphylococcus*	*mprF*- modification of phospholipids with L-lysine (esterification to create LPG)

(*cont.*)

Table 12.1. (*cont.*)

Mechanism of antimicrobial peptide resistance	Pathogens utilizing mechanism[1]	Locus (loci) involved
Membrane or secreted proteases	*E. coli*	*ompT*- outer membrane protease
	Salmonella	*pgtE*[4]- homologous to OmpT

[1]Pathogens include only those described in published manuscripts.
[2]2-CRS, two-component regulatory system.
[3]Although many organisms add Ara4N and pEtN, these additions are not known to be PmrAB-mediated except in *Escherichia coli* and *Salmonella enterica* serovar Typhimurium.
[4]Also activated by PhoPQ in *Salmonella enterica* serovar Typhimurium
[5]Constitutive; modifications have not been shown to be induced by in vivo conditions
[6]Homologous to previously identified genes, but remain unnamed in *Proteus.*

that are a part of the oxygen-independent arsenal of professional phagocytes. Consequently it may be presumed that Salmonellae have evolved mechanisms by which to evade killing by the host innate immune system in order to survive within the host.

12.2.1. Identification of PhoP-PhoQ (PhoPQ) as a Major *Salmonella* Virulence Factor

Fields, Groisman, and Heffron (1989) published the discovery of a *Salmonella typhimurium* gene required for resistance to antimicrobial peptides, intracellular survival, and virulence. It was shown that crude extracts from rabbit peritoneal macrophages and from human neutrophils, as well as purified rabbit defensin NP1, had much greater bactericidal effects on this mutant than on the parent strain. Further analysis indicated that the transposon insertion in this mutant resided in the *phoP* gene, leading to the characterization of the PhoPQ two-component regulatory system, in which PhoQ is the inner membrane sensor and PhoP is the transcriptional activator (Groisman et al., 1989; Miller, Kukral, and Mekalanos, 1989).

The PhoPQ regulatory system, now recognized as a major virulence determinant in *S. typhimurium* (now reclassified as *Salmonella enterica* serovar Typhimurium) and a focal point in pathogenesis research, has been linked to pleiotropic phenotypes. Inactivation of *phoPQ* results in attenuated virulence in mice (Fields et al., 1986), deficiencies in cell invasion (Behlau and Miller, 1993), inability to survive within macrophage phagosomes (Fields et al., 1986), sensitivity to low pH growth conditions (Foster and Hall, 1990), inability to grow with succinate as a sole carbon source (Groisman et al., 1989), increased sensitivity to bile (van Velkinburgh and Gunn, 1999), and increased sensitivity to a number of antimicrobial peptides (e.g., defensins, magainins, cecropins, melittin, mastoparan, polymyxins, and neutrophil peptides CAP-37 and CAP-57) (Fields et al., 1989; Groisman et al., 1989; Miller et al., 1990; Groisman et al., 1992; Gunn and Miller, 1996).

12.2.2. PhoP-Regulated LPS Modification and Antimicrobial Peptide Resistance

As the necessity of PhoPQ in resistance to host antimicrobial peptides became apparent, much research became geared toward characterizing the mechanism(s) by which activation of PhoPQ by specific host cues resulted in increased resistance to antimicrobial peptides. The lipopolysaccharides (LPSs) of Gram-negative bacteria are the most exposed molecules on the cell surface, and are thus the primary interface between the bacterium and antimicrobial peptides. It is believed that the anionic charges that pervade the lipid A and core oligosaccharide fractions of LPS electrostatically attract cationic antimicrobial peptides. The close association between antimicrobial peptide and bacterium then would allow the antimicrobial peptide to disorganize/permeabilize the outer membrane and, ultimately, permit access of the antimicrobial peptide to the cytoplasmic membrane. The permeability and charge of the cell envelope are thus key factors in bacterial susceptibility to antimicrobial peptides.

A number of PhoPQ-regulated genes have been identified that encode proteins that modify LPS, resulting in reduced negative charge of LPS or altered permeability of the outer membrane. Because *phoP* is induced in mice and within macrophage phagosomes (Alpuche-Aranda et al., 1992; Heithoff et al., 1997), and because PhoPQ responds to specific activating factors likely to be encountered in the host (e.g., low Mg^{2+} and Ca^{2+} conditions) (Garcia-Vescovi, Soncini, and Groisman, 1996), the LPS modifications are predicted to be induced in vivo.

Several regulated modifications of lipid A have been associated with antimicrobial peptide resistance in *S. enterica* serovar Typhimurium (Guo et al., 1997). To identify PhoP-regulated genes involved in resistance, antimicrobial peptide resistant PhoP-constitutive (*pho-24*; PhoP^{c}) strains containing mutations in known PhoP-activated genes (*pag*) were screened for susceptibility to C18G, a derivative of the α-helical AP human platelet factor IV. These methods identified a C18G-sensitive mutant with a Tn*phoA* insertion in *pagP*. In a $PhoP^C$ background, the *pagP* mutant showed cross-sensitivity to the α-helical AP magaininlike peptide pGLa and modest sensitivity to the porcine defensin protegrin (Guo et al., 1998). Susceptibility to another, structurally different antimicrobial peptide, polymyxin, remained unchanged. Mass spectrometric [matrix-assisted LASER desorption ionization (MALDI-TOF)] analysis of wild-type and *pagP* mutant lipid A demonstrated that although wild-type lipid A was present in hexa-acylated and hepta-acylated forms, lipid A from a *pagP* mutant (*pagP*::Tn*phoA* or *pagP2*::Tn*10d*) existed in only the hexaacylated form (Guo et al., 1998). These findings suggested a role for *pagP* specifically in palmitoylation of lipid A, as *pagP* did not affect other PhoP-regulated LPS modifications, such as addition of 2-hydroxymyristate.

Although other lauroyl and myristoyl transferases (LpxL and LpxM, respectively) involved in LPS biosynthesis are located in the inner membrane (Brozek and Raetz, 1990; Carty, Sreekumar, and Raetz, 1999), PagP has been identified as a lipid A biosynthetic enzyme localized in the outer membrane (Bishop et al., 2000). The ability of PagP to acylate lipid A in vitro was demonstrated by an assay of palmitoyl transferase activity of membranes from wild-type, $PhoP^C$, and *pagP* mutants with various ^{32}P-labeled lipid A precursors used as acyl acceptors. Experiments evaluating various phospholipids as acyl donors showed a preference of PagP for palmitate from the sn-1 position of a phospholipid (including phosphatidylglycerol, phosphatidylethanolamine, phosphatidylserine, and phosphatidic acid). The current model entails PagP-mediated transfer of the palmitoyl group from the donor phospholipid to the N-linked hydroxymyristate group present on the proximal glucosamine unit of lipid A (at position 2).

A *pagP* mutant evinces antimicrobial peptide sensitivity in part because these mutants show increased outer-membrane permeability (Guo et al., 1998). Experiments were performed in which alkaline phosphatase was expressed in wild-type and *pagP* mutant strains. Because alkaline phosphatase is a periplasmic protein, detection of alkaline phosphatase activity in the extracellular medium suggests disturbance of the outer membrane. Upon exposure to C18G, the *pagP* mutants showed greater release of alkaline phosphatase (and so greater outer-membrane permeabilization by C18G) than the parent strain. Palmitoylation thus could promote resistance to membrane-active antimicrobial peptides by decreasing the permeability (and increasing the stability) of the outer membrane.

12.2.3. *PmrA*-Regulated LPS Modifications and Antimicrobial Peptide Resistance

Although permeability of the outer membrane is an important factor in susceptibility of bacteria to antimicrobial peptides, the charge of the bacterial cell surface molecules plays a significant role in antimicrobial peptide resistance as well. LPS is a source of integrity of the outer membrane (Nikaido and Vaara, 1985). Each LPS molecule contains multiple negative charges, contributed by phosphate groups on lipid A and core, and 3-deoxy-D-manno-octulosonic acid (Kdo) residues of core. The anionically charged LPS molecules are cross-linked by divalent cations, which stabilizes this structure of the cell envelope (Labischinski et al., 1985; Vaara, Plachy, and Nikaido, 1990). Although antimicrobial peptides are diverse in terms of structure, they all contain positive charges (Hancock, 2001), which leads to electrostatic interactions between antimicrobial peptides and LPS, the first step in AP–bacterial interaction.

Polymyxins, a family of antimicrobial peptide produced by *Bacillus polymyxa*, have been extensively studied and have microbicidal effects that follow typical antimicrobial peptide mechanisms of action. Polymyxin-resistant strains of *S. enterica* serovar Typhimurium have been demonstrated to bind polymyxin B (PMB) less effectively than wild-type strains in vitro (Vaara, Varra, and Sarvas, 1979). Moreover, resistance to PMB in *S. enterica* serovar Typhimurium has been shown to confer cross-resistance to polylysine, protamine, and host neutrophil peptides CAP-37 (azurocidin) and CAP-57 (bactericidal/permeability-increasing protein) (Shafer, Casey, and Spitznagel, 1984; Spitznagel, 1990).

Early studies investigating the LPS structure of polymyxin-resistant (*pmr*) mutants of *S. enterica* serovar Typhimurium found that LPS from these strains had increased substitution of lipid A with aminoarabinose (Ara4N) and increased presence of ethanolamine on LPS (Vaara et al., 1981). The particular *pmr* strains tested by Vaara et al. expressed 4-fold–6-fold more Ara4N on lipid A, and bound two times less PMB. ^{31}P-NMR (^{31}P Nuclear Magnetic Resonance) studies performed by Helander and co-workers more precisely analyzed LPS structure for substitutions to phosphate groups present on lipid A and core (Helander, Kilpelainen, and Vaara, 1994). Overall, the data showed that wild-type *S. enterica* serovar Typhimurium LPS was 30% substituted (phosphate monoesters accounted for 65%–74% of all phosphates), whereas *pmr* strains had 62%–68% substitutions (i.e., phosphodiesters). Most of the substitutions occurred in the lipid A, with ~80% phosphodiesters compared with 12%–21% phosphodiesters in wild type. These studies also identified the 4′ phosphate of lipid A as the major site of Ara4N substitution. Additionally, increased substitution of core phosphates with phosphoethanolamine, or pEtN (diphosphate diesters), was detected in *pmr* versus wild type. More recent studies indicated that Ara4N can be detected on the 1-phosphate of lipid A and that pEtN can be found on lipid A as well (Zhou, Ribeiro, and Raetz, 2000). Cumulatively, these data suggest that substitution of lipid A and core phosphates (negatively charged) with Ara4N or pEtN, which have free amino groups (positively charged), reduces the net negative charge of LPS, resulting in decreased binding of PMB.

A single mutation, designated *pmrA*505 (PmrA-constitutive, PmrA^{C}), was identified as the cause of PMB resistance in *pmr* strains; the minimal inhibitory concentration (MIC) of PMB against the mutant was 1000-fold higher than against wild type and MICs of neutrophil peptides were 2-fold–4-fold higher (Roland et al., 1993). Three open reading frames (ORFs) were identified in this region, *pmrC* (also called *pagB*), and *pmrA* and *pmrB*, which show homology to response regulators and sensor kinases, respectively, of two-component regulatory systems. The genetic alteration resulting in PMB resistance was pinpointed to a single G–A transition mutation, resulting in replacement of a histidine residue with an arginine.

The *pmrCAB* locus was demonstrated to be positively autoregulated and also under the control of PhoPQ (Gunn and Miller, 1996). Under Mg^{2+} limitation, PhoPQ activates transcription of *pmrD*, whose product appears to act on PmrA-PmrB (PmrAB) at a posttranscriptional level, perhaps by affecting the phosphorylation state of PmrA (Kox, Wosten, and Groisman, 2000). PmrAB can also be activated independently of PhoPQ by mild acid pH and by high iron concentrations (Soncini and Groisman, 1996; Wosten et al., 2000). Sequence analysis of the sensor protein, PmrB, indicated the presence of an ExxE putative iron-binding motif in the periplasmic region of the protein, suggesting a direct method for detecting extracellular iron (Wosten et al., 2000).

Although a *pmrA*505 (PmrA^{C}) mutant expressed high levels of resistance to PMB and other APs, a PmrA-null mutant was particularly sensitive to these compounds (Gunn et al., 1998). The PmrA-null mutant is deficient in the ability to modify its lipid A with Ara4N or pEtN, and core with pEtN (Gunn et al., 1998). Several PmrA-activated genes were described, including *pmrCAB* itself, *pmrE* (*pagA*, *ugd*; UDP-glucose dehydrogenase), and other undefined genes (Gunn and Miller, 1996; Soncini and Groisman, 1996; Gunn et al., 1998; Tamayo et al., 2002). Gunn and co-workers characterized a seven-gene operon (*pmrHFIJKLM*) regulated by PmrA which encodes gene products that, together with PmrE, are involved in the biosynthesis of the Ara4N and its addition to lipid A (Gunn et al., 1998; Baker, Gunn, and Morona, 1999; Zhou et al., 1999). Mutation of any of the operon genes except *pmrM* prevents the incorporation of Ara4N into lipid A (Gunn et al., 2000; Zhou et al., 2001).

The PmrA-regulated genes involved in production of pEtN substitutions to lipid A and core have not been identified. The specific role of pEtN in PMB resistance, and AP resistance in general, thus remains unknown. Although it is predicted that pEtN substitution of LPS could increase antimicrobial peptide resistance in a mechanism similar to that of Ara4N, evidence suggests that PMB resistance is not solely dependent on alteration of LPS charge (Yethon et al., 2000).

12.2.4. Non-LPS-Mediated Mechanisms of Antimicrobial Peptide Resistance

Much research has been focused on LPS substitution-mediated mechanisms of antimicrobial peptide resistance; however, other components of the outer membrane have been described to play a role in antimicrobial peptide resistance. In addition to upregulating genes involved in LPS modification in response to in vivo signals, PhoPQ have been shown to activate expression of a number of outer-membrane and secreted proteins that potentially play a role in antimicrobial peptide resistance (Belden and Miller, 1994).

By a comparison of outer membranes from PhoP-null and $PhoP^C$ (*pho*-24) strains by high-resolution analysis of two-dimensional polyacrylamide gel electrophoresis, a number of outer-membrane proteins were determined to be PhoP regulated. One of the proteins expressed in the $PhoP^C$ strain (and Mg^{2+}-limiting conditions) but not in a PhoP-null strain (or high Mg^{2+} conditions) was identified by mass spectrometry as PgtE (Guina et al., 2000). Interestingly, PhoP does not directly activate *pgtE* transcription, translation or translocation across the cytoplasmic membrane. Guina et al. (2000) showed that PhoP-activated genes known to be involved in LPS modification (palmitate and Ara4N, previously described) did not affect localization of PgtE to the outer membrane, and the mechanism by which PhoP regulates the presence of PgtE in the outer membrane remains undetermined.

OmpT, with which PgtE shares significant similarity, has been demonstrated to confer increased resistance to protamine in *E. coli* by enzymatically cleaving protamine at specific amino acids (between pairs of basic residues and between a basic residue followed by a nonpolar residue) (Stumpe et al., 1998). In *S. enterica* serovar Typhimurium, mutation of *pgtE* caused increases in susceptibility to human LL-37, mouse cathelin-related antimicrobial peptide (CRAMP), and C18G, which all possess α-helical structure (Guina et al.,

2000). Preliminary evidence indicates that protease activity is the mechanism by which PgtE confers resistance to C18G.

Groisman et al. (1992) identified a number of loci that are necessary for protamine resistance and have not been shown to affect LPS structure. These *sap* (*s*ensitive to *a*ntimicrobial *p*eptides) mutants were attenuated for virulence in the mouse model and could not survive within macrophages in vitro. Five of the *sap* genes are in operonic arrangement, *sapABCDF*, and are predicted to comprise a transporter, based on amino acid similarity to various proteins from other organisms involved in transport of different peptides (Parra-Lopez, Baer, and Groisman, 1993). The *sapG* showed 99% identity to the NAD+ binding protein, TrkA, which participates in low-affinity potassium uptake in *E. coli*, but the sensitivity to protamine conferred by a *sapG* mutation was independent of potassium levels in the medium (Parra-Lopez et al., 1994). A number of other *sap* loci were identified (Groisman et al., 1992), but functions have not been determined. Sap homologs have been identified in other bacteria, in which they have also been shown to play a role in virulence and resistance to antimicrobial peptides (Lopez-Solanilla et al., 1998).

12.2.5. Antimicrobial Peptide Resistance and *Salmonella* Virulence

Hypotheses suggesting that antimicrobial peptide resistance mechanisms are a central component of microbial survival strategies in the host have been based on a number of experiments describing an association between *Salmonella* antimicrobial peptide resistance, ability to survive within macrophages in vitro, and ability to cause disease in the mouse model.

Studies performed by the Heffron group, in which a variety of *S. enterica* serovar Typhimurium mutants were selected for increased susceptibility to killing by macrophages, suggested concomitant virulence defects in the mouse model (Lindgren, Stojiljkovic, and Heffron, 1996). Groisman and co-workers have also shown an association between antimicrobial peptide resistance and virulence in the mouse model of infection; mutants identified for protamine susceptibility had consequent virulence defects in mice (Groisman et al., 1992). PMB-sensitive mutants identified by Tamayo et al. (2002) possessed defects in genes that likely affect outer-membrane structural integrity; these mutants were also less virulent in mice. These findings indicate that bacterial factors that increase AP resistance and/or growth in host cells in vitro are important for survival in the animal, but do not demonstrate a causal relationship.

The PhoPQ mutants have serious defects in virulence in the mouse model, but this cannot conclusively be attributed to antimicrobial peptide sensitivity of these mutants, because of the diverse roles of PhoPQ as a virulence determinant (Miller and Mekalanos, 1990). Antimicrobial peptide sensitive *Salmonella* mutants specifically deficient in producing Ara4N modifications to lipid A (e.g., mutants in *pmrF* or *pmrE*) are attenuated when inoculated by the oral route but not by the intraperitoneal route (Gunn et al., 2000). Although this is perhaps the strongest association of an antimicrobial peptide resistance mechanism and virulence in a host, it is possible that these genes perform other unknown functions that affect virulence. The PhoP-regulated palmitate modification to lipid A by *pagP*, although it plays a role in antimicrobial peptide resistance, does not appear to affect *Salmonella* virulence (Belden and Miller, 1994). However, mutation of a *pagP* homolog in *Legionella pneumophila* affected both antimicrobial peptide resistance and virulence properties of this pathogen (Robey, O'Connell, and Cianciotto, 2001).

12.3. *Staphylococcus* spp.

Staphylococcus aureus are common inhabitants of humans; approximately 30% of humans carry *S. aureus* at any given time, and nearly every person will be colonized by *S. aureus* at some time in their lives. In humans, *S. aureus* is found most commonly on the skin and on mucosal membranes. Carrier status is a major risk factor for infection by *S. aureus*, yet mucosal membranes and skin typically prevent bacterial infection in part through the production of antimicrobial peptides, including the production of human β-defensin 2 (hBD-2) and cathelicidin LL-37 by keratinocytes (Huttner and Bevins, 1999; Nizet et al., 2001). Furthermore, upon breaching the skin, bacteria face antimicrobial peptides (e.g., α-defensins HNP1–HNP3) present in neutrophil granules (Lehrer and Ganz, 1999, 2002). Experiments showing that depletion of neutrophils in mice leads to greater susceptibility to *S. aureus* infection serve as evidence that neutrophils are a crucial component of host innate immunity against *S. aureus* (Verdrengh and Tarkowski, 1997). It has been demonstrated that *S. aureus* is resistant to a variety of antimicrobial peptides (Harder et al., 1997; Peschel et al., 1999), and the mechanisms of resistance are being elucidated.

Although Gram-negative bacteria bear the protection of an outer membrane covered by LPS molecules that can be modified to decrease antimicrobial peptide binding, Gram-positive bacteria possess a cytoplasmic membrane shielded by a thick peptidoglycan layer. The peptidoglycan serves as the initial target of antimicrobial agents, including antimicrobial peptides. Like

Gram-negative bacteria, the cell surface of Gram-positive microbes carries a net negative charge. The anionic charges are contributed by the polar head groups of phospholipids in the lipid bilayer and by the phosphate groups in cell-wall teichoic acids (WTAs) and lipoteichoic acids (LTAs).

As discussed in the previous section for *Salmonella* spp., a common mechanism for resistance to cationic antimicrobial peptides involves modification of the cell envelope to reduce the negative charge, decreasing the electrostatic attraction between the bacterium and the antimicrobial compound. Studies regarding antimicrobial peptide – sensitive mutants of *S. aureus* are bringing to light a similar theme of altering the cell surface to decrease binding of antimicrobial peptides. More importantly, these studies support the correlation between resistance to antimicrobial peptides and the ability to survive within the host.

12.3.1. *DltABCD* and Substitution of Teichoic Acids with D-Alanine

To identify staphylococcal loci involved in resistance to AP and to understand the mechanism(s) of resistance concerned, transposon mutants of naturally antimicrobial peptide–resistant *Staphylococcus xylosus* were screened for the acquisition of susceptibility to gallidermin, a lantibiotic produced by *Staphylococcus gallinarum* with activities similar to host cationic antimicrobial peptides (Peschel et al., 1999). The screen identified a locus with homology to the *Bacillus subtilis dltABCD* operon that participates in the addition of D-alanine to phosphate groups of teichoic acids, which comprise alternating phosphate and alditol residues. Mutations in the *dltABCD* locus in both *S. xylosus* and *S. aureus* reduced the amount of D-alanine incorporated into WTA and LTA. D-alanine substitution of phosphate groups on teichoic acids not only masks the anionic charge of phosphate, but also contributes positive charge provided by the D-alanine free amino group. *dlt* mutants were sensitive to a number of positively charged AP, including gallidermin, HNP1–HNP3, protegrins 3 and 5 isolated from porcine leukocytes, tachyplesin 1 and 3 from horseshoe crab hemocytes, a magainin II derivative from clawed frog skin, and the lantibiotics gallidermin and nisin. Because the D-alanine substitution of teichoic acids appears to provide cross-resistance to a number of antibiotics and many but not all antimicrobial peptides, this mechanism of antimicrobial peptide resistance is relatively nonspecific.

Studies performed by Collins et al. (2002) showed that mutation of the *dlt* locus results in decreased virulence in the mouse model of sepsis and arthritis. Intravenous infection of mice with the *dlt* mutant led to no mortality (compared with 33% mortality resulting from infection with wild type), nearly 5-fold decrease in incidence of arthritis, and decreased bacterial loads in the kidneys. In addition, although both *S. aureus* wild-type and *dlt* null strains were taken up by human neutrophils with similar efficiencies, the Dlt mutant was killed at a much faster rate. The defect that was due to the *dlt* mutation was not limited to susceptibility to defensins, but also included increased sensitivity to killing by neutrophil myeloperoxidase (MPO), which produces oxygen-dependent toxic products present in neutrophil granules.

12.3.2. MprF and Esterification of Membrane Phospholipids with L-Lysine

The importance of cell surface charge with respect to binding by cationic antimicrobial peptides is underscored by the identification of a second determinant in *S. aureus* that affects the net charge of the cell envelope, antimicrobial peptide resistance, and virulence. Transposon mutagenesis of *S. xylosus* also led to the identification of *mprF* (*m*ultiple *p*eptide *r*esistance *f*actor), which resulted in gallidermin sensitivity in *S. xylosus* and *S. aureus* (Peschel et al., 2001). Overall, the *mprF* mutation yielded a profile of phenotypes identical to those for the *dlt* mutants, including (1) increased susceptibility to a variety of antimicrobial peptides, restricted to the larger, cationic antimicrobial peptides previously described for the *dlt* mutant, (2) more efficient killing by human neutrophils, specifically by oxygen-independent mechanisms, and (3) decreased virulence in mice, with no mortality, decreased incidence of arthritis, and fewer bacteria found in the kidneys. Like *dlt* mutations, inactivation of *mprF* affected the net charge of the cell envelope, but involved esterification of a major membrane phospholipid with L-lysine to produce lysylphosphatidylglycerol (LPG). Addition of L-lysine to the phosphatidylglycerol substrate in effect reduces the negative charge present on phospholipid polar head groups and, consequently, antimicrobial peptide binding.

12.3.3. Other Mechanisms of Antimicrobial Peptide Resistance in *S. aureus*

In addition to antimicrobial peptide resistance mechanisms that are dependent on cell surface charge, there are indications that other mechanisms of resistance exist in *S. aureus*. For example, the QacA efflux pump of *S. aureus* has been shown to contribute to resistance to thrombocidin platelet microbicidal protein 1 (tPMP1), a cationic antimicrobial protein released by platelets (Kupferwasser et al., 1999). However, the mechanism by which QacA mediates resistance remains unknown, as the efflux activity of QacA may not be required for resistance.

12.4. RESISTANCE MECHANISMS OF OTHER BACTERIA

12.4.1. *Neisseriae*

The *Neisseria gonorrhoeae* route of entry into the host provides much opportunity for bacterial – antimicrobial peptide interaction; however, *N. gonorrhoeae* is able to survive in the urogenital tract despite innate immune system contact. The gonococcus is highly resistant to the action of antimicrobial peptides such as PMB (>400 μg/ml) and defensin (>200 μg/ml) (Vaara and Viljanen, 1985; Shafer et al., 1998). This resistance appears to be at the level of initial membrane binding, as *N. gonorrhoeae* binds significantly less PMB than other susceptible bacteria (Vaara and Viljanen, 1985). Although this may be true, Shafer et al. (1998) recently identified an efflux pump, Mtr, that plays a role in gonococcal resistance to the antimicrobial peptides LL-37 and protegrin. This Mtr efflux pump also provides resistance to a structurally diverse group of antimicrobial hydrophobic agents including bile salts and fatty acids. However, it played no role in resistance of *N. gonorrhoeae* to the lethal action of PMB or defensin. Therefore the mechanism(s) of resistance to PMB and defensin remain unknown. However, Gunn et al. recently identified several PMB sensitive transposon mutants, some of which also demonstrate increased sensitivity to other antimicrobial peptides (J. S. Gunn, personal communication, 2003). These mutants are currently being characterized to identify the loci involved in antimicrobial peptide resistance.

A homolog of the PhoPQ two-component regulatory system has been identified and characterized in *Neisseria meningitidis* (Johnson et al., 2001). Mutation of this regulatory system resulted in the inability of the strain to grow in low magnesium concentrations and increased sensitivity to defensins. Furthermore, this *phoP* mutant was unable to grow in mouse serum and showed a dramatic decrease in the ability to traverse a layer of human epithelial cells. These data suggest that the *N. meningitidis* and *S.enterica* serovar Typhimurium PhoPQ systems are similar in function and responsiveness and that this regulatory system may play an important role in the pathogenesis of the Neisseriaceae.

12.4.2. *Pseudomonas aeruginosa*

In patients with cystic fibrosis (CF), *Pseudomonas aeruginosa* infection of the lungs is a common occurrence. Infections in younger patients are usually cleared, whereas infection of adults is more severe. This pathology coincides with changes in *P. aeruginosa* that are believed to make the organism more resilient and resistant to host antimicrobial agents such as antimicrobial peptides. Although CF isolates from younger patients have smooth LPS and are nonmucoid, isolates from older patients usually have a rough LPS and are alginate producers (mucoid) (Govan and Deretic, 1996), suggesting that mucoidy might contribute to antimicrobial peptide resistance as well as resistance to other antimicrobials. Furthermore, LPS from primary *P. aeruginosa* isolates from patients with CF contained modifications (Ara4N and palmitate addition to lipid A) consistent with those additions observed to play a role in antimicrobial peptide resistance in other organisms (Ernst et al., 1999). These LPS modifications were lost upon repeated passage of these strains. Additionally, lipid A from patients with brochiectasis, a chronic non-CF lung infection, did not contain these modifications. These data suggest that specific factors present in the CF lung induce lipid A modifications in *P. aeruginosa.*

P. aeruginosa also has *phoPQ* homologs that play a role in resistance to antimicrobial peptides including C18G and PMB (Ernst et al., 1999; Macfarlane, Kwasnicka, and Hancock, 2000). As in *S. enterica* serovar Typhimurium, lipid A from cultures grown in media with low Mg^{2+} contain Ara4N substitutions to the 1- and/or 4′ phosphate as well as palmitate addition to the 3′-ester-linked fatty acid, which are associated with increased resistance to antimicrobial peptides (Ernst et al., 1999). PhoP-null mutants lack these modifications and are more sensitive to PMB. As well as *pmrHFIJKLM* (the seven-gene operon necessary for Ara4N addition, PMB resistance, and oral virulence in *S. enterica* serovar Typhimurium), homologs of PmrAB exist in *P. aeruginosa*; however, their role in *P. aeruginosa* Ara4N addition and PMB resistance has not been elucidated.

12.4.3. *Burkholderia* spp.

Burkholderia cepacia, like *P. aeruginosa*, is a pathogen of the CF lung and is highly resistant to the action of antimicrobial peptides (e.g., MIC >6400 μg/ml for PMB; 400 μg/ml for protegrin; J. S. Gunn, personal communication, 2003). This resistance is largely due to the inability of antimicrobial peptides such as PMB to disrupt and permeabilize the outer membrane, which is likely dependent on differences in LPS structure. The phosphate content of *Burkholderia* LPS is one-third that of *Pseudomonas*, which limits the possible polycation-binding sites of the LPS (Moore and Hancock, 1986; Albrecht et al., 2002). Furthermore, the LPS of *Burkholderia* has an inner core with Ko-$\alpha(2\rightarrow4)$-Kdo disaccharide, with the Ko residue substituted nonstoichiometrically with Ara4N at position 8 (Gronow et al., 2002). Ara4N can also be added to the 4′ phosphate of the lipid A. These Ara4N additions likely play a role in antimicrobial peptide resistance, as has been shown in several other bacteria discussed in this chapter. However, Gunn et al. recently identified four Tn5 insertion mutants of *B. cepacia* that demonstrated increased PMB sensitivity without alterations of the lipid A Ara4N modification (J. S. Gunn, personal communication, 2003). From banding patterns of purified LPS on polyacrylamide gels, LPS from the mutants migrate differently from that of the parent, suggesting a role for other LPS structures (possibly including core Ara4N) in antimicrobial peptide resistance.

Burkholderia pseudomallei is a Gram-negative bacterium that causes the disease melioidosis. Melioidosis most commonly presents as an acute pulmonary infection and occasionally as an acute localized skin infection or septicemia. *B. pseudomallei* is resistant to the killing action of antimicrobial peptides such as protamine, magainin, and PMB, and bacterial replication can still occur in media containing PMB at concentrations of >100 mg/ml (Burtnick and Woods, 1999). By Tn5-OT182 mutagenesis, PMB-susceptible mutants of *B. pseudomallei* have been identified. The genes involved included homologs to *lytB*, *waaF*, and *pmrE* (*pagA*, *ugd*). All of these loci are involved in LPS (*waaF* and *pmrE*) or phospholipid/peptidoglycan synthesis (*lytB*). It is not known if the *pmrE* homolog, which is regulated by PmrAB and is involved in Ara4N addition in the salmonellae, is also involved in Ara4N addition in *B. pseudomallei*.

12.4.4. *Yersinia* spp.

Of the eleven *Yersinia* species, only three are pathogenic to humans: *Yersinia enterocolitica*, *Yersinia pseudotuberculosis*, and *Yersinia pestis*. These organisms likely encounter antimicrobial peptides at sites of mucosal infection (gastrointestinal tract and lung) and within host cells. Two mechanisms of antimicrobial peptide resistance have been identified in the *Yersinia*: LPS modification and efflux pumps.

The three pathogenic species differ in LPS structure, as both *Y. pseudotuberculosis* and *Y. pestis* fail to make extended oligosaccharides whereas *Y. enterocolitica* makes a more classical LPS. The *Y. enterocolitica* serotype O:3 outer core, a hexasaccharide, has been shown to form a branch (Skurnik et al., 1999). Experiments suggest that this branch is necessary for antimicrobial peptide resistance and may function in this respect by sterically hindering antimicrobial peptides from binding the deeper parts of the LPS or by stabilizing the outer membrane by interacting with adjacent LPS molecules. Temperature has been shown to affect the ability of *Yersinia* (and LPS from these organisms) to bind antimicrobial peptides. For example, when grown at 37 °C, *Y. pseudotuberculosis* and *Y. pestis* are more resistant to and bind less PMB than *Y. enterocolitica* (Bengoechea and Skurnik, 2000).

Y. pestis possesses a PhoPQ two-component system, which, when mutated, does not affect the lipid A structure as seen in *Salmonella*, but affects the ability to add a terminal galactose to the core (Hitchen et al., 2002). The *Y. pestis phoP* mutant was more sensitive to PMB, but it is not clear if the lipooligosaccharide (LOS) defect is directly responsible for the observed sensitivity. Additionally, *Yersinia* LPS can be modified with Ara4N and ethanolamine. However, it is not clear if these additions are regulated by the PhoPQ or PmrAB homologs or whether these additions affect antimicrobial peptide resistance in the Yersiniae.

The *Yersinia rosAB* locus encodes an efflux pump coupled to a potassium antiporter. This system is activated by a temperature shift to 37 °C or by the presence of antimicrobial peptide (Bengoechea and Skurnik, 2000). The cytoplasmic membrane protein, RosA, pumps peptides out of the cell likely by interacting with TolC, which acts as a bridge between the cytoplasmic membrane and the outer membrane. RosB aids in the process by inducing acidification of the cytoplasm. The lowered pH could act as a positive signal to activate other antimicrobial peptide resistance mechanisms and also may inhibit the action of some peptides sensitive to pH.

12.4.5. *Proteus* spp.

Proteus mirabilis is one of the most common causes of urinary tract infections and also can cause nosocomial infections. This organism is highly resistant to antimicrobial peptides; for example, the MIC of PMB is >6400 μg/ml and for protegrin it is >400 μg/ml (Kaca, Radziejewska-Lebrecht, and Bhat, 1990; McCoy et al., 2001). Recently, McCoy et al. (2001) identified three loci that play a role in resistance to PMB, one of which is also necessary for resistance to β-sheet antimicrobial peptides such as protegrin. These loci include a gene with homology to a member of the *sap* locus in *S. enterica* serovar Typhimurium, a gene with homology to the A subunit of the F_1F_0ATPase, and a locus encoding a putative O-acetyltransferase. All mutations eliminated the addition of Ara4N to the LPS, but it is not clear if this effect is direct or indirect. In addition, the swarming motility of the PMB- and protegrin-sensitive mutants on an agar-containing medium was dramatically altered, demonstrating that surface modifications affecting AP resistance can also affect other virulence phenotypes.

Other studies have also implicated LPS as the major factor in *Proteus* resistance to antimicrobial peptides. Boll et al. (1994) and St Swierzko et al. (2000) have shown that the two Ara4N residues in the Kdo/lipidA region are essential for resistance to PMB, but not for resistance to the terminal portion of a cationic antimicrobial peptide, $CAP18_{109-135}$.

12.4.6. *Shigella flexneri*

Shigella flexneri contains a PhoPQ system that, much like the *Salmonella* system, is activated by low Mg^{2+} to increase resistance to purified antimicrobial peptides and extracts from polymorphonuclear (PMN) cells (Moss et al., 2000). A *S. flexneri* PhoPQ mutant was also impaired for survival within PMNs. Although several virulence properties of the organism, such as epithelial cell invasion, intracellular spread, and resistance to low pH were unaffected in the mutant, other outcomes of infection, such as in the Sereny test, showed that the *phoP* mutant was unable to cause significant keratoconjunctivitis. These data support the hypothesis that PhoPQ plays an important role in pathogenesis of *Shigella*.

12.5. SUMMARY

Antimicrobial peptides are an important part of the host defense mechanism of insects, animals, and plants. However, the expression of these antimicrobial peptides is clearly not able to inhibit the invasion, dissemination, and virulence of all pathogenic microbes, as many of these organisms (including the bacterial species discussed in this chapter) have developed several unique means of avoiding the threat of antimicrobial peptide killing. Most involve the elimination of antimicrobial peptide binding to the cell surface (e.g., LPS modification, LTA modification, the loss of specific phospholipids, secreted proteases) or expulsion of the antimicrobial peptide from the cell after it has entered (e.g., efflux pumps) (Fig. 12.1). Furthermore, the PhoPQ system has been implicated in membrane (usually LPS) modifications in a number of organisms, which often affect antimicrobial peptide resistance. Although these generalities hold true for the identified resistance mechanisms uncovered in recent years, those specific mechanisms elicited by the individual bacteria discussed in this chapter are often genetically distinct and unique in nature (Table 12.1). For example, the PhoPQ-mediated resistance mechanisms can range from modification of LPS phosphate groups with Ara4N or pEtN to alteration of the acylation status of LPS or other membrane components or to LPS sugar modifications.

Most mechanisms of antimicrobial peptide resistance utilized by bacteria do not function in conferring resistance to a single antimicrobial peptide, or even necessarily to a specific structural class of antimicrobial peptide, particularly those mechanisms involving modification of the cell envelope. For example, *Salmonella* resistance to PMB by PmrA has been shown to confer cross-resistance to polylysine, protamine, and host neutrophil antimicrobial proteins, CAP-37 and CAP-57, which differ significantly in structure (Shafer et al., 1984; Spitznagel, 1990). The antimicrobial peptide resistance mechanism involving acylation of *Salmonella* lipid A by PagP results in increased resistance to structurally related protegrin and the magaininlike peptide pGLa, but does not affect PMB resistance (Guo et al., 1998). Analogously, *S. aureus* Dlt-mediated addition of D-ala to teichoic acids confers increased resistance to HNP1–HNP3, protegrins 3 and 5, tachyplesin 1 and 3, a magainin II derivative, and the lantibiotics gallidermin and nisin, which vary considerably in structure. More specific antimicrobial peptide resistance mechanisms involve enzymatic activity directed against the AP itself, such as the *Salmonella* outer-membrane protease PgtE, which acts preferentially against α-helical antimicrobial peptides, such as human LL-37, mouse CRAMP, and C18G. Application of a single resistance mechanism to a variety of antimicrobial compounds allows for versatile "strategic" approaches by the bacterium for survival in diverse environments in vitro and within the host.

Many biopharmaceutical companies are investigating the use of antimicrobial peptides as a new class of "antibiotics." However, we are just beginning to understand the range of bacterial-resistance mechanisms that exist and the likelihood of the genetic transfer of antimicrobial peptide resistance mechanisms between bacteria. Interestingly, serial passage with sublethal antimicrobial peptide concentrations can result in stable resistant mutants with some antimicrobial peptides, but cannot be easily selected with other types of antimicrobial peptides. Lessons, such as those learned from the acquisition of vancomycin resistance in bacteria, demonstrate that complex resistances can occur and remind us that, given the time and the opportunity, the resilience of microbes can be proven time and time again. The various mechanisms described in this chapter are at the forefront of understanding bacterial antimicrobial peptide resistance and therefore will likely produce important information relevant to the therapeutic use of these peptides in humans.

ACKNOWLEDGMENTS

We regret that, because of space constraints, all manuscripts associated with antimicrobial peptide resistance mechanisms could not be discussed or referenced. We apologize to those individuals who were excluded.

REFERENCES

Albrecht, M. T., Wang, W., Shamova, O., Lehrer, R. I., and Schiller, N. L. (2002). Binding of protegrin-1 to *Pseudomonas aeruginosa* and *Burkholderia cepacia*. *Respiratory Research*, 3, 18.

Alpuche-Aranda, C. M., Swanson, J. A., Loomis, W. P., and Miller, S. I. (1992). *Salmonella typhimurium* activates virulence gene transcription within acidified macrophage phagosomes. *Proceedings of the National Academy of Sciences USA*, 89, 10079–83.

Baker, S. J., Gunn, J. S., and Morona, R. (1999). The *Salmonella typhi* melittin resistance gene *pqaB* affects intracellular growth in PMA-differentiated U937 cells, polymyxin B resistance and lipopolysaccharide. *Microbiology*, 145, 367–78.

Behlau, I. and Miller, S. I. (1993). A PhoP-repressed gene promotes *Salmonella typhimurium* invasion of epithelial cells. *Journal of Bacteriology*, 175, 4475–84.

Belden, W. J. and Miller, S. I. (1994). Further characterization of the PhoP regulon: identification of new PhoP-activated virulence loci. *Infection and Immunity*, 62, 5095–01.

Bengoechea, J. A. and Skurnik, M. (2000). Temperature-regulated efflux pump/potassium antiporter system mediates resistance to cationic antimicrobial peptides in *Yersinia*. *Molecular Microbiology*, 37, 67–80.

Bishop, R. E., Gibbons, H. S., Guina, T., Trent, M. S., Miller, S. I., and Raetz, C. R. (2000). Transfer of palmitate from phospholipids to lipid A in outer membranes of gram-negative bacteria. *EMBO Journal*, 19, 5071–80.

Boll, M., Radziejewska-Lebrecht, J., Warth, C., Krajewska-Pietrasik, D., and Mayer, H. (1994). 4-amino-4-deoxy-L-arabinose in LPS of enterobacterial R-mutants and its possible role for their polymyxin reactivity. *FEMS Immunology and Medical Microbiology*, 8, 329–41.

Brozek, K. A. and Raetz, C. R. (1990). Biosynthesis of lipid A in *Escherichia coli*. Acyl carrier protein-dependent incorporation of laurate and myristate. *Journal of Biological Chemistry*, 265, 15410–17.

Burtnick, M. N. and Woods, D. E. (1999). Isolation of polymyxin B-susceptible mutants of *Burkholderia pseudomallei* and molecular characterization of genetic loci involved in polymyxin B resistance. *Antimicrobial Agents and Chemotherapy*, 43, 2648–56.

Carty, S. M., Sreekumar, K. R., and Raetz, C. R. (1999). Effect of cold shock on lipid A biosynthesis in *Escherichia coli*. Induction At 12 degrees C of an acyltransferase specific for palmitoleoyl-acyl carrier protein. *Journal of Biological Chemistry*, 274, 9677–85.

Collins, L. V., Kristian, S. A., Weidenmaier, C., Faigle, M., Van Kessel, K. P., Van Strijp, J. A., Gotz, F., Neumeister, B., and Peschel, A. (2002). *Staphylococcus aureus* strains lacking D-alanine modifications of teichoic acids are highly susceptible to human neutrophil killing and are virulence attenuated in mice. *Journal of Infectious Disease*, 186, 214–19.

Ernst, R. K., Yi, E. C., Guo, L., Lim, K. B., Burns, J. L., Hackett, M., and Miller, S. I. (1999). Specific lipopolysaccharide found in cystic fibrosis airway *Pseudomonas aeruginosa. Science*, 286, 1561–5.

Fields, P. I., Groisman, E. A., and Heffron, F. (1989). A *Salmonella* locus that controls resistance to microbicidal proteins from phagocytic cells, *Science*, 243, 1059–62.

Fields, P. I., Swanson, R. V., Haidaris, C. G., and Heffron, F. (1986). Mutants of *Salmonella typhimurium* that cannot survive within the macrophage are avirulent. *Proceedings of the National Academy of Sciences USA*, 83, 5189–93.

Foster, J. W. and Hall, H. K. (1990). Adaptive acidification tolerance response of *Salmonella typhimurium. Journal of Bacteriology*, 172, 771–8.

Garcia-Vescovi, E., Soncini, F. C., and Groisman, E. A. (1996). Mg^{2+} as an extracellular signal: Environmental regulation of *Salmonella* virulence. *Cell*, 84, 165–74.

Govan, J. R. and Deretic, V. (1996). Microbial pathogenesis in cystic fibrosis: Mucoid *Pseudomonas aeruginosa* and *Burkholderia cepacia. Microbiological Reviews*, 60, 539–74.

Groisman, E. A., Chiao, E., Lipps, C. J., and Heffron, F. (1989). *Salmonella typhimurium phoP* virulence gene is a transcriptional regulator. *Proceedings of the National Academy of Sciences USA*, 86, 7077–81.

Groisman, E. A., Parra-Lopez, C., Salcedo, M., Lipps, C. J., and Heffron, F. (1992). Resistance to host antimicrobial peptides is necessary for *Salmonella* virulence. *Proceedings of the National Academy of Sciences USA*, 89, 11939–43.

Gronow, S., Noah, C., Blumenthal, A., Lindner, B., and Brade, H. (2002). Construction of a deep-rough mutant of *Burkholderia cepacia* ATCC 25416 and characterization of its chemical and biological properties. *Journal of Biological Chemistry*, 278, 1647–55.

Guina, T., Yi, E. C., Wang, H., Hackett, M., and Miller, S. I. (2000). A PhoP-regulated outer membrane protease of *Salmonella enterica* serovar Typhimurium promotes resistance to alpha-helical antimicrobial peptides. *Journal of Bacteriology*, 182, 4077–86.

Gunn, J. S., Lim, K. B., Krueger, J., Kim, K., Guo, L., Hackett, M., and Miller, S. I. (1998). PmrA-PmrB regulated genes necessary for 4-aminoarabinose lipid A modification and polymyxin resistance. *Molecular Microbiology*, 27, 1171–82.

Gunn, J. S. and Miller, S. I. (1996). PhoP/PhoQ activates transcription of *pmrA/B*, encoding a two-component system involved in *Salmonella typhimurium* antimicrobial peptide resistance. *Journal of Bacteriology*, 178, 6857–64.

Gunn, J. S., Ryan, S. S., Van Velkinburgh, J. C., Ernst, R. K., and Miller, S. I. (2000). Genetic and functional analysis of a PmrA-PmrB-regulated locus necessary for lipopolysaccharide modification, antimicrobial peptide resistance, and oral virulence of *Salmonella enterica* serovar Typhimurium. *Infection and Immunity*, 68, 6139–46.

Guo, L., Lim, K., Gunn, J. S., Bainbridge, B., Darveau, R., Hackett, M., and Miller, S. I. (1997). Regulation of lipid A modifications by *Salmonella typhimurium* virulence genes *phoP-phoQ*. *Science*, 276, 250–3.

Guo, L., Lim, K. B., Poduje, C. M., Daniel, M., Gunn, J. S., Hackett, M., and Miller, S. I. (1998). Lipid A acylation and bacterial resistance against vertebrate antimicrobial peptides. *Cell*, 95, 189–98.

Hancock, R. E. (2001). Cationic peptides: Effectors in innate immunity and novel antimicrobials. *Lancet Infectious Disease*, 1, 156–64.

Harder, J., Bartels, J., Christophers, E., and Schroder, J. M. (1997). A peptide antibiotic from human skin. *Nature (London)*, 387, 861.

Heithoff, D. M., Conner, C. P., Hanna, P. C., Julio, S. M., Hentschel, U., and Mahan, M. J. (1997). Bacterial infection as assessed by in vivo gene expression. *Proceedings of the National Academy of Sciences USA*, 94, 934–9.

Helander, I. M., Kilpelainen, I., and Vaara, M. (1994). Increased substitution of phosphate groups in lipopolysaccharides and lipid A of the polymyxin-resistant *pmrA* mutants of *Salmonella typhimurium*: a ^{31}P-NMR study. *Molecular Microbiology*, 11, 481–7.

Hitchen, P. G., Prior, J. L., Oyston, P. C., Panico, M., Wren, B. W., Titball, R. W., Morris, H. R., and Dell, A. (2002). Structural characterization of lipo-oligosaccharide (LOS) from *Yersinia pestis*: Regulation of LOS structure by the PhoPQ system. *Molecular Microbiology*, 44, 1637–50.

Huttner, K. M. and Bevins, C. L. (1999). Antimicrobial peptides as mediators of epithelial host defense. *Pediatric Research*, 45, 785–94.

Johnson, C. R., Newcombe, J., Thorne, S., Borde, H. A., Eales-Reynolds, L. J., Gorringe, A. R., Funnell, S. G., and McFadden, J. J. (2001). Generation and characterization of a PhoP homologue mutant of *Neisseria meningitidis*. *Molecular Microbiology*, 39, 1345–55.

Kaca, W., Radziejewska-Lebrecht, J., and Bhat, U. R. (1990). Effect of polymyxins on the lipopolysaccharide-defective mutants of *Proteus mirabilis*. *Microbios*, 61, 23–32.

Kox, L. F., Wosten, M. M., and Groisman, E. A. (2000). A small protein that mediates the activation of a two-component system by another two-component system. *EMBO Journal*, 19, 1861–72.

Kupferwasser, L. I., Skurray, R. A., Brown, M. H., Firth, N., Yeaman, M. R., and Bayer, A. S. (1999). Plasmid-mediated resistance to thrombin-induced platelet microbicidal protein in staphylococci: Role of the *qacA* locus. *Antimicrobial Agents and Chemotherapy*, 43, 2395–9.

Labischinski, H., Barnickel, G., Bradaczek, H., Naumann, D., Rietschel, E. T., and Giesbrecht, P. (1985). High state of order of isolated bacterial lipopolysaccharide and its possible contribution to the permeation barrier property of the outer membrane. *Journal of Bacteriology*, 162, 9–20.

Lehrer, R. I. and Ganz, T. (1999). Antimicrobial peptides in mammalian and insect host defence. *Current Opinion in Immunology*, 11, 23–27.

Lehrer, R. I. and Ganz, T. (2002). Defensins of vertebrate animals. *Current Opinion in Immunology*, 14, 96–102.

Lindgren, S. W., Stojiljkovic, I., and Heffron, F. (1996). Macrophage killing is an essential virulence mechanism of *Salmonella typhimurium. Proceedings of the National Academy of Sciences USA*, 93, 4197–201.

Lopez-Solanilla, E., Garcia-Olmedo, F., and Rodriguez-Palenzuela, P. (1998). Inactivation of the *sapA* to *sapF* locus of *Erwinia chrysanthemi* reveals common features in plant and animal bacterial pathogenesis. *Plant Cell*, 10, 917–24.

Macfarlane, E. L., Kwasnicka, A., and Hancock, R. E. (2000). Role of *Pseudomonas aeruginosa* PhoP-phoQ in resistance to antimicrobial cationic peptides and aminoglycosides. *Microbiology*, 146, 2543–54.

McCoy, A. J., Liu, H., Falla, T. J., and Gunn, J. S. (2001). Identification of *Proteus mirabilis* mutants with increased sensitivity to antimicrobial peptides. *Antimicrobial Agents and Chemotherapy*, 45, 2030–7.

Miller, S. I., Kukral, A. M., and Mekalanos, J. J. (1989). A two-component regulatory system (*phoP phoQ*) controls *Salmonella typhimurium* virulence. *Proceedings of the National Academy of Sciences USA*, 86, 5054–8.

Miller, S. I. and Mekalanos, J. J. (1990). Constitutive expression of the PhoP regulon attenuates *Salmonella* virulence and survival within macrophages. *Journal of Bacteriology*, 172, 2485–90.

Miller, S. I., Pulkkinen, W. S., Selsted, M. E., and Mekalanos, J. J. (1990). Characterization of defensin resistance phenotypes associated with mutations in the *phoP* virulence regulon of *Salmonella typhimurium. Infection and Immunity*, 58, 3706–10.

Moore, R. A. and Hancock, R. E. (1986). Involvement of outer membrane of *Pseudomonas cepacia* in aminoglycoside and polymyxin resistance. *Antimicrobial Agents and Chemotherapy*, 30, 923–6.

Moss, J. E., Fisher, P. E., Vick, B., Groisman, E. A., and Zychlinsky, A. (2000). The regulatory protein PhoP controls susceptibility to the host inflammatory response in *Shigella flexneri. Cell Microbiology*, 2, 443–52.

Nikaido, H. and Vaara, M. (1985). Molecular basis of bacterial outer membrane permeability. *Microbiology Reviews*, 49, 1–32.

Nizet, V., Ohtake, T., Lauth, X., Trowbridge, J., Rudisill, J., Dorschner, R. A., Pestonjamasp, V., Piraino, J., Huttner, K., and Gallo, R. L. (2001). Innate antimicrobial peptide protects the skin from invasive bacterial infection. *Nature (London)*, 414, 454–7.

Parra-Lopez, C., Baer, M. T., and Groisman, E. A. (1993). Molecular genetic analysis of a locus required for resistance to antimicrobial peptides in *Salmonella typhmurium*. *EMBO Journal*, 12, 4053–62.

Parra-Lopez, C., Lin, R., Aspedon, A., and Groisman, E. A. (1994). A *Salmonella* protein that is required for resistance to antimicrobial peptides and transport of potassium. *EMBO Journal*, 13, 3964–72.

Peschel, A., Jack, R. W., Otto, M., Collins, L. V., Staubitz, P., Nicholson, G., Kalbacher, H., Nieuwenhuizen, W. F., Jung, G., Tarkowski, A., van Kessel, K. P., and van Strijp, J. A. (2001). *Staphylococcus aureus* resistance to human defensins and evasion of neutrophil killing via the novel virulence factor MprF is based on modification of membrane lipids with L-lysine. *Journal of Experimental Medicine*, 193, 1067–76.

Peschel, A., Otto, M., Jack, R. W., Kalbacher, H., Jung, G., and Gotz, F. (1999). Inactivation of the *dlt* operon in *Staphylococcus aureus* confers sensitivity to defensins, protegrins, and other antimicrobial peptides. *Journal of Biological Chemistry*, 274, 8405–10.

Robey, M., O'Connell, W., and Cianciotto, N. P. (2001). Identification of *Legionella pneumophila rcp*, a pagP-like gene that confers resistance to cationic antimicrobial peptides and promotes intracellular infection. *Infection and Immunity*, 69, 4276–86.

Roland, K. L., Martin, L. E., Esther, C. R., and Spitznagel, J. K. (1993). Spontaneous *pmrA* mutants of *Salmonella typhimurium* LT2 define a new two-component regulatory system with a possible role in virulence. *Journal of Bacteriology*, 175, 4154–64.

Shafer, W. M., Casey, S. G., and Spitznagel, J. K. (1984). Lipid A and resistance of *Salmonella typhimurium* to antimicrobial granule proteins of human neutrophils. *Infection and Immunity*, 43, 834–8.

Shafer, W. M., Qu, X., Waring, A. J., and Lehrer, R. I. (1998). Modulation of *Neisseria gonorrhoeae* susceptibility to vertebrate antibacterial peptides due to a member of the resistance/nodulation/division efflux pump family. *Proceedings of the National Academy of Sciences USA*, 95, 1829–33.

Skurnik, M., Venho, R., Bengoechea, J. A., and Moriyon, I. (1999). The lipopolysaccharide outer core of *Yersinia enterocolitica* serotype O:3 is required for virulence and plays a role in outer membrane integrity. *Molecular Microbiology*, 31, 1443–62.

Soncini, F. C. and Groisman, E. A. (1996). Two-component regulatory systems can interact to process multiple environmental signals. *Journal of Bacteriology*, 178, 6796–801.

Spitznagel, J. K. (1990). Antibiotic proteins of human neutrophils. *Journal of Clinical Investigation*, 86, 1381–6.

St Swierzko, A., Kirikae, T., Kirikae, F., Hirata, M., Cedzynski, M., Ziolkowski, A., Hirai, Y., Kusumoto, S., Yokochi, T., and Nakano, M. (2000). Biological activities of lipopolysaccharides of *Proteus* spp. and their interactions with polymyxin B and an 18-kDa cationic antimicrobial protein (CAP18)-derived peptide. *Journal of Medical Microbiology*, 49, 127–38.

Stumpe, S., Schmid, R., Stephens, D. L., Georgiou, G., and Bakker, E. P. (1998). Identification of OmpT as the protease that hydrolyzes the antimicrobial peptide protamine before it enters growing cells of *Escherichia coli*. *Journal of Bacteriology*, 180, 4002–6.

Tamayo, R., Ryan, S. S., McCoy, A. J., and Gunn, J. S. (2002). Identification and genetic characterization of pmra-regulated genes and genes involved in polymyxin B resistance in *Salmonella enterica* serovar Typhimurium. *Infection and Immunity*, 70, 6770–8.

Vaara, M., Plachy, W. Z., and Nikaido, H. (1990). Partitioning of hydrophobic probes into lipopolysaccharide bilayers. *Biochimica et Biophysica Acta*, 1024, 152–8.

Vaara, M., Vaara, T., Jenson, M., Helander, I., Nurminen, M., Rietschel, E. T., and Makela, P. H. (1981). Characterization of the lipopolysacharride from the polymyxin-resistant *pmrA* mutants of *Salmonella typhimurium*. *FEBS Letters*, 129, 145–9.

Vaara, M., Vaara, T., and Sarvas, M. (1979). Decreased binding of polymyxin by polymyxin-resistant mutants of *Salmonella typhimurium*. *Journal of Bacteriology*, 139, 664–7.

Vaara, M. and Viljanen, P. (1985). Binding of polymyxin B nonapeptide to gram-negative bacteria. *Antimicrobial Agents and Chemotherapy*, 27, 548–54.

van Velkinburgh, J. C., and Gunn, J. S. (1999). PhoP-PhoQ-regulated loci are required for enhanced bile resistance in *Salmonella* spp. *Infection and Immunity*, 67, 1614–22.

Verdrengh, M. and Tarkowski, A. (1997). Role of neutrophils in experimental septicemia and septic arthritis induced by *Staphylococcus aureus*. *Infection and Immunity*, 65, 2517–21.

Wosten, M. M., Kox, L. F., Chamnongpol, S., Soncini, F. C., and Groisman, E. A. (2000). A signal transduction system that responds to extracellular iron. *Cell*, 103, 113–25.

Yethon, J. A., Gunn, J. S., Ernst, R. K., Miller, S. I., Laroche, L., Malo, D., and Whitfield, C. (2000). *Salmonella enterica* serovar typhimurium waaP mutants show increased susceptibility to polymyxin and loss of virulence in vivo. *Infection and Immunity*, 68, 4485–91.

Zhou, Z., Lin, S., Cotter, R. J., and Raetz, C. R. (1999). Lipid A modifications characteristic of *Salmonella typhimurium* are induced by NH_4VO_3 in *Escherichia coli* K12. Detection of 4-amino-4-deoxy-L-arabinose, phosphoethanolamine and palmitate. *Journal of Biological Chemistry*, 274, 18503–14.

Zhou, Z., Ribeiro, A. A., Lin, S., Cotter, R. J., Miller, S. I., and Raetz, C. R. (2001). Lipid A modifications in polymyxin-resistant *Salmonella typhimurium*: PmrA-dependent 4-amino-4-deoxy-L-arabinose, and phosphoethanolamine incorporation. *Journal of Biological Chemistry*, 276, 43111–21.

Zhou, Z., Ribeiro, A. A., and Raetz, C. R. (2000). High-resolution NMR spectroscopy of lipid A molecules containing 4-amino-4-deoxy-L-arabinose and phosphoethanolamine substituents. Different attachment sites on lipid A molecules from NH_4VO_3-treated *Escherichia coli* versus *kdsA* mutants of *Salmonella typhimurium*. *Journal of Biological Chemistry*, 275, 13542–51.

CHAPTER 13

Roles of antimicrobial peptides in pulmonary disease

Robert Bals

13.1. INTRODUCTION

The respiratory tract has been one focus of research on antimicrobial peptides in recent years. Findings in 1996 that airway epithelia from patients with cystic fibrosis (CF) failed to kill bacteria led to increased interest in antimicrobial factors in airway secretions and the role of antimicrobial peptides in the lung (Smith et al., 1996; Schutte and McCray, 2002). These observations and the large body of data on antimicrobial peptides in other fields have resulted in important insights into the biological role of these substances in the pulmonary tract (van Wetering et al., 1999; Aarbiou, Rabe, and Hiemstra et al., 2002). Two aspects have made antimicrobial peptides a focus of scientific and commercial interest during the past years: (1) Recent insight into the basic biology of antimicrobial peptides showed that these molecules have various functions in host defense, inflammation, and tissue regeneration and are likely involved in the pathogenesis of several diseases; and (2) antimicrobial peptides might qualify as candidates for drug development. The evolution and rapid spread of resistant microorganisms are significant problems in nosocomial infections and are of increasing importance in community-acquired diseases. It is the aim of this chapter to summarize current knowledge of antimicrobial peptides in the respiratory tract and their role in pulmonary disease.

13.2. ANTIMICROBIAL PEPTIDES EXPRESSED IN THE RESPIRATORY TRACT

13.2.1. Host Defense in the Airways

The respiratory tract is shielded by a multicomponent host defense system that involves structural, physical, and functional mechanisms

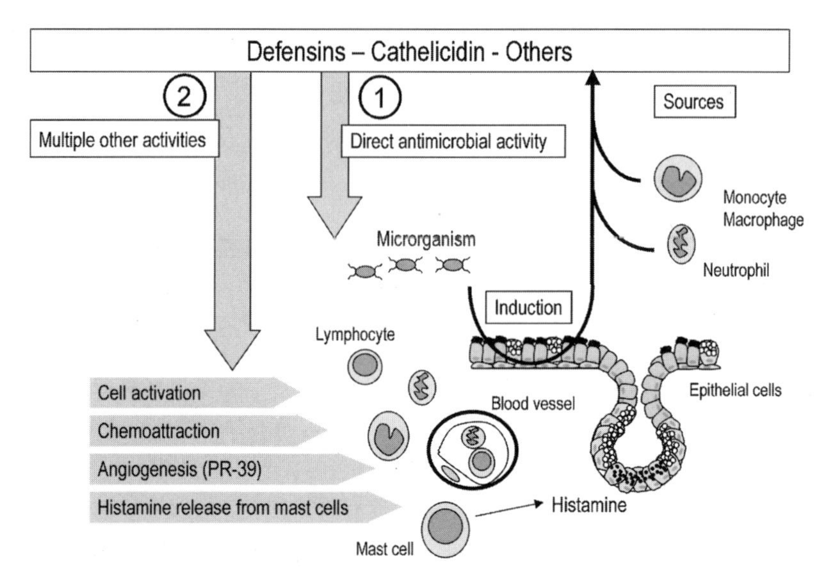

Figure 13.1. Sources and functions of antimicrobial peptides in the respiratory tract. The figure schematically illustrates the sources of the peptides. Besides their direct antimicrobial function, the peptide act as mediators by binding to specific receptors.

(Whitsett, 2002). Various cell types provide the structural base of the pulmonary host defense system. The pseudostratified epithelium of the larger airways is composed of ciliated, basal, goblet, brush, and small-granule cells (Diamond, Legarda, and Ryan, 2000; Davies, 2001; Hamilton et al., 2001). The epithelium of the smaller bronchioli consists mainly of ciliated and Clara cells. The walls of the larger airway contain glands composed of two secretory cell types, serous and mucous. The lung parenchyma comprises alveolar epithelial cells (types I and II), endothelial, and interstitial cells. The goblet, Clara, gland, and type II alveolar epithelial cells are parts of the secretory apparatus of the lung. Different cell types of the innate and adaptive immune system have been described in the lungs: alveolar macrophages, neutrophils, eosinophils, dendritic cells, mast cells, natural-killer (NK) cells, and lymphocytes (Bals, Weiner, and Wilson et al., 1999; Whitsett, 2002). Antimicrobial peptides in the lung have several cellular sources: (1) Epithelial cells of the airways, including serous gland cells that seem to contribute a significant amount, (2) host defense cells such as neutrophils, macrophages, NK cells, and others (Fig. 13.1).

It has become evident that the respiratory epithelium is actively involved in inflammation and host defense in multiple ways: (1) providing a physical barrier; (2) being one component of mucociliary clearance; (3) detecting microorganisms, through pattern-recognition receptors (PRRs) expressed on epithelial cells recognizing pathogen-associated molecular patterns (PAMPs); (4) secreting a variety of proinflammatory and anti-inflammatory mediators; and (5) secreting a variety of antimicrobial substances including antimicrobial peptides (Diamond et al., 2000). Mucociliary clearance provides an efficient system to clear pathogenic particles and is assisted by aerodynamic filtering and airway reflexes, such as coughing and sneezing. Mucus secreted by mucous gland and goblet cells of the large airways entraps particles that are then propelled by the movement of cilia. Secretions of the airways contain proteins and peptides that directly kill or inhibit pathogens or modulate the inflammatory response (Fig. 13.1). Classical components of the airway surface fluid (ASF) that have antimicrobial activity are lysozyme, lactoferrin, secretory phospholipase A2, and secretory leukocyte protease inhibitor (SLPI). Other substances, such as complement, surfactant proteins, and Clara-cell proteins (CC10, CCSP) likely contribute to host defense (Wilmott, Fiedler, and Stark, 1998). Antimicrobial peptides have emerged as potential players in host defense at mucosal surfaces such as the airways.

An assessment of the relative contribution of individual proteins or peptides to innate immune responses is quite challenging. Biochemical methods have been used to isolate and detect the molecules from biological samples. Functional studies primarily have been restricted to in vitro experiments using purified components. The actual concentration of antimicrobial peptides and proteins at the site of action (e.g., in the gel or sol layer of the ASF) is difficult to determine because of problems in sampling the ASF.

13.2.2. Systematic Overview

Antimicrobial peptides can be grouped according to their size, conformational structure, or predominant amino acid structure; however, the diversity of the molecules is so great that it is difficult to categorize them in a generally accepted classification. Based on their gross composition and three-dimensional (3D) structure, peptide antibiotics can be divided in four main classes (Hancock, 1997; Andreu and Rivas, 1998; van 't Hof et al., 2001):

Group I: linear peptides with an α-helical structure
Group II: β-sheet structures stabilized by two or three disulphide bridges
Group III: extended structures with a predominance of one or more amino acids
Group IV: peptides with loop structures

Conversely different classifications of antimicrobial peptides families can be generated. For example, cathelicidins are characterized by the conserved sequence of their propeptide and contain mature peptides of various structures.

The next subsection gives an overview of antimicrobial peptides of the various families that are present in the respiratory tract (Table 13.1). A more detailed description of their biology is provided in other chapters of this book.

Defensins

Mammalian defensins are cationic, relatively arginine-rich non-glycosylated peptides with molecular weights of 3.5–4.5 kd and contain six cysteines that form three characteristic intramolecular disulfide bridges (Lehrer, Ganz, and Selsted, 1991). According to the spacing of the cysteines, the alignment of the disulfide bridges, and the overall molecular structure, defensins can be divided into three classes: α-defensins, β-defensins, and θ-defensins. The genes of α- and β-defensins are located in a cluster on chromosome 8p23 (Linzmeier et al., 1999).

(1) α-defensins: At the present time, six α-defensins have been identified from humans. Human neutrophil peptides 1–4 (HNP1–HNP4) are localized in azurophilic granules of neutrophil granulocytes where they represent the principal protein and contribute to the oxygen-independent killing of phagocytized microorganisms (Ganz et al., 1985; Selsted et al., 1985). The two other α-defensins, human defensins 5 and 6 (HD5 and HD6), are primarily found in Paneth cells of the small intestine. HD5 transcripts were also found in airway epithelial cells (Frye et al., 2000). HNP1–HNP4 are present in airway secretions, originating from neutrophils that invaded the airway or alveolar lumen.

(2) β-defensins: In 1991, an antimicrobial peptide from cow, called tracheal antimicrobial peptide (TAP) (Diamond et al., 1991), was isolated. TAP contains six cysteines connected by three disulfide bridges, however, connected in a different way compared with α-defensins. Therefore members of this new family of antimicrobial peptides were named β-defensins. The first human β-defensin, called human β-defensin 1 (HBD1), was isolated from large volumes of hemofiltrate (Bensch et al., 1995) and is expressed constitutively in epithelial cells of the urinary and respiratory tracts (Goldman et al., 1997; McCray and Bentley, 1997; Valore et al., 1998). Human β-defensin 2 (HBD2) was isolated from psoriatic skin by an affinity chromatography procedure, applying columns coated with components of *Escherichia coli* (Harder et al., 1997) and from airway epithelium by a strategy based on screening of databases (Bals, Wang, and Zasloff et al., 1998). HBD2 was found to be

Table 13.1. *Antimicrobial peptides of the human respiratory tract. A detailed discussion of peptide function is given in the text*

Peptide family	Peptide name	Cellular source	Receptor	Function
α-defensin	HNP1–HNP4	Neutrophil	Not identified	Antibiotic; IL-8 secretion from epithelial cells; chemoattraction of T cells, dendritic cells, monocytes; increased proliferation of epithelial cells; activation of mast cells
	HD5	Airway epithelial cell		Antibiotic
β-defensin	HBD1–HBD4	Airway epithelial cell, macrophage, type II pneumocyte	CCR6 (HBD1, HBD2), unknown (HBD3, HBD4)	Antibiotic; chemoattractant for T cells, dendritic cells, and macrophages; activation of mast cells
Cathelicidin	LL-37/hCAP-18	Airway epithelial cell, macrophage, type II pneumocyte, neutrophil	FPRL-1	Antibiotic; chemoattractant for neutrophils, macrophages, and lymphocytes; activation of mast cells
—	Granulysin	Cytolytic T lymphocytes and NK cells	—	Antibiotic

expressed in epithelia of the inner or outer surfaces of the human body, such as skin and the respiratory and gastrointestinal tracts (Bals, Wang, and Zasloff et al., 1998; Singh et al., 1998). Additionally, HBD1 and HBD2 are expressed by monocytes, macrophages, and dendritic cells (Duits et al., 2002). Both peptides have been detected in airway secretions and concentrations have been found in the microgram per milliliter range (Bals, Wang, and Zasloff et al., 1998; Singh et al., 1998). HBD3 was identified in parallel by a bioscreening and a computational approach (Garcia, Jaumann et al., 2001; Harder et al., 2001; Jia et al., 2001), whereas HBD4 was identified solely by searches of genomic databases (Garcia, Krause et al., 2001). Expression of HBD3 and HBD4 in respiratory epithelial cells is regulated by different stimuli. Recently up to 28 new human and 43 new mouse β-defensin genes in five syntenic chromosomal regions have been identified by bioinformatic screening of the human and murine genomes (Schutte et al., 2002).

(3) θ-defensins: A novel class of defensins has been isolated from rhesus monkey neutrophils and named θ-defensins according to their circular molecular structure (Tang et al., 1999). The peptide rhesus θ-defensin 1 (rTD-1) is produced by the posttranslational ligation of two truncated α-defensins and demonstrated salt-independent antimicrobial activity. rTD-2 and rTD-3 are formed by tandem nonapeptide repeats derived from only one of the rTD-1 precursors (Leonova et al., 2001). No data about the presence of these molecules in the respiratory tract are available at this time.

Cathelicidins

Peptide antibiotics of the cathelicidin family contain a highly conserved signal sequence and proregion (termed "cathelin" for cathepsin L inhibitor) but show substantial heterogeneity in the C-terminal domain that encodes the mature peptide, which can range in size from 12 to 80 or more amino acids (Zanetti, Gennaro, and Romeo, 1995). The only human cathelicidin, LL-37/hCAP-18, was isolated from human bone marrow (Cowland, Johnsen, and Borregaard, 1995; Gudmundsson et al., 1996). LL-37/hCAP-18 is expressed in myeloid cells where it resides in granules but is also found in inflamed skin. LL-37/hCAP-18 has been found to be regulated by inflammatory stimuli (Agerberth et al., 1995; Cowland et al., 1995). In the airways, the peptide is produced by epithelial cells, macrophages, and neutrophils and secreted into the ASF (Bals, Wang, and Zasloff et al., 1998). LL-37 has been detected in tissue culture supernatants of respiratory epithelial cells as well as in lung washings from patients (Bals, Wang, and Zasloff et al., 1998; Agerberth et al., 1999). At this time there are no details known about the processing of LL-37/hCAP-18 in the pulmonary tract. In neutrophils, in which

LL-37/hCAP-18 is localized to specific granules, the peptide is stored in its propeptide form and cleaved after secretion by the activity of protease 3 (Sorensen et al., 2001). Cathelicidins have been isolated from mice (CRAMP) (cathelin-related antimicrobial peptide), rat (rCRAMP), pigs (protegrin, PMAP-23, PR-39), monkeys (rhLL-37, RL-37), rabbit (CAP-18), and sheep (SMAP 29, SMAP34) (Zanetti et al., 1995). Mice deficient in CRAMP were found to be more susceptible to bacterial infections of the skin (Nizet et al., 2001), highlighting the role of cathelicidin peptides in host defense.

Granulysin

Granulysin is an antimicrobial peptide produced by human cytolytic T lymphocytes and NK cells. It is active against a broad range of microbes, including Gram-positive and Gram-negative bacteria, fungi, and parasites. Like porcine NK lysin and amoebapores made by *Entamoeba histolytica*, granulysin is related to saposins, small lipid-associated proteins present in the central nervous system. The presence of this molecule indicates a broader and, perhaps, more significant role for T lymphocytes in both innate and acquired antimicrobial defenses (Stenger et al., 1998; Ochoa et al., 2001). The expression pattern of this peptide indicates it is also present in the respiratory tract. Its biological role is unclear at this time.

Histatins

Histatins are a family of histidine-rich peptides present in human saliva. Their presence in airway secretions has not been investigated. The primary structures of the major family members (histatins 1 and 3) have been determined and revealed lengths of 38- and 32-amino-acid residues. Smaller members of the histatin family, including histatin 5 (24 residues), originate from histatin 1 and 3 by posttranslational processing. The genes that encode histatins 1 and 3 consist of several exons and have been mapped to human chromosome 4q13. The antimicrobial activity includes especially strong antifungal effects. Their biological role in the lung is unclear at this time.

Dermcidin

Dermcidin is an antimicrobial peptide that has a broad spectrum of activity and no homology to other known antimicrobial peptides (Schittek et al., 2001). This protein is expressed in sweat glands of the skin, secreted into sweat and transported to the epidermal surface. In sweat, a proteolytically processed 47-amino-acid peptide is generated that showed antimicrobial activity against a variety of pathogenic microorganisms. Whether dermcidin is expressed in the respiratory tract is unknown.

Regulation

Human airway epithelial cells produce a variety of antimicrobial peptides, including β-defensins, HD5, and LL-37. Several of the peptides are upregulated by inflammatory mediators or microbial structures (Singh et al., 1998; Harder et al., 2000). HBD2 signaling pathways involve mitogen-activated protein kinase (Krisanaprakornkit, Kimball, and Dale, 2002) including Src-dependent Raf-MEK1/2-ERK activation (Moon et al., 2002) and NF-κB (Tsutsumi-Ishii and Nagaoka, 2002). Human Toll-like receptor 2 mediates induction of the antimicrobial peptide hBD-2 in response to bacterial lipoprotein (Birchler et al., 2001) and possibly lipopolysaccharide (LPS) (Becker et al., 2000). In contrast, β-defensins seem also to act as endogenous ligands for Toll-like receptors (Biragyn et al., 2002).

13.3. FUNCTIONS OF ANTIMICROBIAL PEPTIDES IN THE RESPIRATORY TRACT

13.3.1. Antimicrobial Activity of Antimicrobial Peptides

The antimicrobial activity of peptide antibiotics was deduced from in vitro tests assaying purified substances against microorganisms. Antimicrobial peptides have a broad spectrum activity against Gram-positive and Gram-negative bacteria, as well as against fungi and enveloped viruses. Minimum inhibitory concentrations of the peptides are in the range from 0.1 to 10 μg/ml. The antimicrobial activity is based on several mechanisms that are described in more detail in other chapters of this book. In most cases, interactions between the peptide and surface membranes of the target organisms are considered to be responsible for activity. The initial binding is thought to depend on electrostatic interactions between the positively charged peptides and the negatively charged molecules at the surface of the target. A secondary step results in the modification of the biophysical properties of the membrane caused by direct interactions with the peptide. This interaction with the membrane finally leads to loss of membrane function, including breakdown of membrane potential, leakage of metabolites and ions, and alteration of membrane permeability. Other mechanisms of antimicrobial activity include the inhibition of protein and RNA synthesis by Bac5 and Bac7 (Skerlavaj, Romeo, and Gennaro, 1990) or interference with protein production by PR-39 (Cabiaux et al., 1994). The antimicrobial spectrum of individual peptides depends on their structure and amino acid sequence. Peptides rich in Pro or Arg residues are usually more active against Gram-negative than Gram-positive bacteria. Other linear (helical or Trp-rich) and Cys-containing peptides are generally active against Gram-positive and Gram-negative organisms. Individual antimicrobial components of the airway surface act synergistically against microorganisms (Bals, Wang, and Zasloff et al., 1998; Bals, Wang, and Zasloff et al., 1998; Singh et al., 2000).

In recent months several groups have published results that provide proof of the host defense function of antimicrobial peptides in living organisms. Indirect in vivo evidence for the host defense function of antimicrobial peptides came from a study on mice with a disrupted gene for matrilysin (metalloprotease 7). Mice with missing matrilysin were more susceptible to infection with enteropathogens (Wilson et al., 1999). Studies in a human bronchial xenograft model revealed decreased antimicrobial activity of ASF after inhibition of HBD1 transcription by antisense oligonucleotides (Goldman et al., 1997). Mice deficient in an antimicrobial peptide, mouse β-defensin-1 (mBD-1), revealed delayed clearance of *Haemophilus influenzae* from lung (Moser et al., 2002). Mice with deleted CRAMP, the murine homologue of LL-37, showed more prominent infection after cutaneous inoculation of bacteria (Nizet et al., 2001). Drosophila mutant for both the *immune deficiency gene* (imd) and the *toll* (Tl) signaling pathways fail to express most of the antimicrobial genes and rapidly succumb to either fungal or bacterial infections, indicating that these pathways, and likely the involved antimicrobial peptides, are essential for antimicrobial resistance in insects (Lemaitre et al., 1996). In reverse, constitutive expression of a single antimicrobial peptide can restore wild-type resistance to infection in immunodeficient Drosophila imd and spätzle double mutants that do not express any known endogenous antimicrobial peptide gene (Tzou, Reichhart, and Lemaitre, 2002). Also, the overexpression of LL-37 by viral gene transfer resulted in augmentation of innate host defense in a bronchial xenograft model of CF and in murine animal models of pneumonia and septic shock (Bals, Wang, and Zasloff et al., 1998; Bals, Wiener, and Wilson et al., 1999).

The development of microbial resistance against antimicrobial peptides is a rare event. Increase of the phosphocholine content of cell walls of *Haemophilus influenzae* decreased the susceptibility to LL-37 (Lysenko et al., 2000). Inactivation of the *dlt* operon in *Staphylococcus aureus* confers sensitivity to defensins, protegrins, and other antimicrobial peptides (Peschel et al., 1999). When exposed to the environment found in the airways of patients with CF, *Pseudomonas aeruginosa* is able to modify the structure of the LPS of the outer membrane (Ernst et al., 1999). These changes to endotoxin decrease the susceptibility of these bacteria to cationic antimicrobial peptides. Infection or chronic inflammation seems also to be responsible for the transcriptional downregulation of antimicrobial peptides. Gastrointestinal *Shigella dysenteriae* infections in humans are associated with reduced expression of LL-37 in epithelial cells (Islam et al., 2001). Proteases secreted by common pathogenic bacteria degrade and inactivate the antimicrobial peptide LL-37 (Schmidtchen et al., 2002).

13.3.2. Role of Antimicrobial Peptides in Inflammation, Angiogenesis, and Cell Function

Antimicrobial peptides have a variety of other biological effects beside their antimicrobial activity. Based on their membrane activity, antimicrobial peptides have a concentration dependent toxicity toward eukaryotic cells. High concentrations of α-defensins have been described in the secretions of patients with CF (Soong et al., 1997) and chronic bronchitis (Panyutich et al., 1993), in which these substances likely contribute to overwhelming inflammation. The α-defensins induce IL-8 production by lung epithelial cells (van Wetering, Mannesse-Lazeroms, Dijkman et al., 1997). The cellular damage by α-defensins may be augmented by defensin-induced lysis of epithelial cells (van Wetering, Mannesse-Lazeroms, van Sterkenburg et al., 1997) or binding of α-defensins to protease inhibitors of the serpin family such as alpha-1-antitrypsin (Panyutich et al., 1995).

Other than this unspecific toxicity, some antimicrobial peptides bind to specific receptors at low concentrations, activate specific intracellular signaling pathways, and stimulate various cellular functions. The α-defensins are able to stimulate a variety of cells by mechanisms not yet identified. They attract human $CD4^+/CD45RA^+$ (naive) or $CD8^+$ T cells (Chertov et al., 1996; Alizadeh et al., 2000), immature dendritic cells (Alizadeh et al., 2000), and monocytes (Territo et al., 1989; Chertov et al., 1996), and they induce proliferation of lung epithelial cells (Aarbiou, Rabe, and Hiemstra et al., 2002). They also induce secretion of histamine from mast cells through a G-protein coupled receptor (Befus et al., 1999; Niyonsaba et al., 2001) and the release of inflammatory mediators from T cells (Lillard et al., 1999) and epithelial cells (van Wetering et al., 2002). When instilled into the respiratory tract of animals, they mediate an acute inflammatory response (Zhang et al., 2001). α-defensins also inhibit adrenocorticotropic hormone-(ACTH-) stimulated cortisol production (Zhu et al., 1988). In addition, HNP-1 blocks adenoviral infection in vitro (Bastian and Schafer, 2001).

HBD1 and HBD2 were found to bind to a chemokine receptor known as CCR-6 (Yang et al., 1999). This receptor is found on immature dendritic and memory T cells ($CD4^+/CD45RO^+$), and consequently these findings are interpreted as a link between innate and adaptive immune mechanisms mediated by defensins. HBD3 and HBD4 chemoattract monocytes by mechanisms that have not yet been clarified (Garcia, Jaumann et al., 2001). Furthermore, HBD2 acts as a chemotaxin for mast cells through a pertussis toxin-sensitive pathway (Niyonsaba et al., 2001). LL-37 was found to bind to formyl peptide receptor-like 1 (FPRL1), a promiscuous receptor expressed on a variety of cells including neutrophils, monocytes, and lymphocytes (Yang et al., 2000). By activation of this G-protein coupled receptor, LL-37 attracts neutrophils, monocytes, CD4 T cells, and mast cells (Nobuhara et al., 1998; Niyonsaba et al., 2001). Analysis of global gene expression by gene array technology revealed that LL-37 has a significant impact on the gene expression and biology of macrophages (Scott et al., 2002). LL-37 also binds to apolipoprotein A-I (Wang et al., 1998). Interestingly, it has been found that IFN-inducible ELR$^-$ CXC chemokines display antimicrobial activity highlighting a structure–function relationship between antimicrobial peptides and chemokines (Cole et al., 2001).

Taken together, vertebrate antimicrobial peptides have a variety of additional functions besides their microbicidal function. The impact of these nonmicrobicidal functions on the pathogenesis of diseases is completely unknown. The nonmicrobicidal functions offer interesting opportunities to investigate the roles of antimicrobial peptides in inflammatory diseases; however, they might also cause adverse side effects when antimicrobial peptides are used as innovative therapeutics.

13.4. ROLE OF ANTIMICROBIAL PEPTIDES IN PULMONARY DISEASE

Antimicrobial peptides may have a role in a variety of infectious and inflammatory diseases of the lung. Based on their functions, several pathogenetic models are relevant.

1. *Inborn or acquired deficiencies* of antimicrobial peptides result in loss of function and subsequent increased susceptibility to infections. Only very limited data on states of decreased activity of antimicrobial peptides are available. In morbus Kostmann (Zeidler and Welte, 2002), a severe congenital neutropenia, periodontal disease has been linked with the deficiency of antimicrobial peptides in neutrophils (Putsep et al., 2002). A deficiency in the expression of antimicrobial peptides may account for the susceptibility of patients with atopic dermatitis to skin infection with *S. aureus* (Ong et al., 2002).

2. Several pulmonary diseases are associated with inflammation that results in many cases in *overexpression of antimicrobial peptide* genes. Based on the receptor-mediated functions of antimicrobial peptides, increased concentrations have a proinflammatory effect. Several examples are given in the next subsection.
3. *Polymorphisms of genes* of antimicrobial peptides might predispose to the development of pulmonary diseases. Several polymorphisms have been found in the HBD1 gene (Dork and Stuhrmann, 1998; Vatta et al., 2000; Circo et al., 2002; Matsushita et al., 2002).

13.4.1. Pneumonia and Tuberculosis

Infections of the respiratory tract are one of the most common disease groups. High numbers of hospital-acquired pneumonias and increasing numbers of infections with multiply resistant bacteria are prominent problems. Several studies found increased concentrations of defensins during infectious pulmonary diseases such as neonatal and adult pneumonia (Hiratsuka et al., 1998; Schaller-Bals, Schulze, and Bals, et al., 2002). Elevated levels of α-defensins were found in patients with empyema (Ashitani, Mukae, Nakazato, Taniguchi, et al., 1998). Tuberculosis is an infectious diseases that in most cases involves the lung. The numbers of pan-resistant *Mycobacterium tuberculosis* strains are increasing and are a significant clinical problem. Levels of defensins are increased in the plasma or bronchioalveolar lavage fluid in pulmonary tuberculosis (Ashitani et al., 2002) and infections with *Mycobacterium avium-intracellulare* (Ashitani et al., 2001).

13.4.2. Cystic Fibrosis and Diffuse Panbronchiolitis

Both diseases are characterized by chronic infection associated with overwhelming inflammation. CF is caused by a genetic defect of the CF transmembrane conductance regulator (CFTR) (Davis, Drumm, and Konstan, 1996). CF represents a model disease for defects of the innate host defense, and research on this disease attracted significant attention to the field of antimicrobial peptides in the late 1990s. Several hypotheses to explain the pathogenesis of CF lung disease have been proposed. The "hypotonic airway surface fluid/antimicrobial substance" hypothesis proposes that the defects in CFTR result in elevated salt concentrations that inactivate antimicrobial substances (Smith et al., 1996; Goldman et al., 1997). The core of this theory is the elevated NaCl content in CF ASF, an observation that is increasingly controversial. In contrast, the biogenesis or secretion of functional antimicrobial substances may be altered by intracellular defects in airway epithelial cells, as suggested by a salt-independent decrease of antimicrobial activity of CF airway secretions (Bals et al., 2001). CF lung disease is clearly caused by a defect of the local innate immune system (Bals, Weiner, and Wilson et al., 1999). Whether antimicrobial peptides have a direct role in the initial processes that link the defective CFTR with impaired host defense is unclear. In contrast, it appears that antimicrobial peptides contribute to the overwhelming inflammatory activity. Several reports found increased concentrations of β-defensins in CF-airways (Singh et al., 1998; Bals et al., 2001). Also, α-defensins are found at increased levels (Soong et al., 1997). One study found relative inhibition of expression of epithelial defensins in CF cells (Dauletbaev et al., 2002). Infections in CF and other lung diseases are characterized by the formation of a microbial biofilm, a specific state of bacterial growth that is characterized by secretion of viscous materials and decreased susceptibility to exogenous antibiotics (Costerton, Stewart, and Greenberg, 1999). The interaction between antimicrobial peptides and bacteria in biofilms has not been investigated. Diffuse panbronchiolitis is a chronic inflammatory lung disease of unknown origin that is phenotypically related to CF (Yanagihara, Kadoto, and Kohno, 2001). Neutrophil-derived defensins are elevated in the airways in diffuse panbronchiolitis and may be a marker of neutrophil activity in this disease (Ashitani, Mukae, Nakazato, Taniguchi, et al., 1998).

13.4.3. Asthma and Chronic Obstructive Pulmonary Disease

Both diseases are obstructive pulmonary diseases that are characterized by airflow limitation and a chronic inflammatory process of the airways, and both are of outstanding medical and economic importance. Inflammation in asthma is characterized by a Th2 orchestrated inflammation (Busse and Lemanske, 2001; Kay, 2001), whereas, in chronic obstructive pulmonary disease (COPD), chronic smoke exposure (or other rare causes) results in neutrophil influx and activation of proteases (Barnes, 2000). Based on their function as inflammatory mediators, antimicrobial peptides likely are involved in the pathogenesis of these diseases. Relatively little is known about concentrations of antimicrobial peptides in asthma and COPD. A polymorphism of the HBD1 gene is found at higher frequency in patients with COPD (Matsushita et al., 2002).

13.4.4. Adult Respiratory Distress Syndrome

Adult respiratory distress syndrome (ARDS) is a catastrophic inflammatory pulmonary disease that can be caused by a variety of conditions, such as trauma, hypoxia, infection, or intoxication (Bellingan, 2002). Neutrophil defensins are elevated in plasma and in bronchoalveolar lavage fluid from patients with ARDS (Ashitani et al., 1996).

13.4.5. Pulmonary Fibrosis and Sarcoidosis

Pulmonary fibrosis is a descriptive term for a group of lung diseases characterized by various amounts of inflammation, destruction of lung parenchyma, and replacement with fibrous materials accompanied by loss of pulmonary function (Green, 2002). Plasma concentrations of α-defensins are raised in patients with pulmonary fibrosis (Mukae et al., 2002). Sarcoidosis is an inflammatory disease that mainly involves the pulmonary system (Sharma, 2002). LL-37 is found to be upregulated in lungs of patients with this disease (Agerberth et al., 1999).

13.5. FUTURE ASPECTS

The broad spectrum of activity and the low incidence of bacterial resistance are attractive features of antimicrobial peptides. The specific mode of action that can involve pore formation in biomembranes makes antimicrobial peptides a new class of potential antibiotic drugs. Various attempts have been made to develop antimicrobial peptides as innovative antimicrobial or anti-LPS drugs. The coding sequence of antimicrobial peptides can also be transferred into target cells by means of gene transfer, as shown by the overexpression of LL-37 in animal models (Bals, Wang, and Zasloff et al., 1998; Bals, Weiner, and Wilson et al., 1999) or the transfer of histatin 3 to salivary glands (O'Connell et al., 1996). Another interesting approach is to stimulate the expression of antimicrobial peptides by the application of small chemicals compounds. The application of the essential amino acid L-isoleucine to airway epithelial cells upregulates the expression of epithelial defensins (Fehlbaum et al., 2000). In most preclinical and clinical studies antimicrobial peptides have been applied topically. Animal studies have used various disease models and application routes. Peptides of various classes were applied for models of pneumonia (Steinberg et al., 1997), septic shock (Kirikae et al., 1998), and oral mucosaitis (Loury et al., 1999). Other applications in animal models include fusion of mBDs to a nonimmunogenic tumor antigen, lymphoma idiotype, resulting in a protective immune response against the tumor (Biragyn et al., 2001). Human studies in which antimicrobial peptides were used were carried out by small biotech companies, in association with larger pharmaceutical companies, forming strategic alliances. Despite several studies, there are unanswered concerns about production costs, susceptibility to proteases in vivo, and unknown toxicities.

Based on the functions of antimicrobial peptides as mediators, future research will focus on the role of these substances in the pathogenesis of inflammatory diseases and will reveal whether the peptides are involved in the development of inflammatory diseases. Furthermore, the in vivo function of antimicrobial peptides will be characterized by use of animal models.

Antimicrobial peptides have emerged as effector substances of the pulmonary innate immune system involving not only activities as endogenous antibiotics but also as mediators of inflammation. Studying the biology of antimicrobial peptides should allow the development of novel therapeutics for pulmonary diseases.

REFERENCES

Aarbiou, J., Ertmann, M., van Wetering, S., van Noort, P., Rook, D., Rabe, K. F., Litvinov, S. V., van Krieken, J. H., de Boer, W. I., and Hiemstra, P. S. (2002). Human neutrophil defensins induce lung epithelial cell proliferation in vitro. *Journal of Leukocyte Biology*, 72, 167–74.

Aarbiou, J., Rabe, K. F., and Hiemstra, P. S. (2002). Role of defensins in inflammatory lung disease. *Annals of Medicine*, 34, 96–101.

Agerberth, B., Grunewald, J., Castanos-Velez, E., Olsson, B., Jornvall, H., Wigzell, H., Eklund, A., and Gudmundsson, G. H. (1999). Antibacterial components in bronchoalveolar lavage fluid from healthy individuals and sarcoidosis patients. *American Journal of Respiratory and Critical Care Medicine*, 160, 283–90.

Agerberth, B., Gunne, H., Odeberg, J., Kogner, P., Boman, H. G., and Gudmundsson, G. H. (1995). FALL-39, a putative human peptide antibiotic, is cysteine-free and expressed in bone marrow and testis. *Proceedings of the National Academy of Sciences USA*, 92, 195–9.

Alizadeh, A. A., Eisen, M. B., Davis, R. E., Ma, C., Lossos, I. S., Rosenwald, A., Boldrick, J. C., Sabet, H., Tran, T., Yu, X., Powell, J. I., Yang, L., Marti, G. E., Moore, T., Hudson, J., Jr., Lu, L., Lewis, D. B., Tibshirani, R., Sherlock, G., Chan, W. C., Greiner, T. C., Weisenburger, D. D., Armitage, J. O., Warnke, R., Levy, R., Wilson, W., Grever, M. R., Byrd, J. C., Botstein, D., Brown, P. O., and Staudt, L. M. (2000). Distinct types of diffuse large B-cell lymphoma identified by gene expression profiling. *Nature (London)*, 403, 503–11.

Andreu, D. and Rivas, L. (1998). Animal antimicrobial peptides: An overview. *Biopolymers*, 47, 415–33.

Ashitani, J., Mukae, H., Hiratsuka, T., Nakazato, M., Kumamoto, K., and Matsukura, S. (2001). Plasma and BAL fluid concentrations of antimicrobial peptides in patients with Mycobacterium avium-intracellulare infection. *Chest*, 119, 1131–7.

Ashitani, J., Mukae, H., Hiratsuka, T., Nakazato, M., Kumamoto, K., and Matsukura, S. (2002). Elevated levels of alpha-defensins in plasma and BAL fluid of patients with active pulmonary tuberculosis. *Chest*, 121, 519–26.

Ashitani, J., Mukae, H., Ihiboshi, H., Taniguchi, H., Mashimoto, H., Nakazato, M., and Matsukura, S. (1996). Defensin in plasma and in bronchoalveolar lavage fluid from patients with acute respiratory distress syndrome [in Japanese]. *Nihon Kyobu Shikkan Gakkai Zasshi*, 34, 1349–53.

Ashitani, J., Mukae, H., Nakazato, M., Ihi, T., Mashimoto, H., Kadota, J., Kohno, S., and Matsukura, S. (1998). Elevated concentrations of defensins in bronchoalveolar lavage fluid in diffuse panbronchiolitis. *European Respiratory Journal*, 11, 104–11.

Ashitani, J., Mukae, H., Nakazato, M., Taniguchi, H., Ogawa, K., Kohno, S., and Matsukura, S. (1998). Elevated pleural fluid levels of defensins in patients with empyema. *Chest*, 113, 788–94.

Bals, R., Wang, X., Wu, Z., Freeman, Banfa, V., Zasloff, M., and Wilson, J. (1998). Human beta-defensin 2 is a salt-sensitive peptide antibiotic expressed in human lung. *Journal of Clinical Investigation*, 102, 874–880.

Bals, R., Wang, X., Zasloff, M., and Wilson, J. M. (1998). The peptide antibiotic LL-37/hCAP-18 is expressed in epithelia of the human lung where it has broad antimicrobial activity at the airway surface. *Proceedings of the National Academy of Sciences USA.*, 95, 9541–6.

Bals, R., Weiner, D., Moscioni, A., Meegalla, R., and Wilson, J. (1999). Augmentation of innate host defense by expression of a cathelicidin antimicrobial peptide. *Infection and Immunity*, 67, 6084–9.

Bals, R., Weiner, D., and Wilson, J. (1999). The innate immune system in cystic fibrosis lung disease. *Journal of Clinical Investigation*, 103, 303–7.

Bals, R., Weiner, D. J., Meegalla, R. L., Accurso, F., and Wilson, J. M. (2001). Salt-independent abnormality of antimicrobial activity in cystic fibrosis airway surface fluid. *American Journal of Respiratory Cellular and Molecular Biology*, 25, 21–25.

Bals, R., Weiner, D. J., Meegalla, R. L., and Wilson, J. M. (1998). Transfer of a cathelicidin peptide antibiotic gene restores bacterial killing in a cystic fibrosis xenograft model. *Journal of Clinical Investigation*, 103, 1113–7.

Barnes, P. J. (2000). Chronic obstructive pulmonary disease. *New England Journal of Medicine*, 343, 269–80.

Bastian, A. and Schafer, H. (2001). Human alpha-defensin 1 (HNP-1) inhibits adenoviral infection in vitro. *Regulatory Peptide*, 101, 157–61.

Becker, M. N., Diamond, G., Verghese, M. W., and Randell, S. H. (2000). CD14-dependent lipopolysaccharide-induced beta-defensin-2 expression in human tracheobronchial epithelium. *Journal of Biological Chemistry*, 275, 29731–6.

Befus, A. D., Mowat, C., Gilchrist, M., Hu, J., Solomon, S., and Bateman, A. (1999). Neutrophil defensins induce histamine secretion from mast cells: mechanisms of action. *Journal of Immunology*, 163, 947–53.

Bellingan, G. J. (2002). The pulmonary physician in critical care * 6: The pathogenesis of ALI/ARDS. *Thorax*, 57, 540–6.

Bensch, K., Raida, M., Magert, H.-J., Schulz-Knappe, P., and Forssmann, W.-G. (1995). hBD-1: A novel b-defensin from human plasma. *FEBS Letters*, 368, 331–5.

Biragyn, A., Ruffini, P. A., Leifer, C. A., Klyushnenkova, E., Shakhov, A., Chertov, O., Shirakawa, A. K., Farber, J. M., Segal, D. M., Oppenheim, J. J., and Kwak, L. W. (2002). Toll-like receptor 4-dependent activation of dendritic cells by beta-defensin 2. *Science*, 298, 1025–9.

Biragyn, A., Surenhu, M., Yang, D., Ruffini, P. A., Haines, B. A., Klyushnenkova, E., Oppenheim, J. J., and Kwak, L. W. (2001). Mediators of innate immunity that target immature, but not mature, dendritic cells induce antitumor immunity when genetically fused with nonimmunogenic tumor antigens. *Journal of Immunology*, 167, 6644–53.

Birchler, T., Seibl, R., Buchner, K., Loeliger, S., Seger, R., Hossle, J. P., Aguzzi, A., and Lauener, R. P. (2001). Human Toll-like receptor 2 mediates induction of the antimicrobial peptide human beta-defensin 2 in response to bacterial lipoprotein. *European Journal of Immunology*, 31, 3131–7.

Busse, W. W. and Lemanske Jr., R. F. (2001). Asthma. *New England Journal of Medicine*, 344, 350–62.

Cabiaux, V., Agerberth, B., Johansson, J., Homble, F., Goormaghtigh, E., and Ruysschaert, J. M. (1994). Secondary structure and membrane interaction of PR-39, a Pro+Arg-rich antibacterial peptide. *European Journal of Biochemistry*, 224, 1019–27.

Chertov, O., Michiel, D. F., Xu, L., Wang, J. M., Tani, K., Murphy, W. J., Longo, D. L., Taub, D. D., and Oppenheim, J. J. (1996). Identification of defensin-1, defensin-2, and CAP37/azurocidin as T-cell chemoattractant proteins released from interleukin-8-stimulated neutrophils. *Journal of Biological Chemistry*, 271, 2935–40.

Circo, R., Skerlavaj, B., Gennaro, R., Amoroso, A., and Zanetti, M. (2002). Structural and functional characterization of hBD-1(Ser35), a peptide deduced from a DEFB1 polymorphism. *Biochemical and Biophysical Research Communications*, 293, 586–92.

Cole, A. M., Ganz, T., Liese, A. M., Burdick, M. D., Liu, L., and Strieter, R. M. (2001). Cutting edge: IFN-inducible ELR- CXC chemokines display defensin-like antimicrobial activity. *Journal of Immunology*, 167, 623–7.

Costerton, J., Stewart, P., and Greenberg, E. (1999). Bacterial biofilms: a common cause of persistent infection. *Science*, 284, 1318–1322.

Cowland, J., Johnsen, A., and Borregaard, N. (1995). hCAP-18, a cathelin/pro-bactenecin-like protein of human neutrophil specific granules. *FEBS Letters*, 368, 173–6.

Dauletbaev, N., Gropp, R., Frye, M., Loitsch, S., Wagner, T. O., and Bargon, J. (2002). Expression of human beta defensin (HBD-1 and HBD-2) mRNA in nasal epithelia of adult cystic fibrosis patients, healthy individuals, and individuals with acute cold. *Respiration*, 69, 46–51.

Davies, D. E. (2001). The bronchial epithelium in chronic and severe asthma. *Current Allergy and Asthma Reports*, 1, 127–33.

Davis, P. B., Drumm, M., and Konstan, M. W. (1996). Cystic fibrosis. *American Journal of Respiratory and Critical Care Medicine*, 154, 1229–56.

Diamond, G., Legarda, D., and Ryan, L. K. (2000). The innate immune response of the respiratory epithelium. *Immunological Reviews*, 173, 27–38.

Diamond, G., Zasloff, M., Eck, H., Brasseur, M., Maloy, W. L., and Bevins, C. L. (1991). Tracheal antimicrobial peptide, a cysteine-rich peptide from mammalian tracheal mucosa: peptide isolation and cloning of a cDNA. *Proceedings of the National Academy of Sciences USA.*, 88, 3952–3956.

Dörk, T. and Stuhrmann, M. (1998). Polymorphisms of the human beta-defensin-1 gene. *Molecular and Cellular Probes*, 12, 171–3.

Duits, L. A., Ravensbergen, B., Rademaker, M., Hiemstra, P. S., and Nibbering, P. H. (2002). Expression of beta-defensin 1 and 2 mRNA by human monocytes, macrophages and dendritic cells. *Immunology*, 106, 517–25.

Ernst, R. K., Yi, E. C., Guo, L., Lim, K. B., Burns, J. L., Hackett, M., and Miller, S. I. (1999). Specific lipopolysaccharide found in cystic fibrosis airway Pseudomonas aeruginosa. *Science*, 286, 1561–5.

Fehlbaum, P., Rao, M., Zasloff, M., and Anderson, G. M. (2000). An essential amino acid induces epithelial beta-defensin expression. *Proceedings of the National Academy of Sciences USA*, 97, 12723–8.

Frye, M., Bargon, J., Dauletbaev, N., Weber, A., Wagner, T. O., and Gropp, R. (2000). Expression of human alpha-defensin 5 (HD5) mRNA in nasal and bronchial epithelial cells. *Journal of Clinical Pathology*, 53, 770–3.

Ganz, T., Selsted, M. E., Szklarek, D., Harwig, S. S., Daher, K., Bainton, D. F., and Lehrer, R. I. (1985). Defensins. Natural peptide antibiotics of human neutrophils. *Journal of Clinical Investigation*, 76, 1427–35.

Garcia, J. R., Jaumann, F., Schulz, S., Krause, A., Rodriguez-Jimenez, J., Forssmann, U., Adermann, K., Kluver, E., Vogelmeier, C., Becker, D., Hedrich, R., Forssmann, W. G., and Bals, R. (2001). Identification of a novel, multifunctional beta-defensin (human beta-defensin 3) with specific antimicrobial activity. Its interaction with plasma membranes of Xenopus oocytes and the induction of macrophage chemoattraction. *Cell and Tissue Research*, 306, 257–64.

Garcia, J. R., Krause, A., Schulz, S., Rodriguez-Jimenez, F. J., Kluver, E., Adermann, K., Forssmann, U., Frimpong-Boateng, A., Bals, R., and Forssmann, W. G. (2001). Human beta-defensin 4: A novel inducible peptide with a specific salt-sensitive spectrum of antimicrobial activity. *FASEB Journal*, 15, 1819–21.

Goldman, M. J., Anderson, G. M., Stolzenberg, E. D., Kari, U. P., Zasloff, M., and Wilson, J. M. (1997). Human beta-defensin-1 is a salt-sensitive antibiotic in lung that is inactivated in cystic fibrosis. *Cell*, 88, 553–60.

Green, F. H. (2002). Overview of pulmonary fibrosis. *Chest*, 122, 334S-9S.

Gudmundsson, G. H., Agerberth, B., Odeberg, J., Bergman, T., Olsson, B., and Salcedo, R. (1996). The human gene FALL39 and processing of the cathelin precursor to the antibacterial peptide LL-37 in granulocytes. *European Journal of Biochemistry*, 238, 325–32.

Hamilton, L. M., Davies, D. E., Wilson, S. J., Kimber, I., Dearman, R. J., and Holgate, S. T. (2001). The bronchial epithelium in asthma–much more than a passive barrier. *Monaldi Archive of Chest Diseases*, 56, 48–54.

Hancock, R. E. (1997). Peptide antibiotics. *Lancet*, 349, 418–22.

Harder, J., Bartels, J., Christophers, E., and Schroder, J. M. (2001). Isolation and characterization of human beta-defensin-3, a novel human inducible peptide antibiotic. *Journal of Biological Chemistry*, 276, 5707–13.

Harder, J., Bartels, J., Christophers, E., and Schroeder, J.-M. (1997). A peptide antibiotic from human skin. *Nature (London)*, 387, 861.

Harder, J., Meyer-Hoffert, U., Teran, L. M., Schwichtenberg, L., Bartels, J., Maune, S., and Schroder, J. M. (2000). Mucoid Pseudomonas aeruginosa, TNF-alpha, and IL-1beta, but not IL-6, induce human beta-defensin-2 in respiratory epithelia. *American Journal of Respiratory Cellular and Molecular Biology*, 22, 714–21.

Hiratsuka, T., Nakazato, M., Date, Y., Ashitani, J., Minematsu, T., Chino, N., and Matsukura, S. (1998). Identification of human beta-defensin-2 in respiratory tract and plasma and its increase in bacterial pneumonia. *Biochemical and Biophysical Research Communications*, 249, 943–7.

Islam, D., Bandholtz, L., Nilsson, J., Wigzell, H., Christensson, B., Agerberth, B., and Gudmundsson, G. (2001). Downregulation of bactericidal peptides in enteric infections: A novel immune escape mechanism with bacterial DNA as a potential regulator. *Nature Medicine*, 7, 180–5.

Jia, H. P., Schutte, B. C., Schudy, A., Linzmeier, R., Guthmiller, J. M., Johnson, G. K., Tack, B. F., Mitros, J. P., Rosenthal, A., Ganz, T., and McCray Jr., P. B. (2001). Discovery of new human beta-defensins using a genomics-based approach. *Gene*, 263, 211–8.

Kay, A. B. (2001). Allergy and allergic diseases. Second of two parts. *New England Journal of Medicine*, 344, 109–13.

Kirikae, T., Hirata, M., Yamasu, H., Kirikae, F., Tamura, H., Kayama, F., Nakatsuka, K., Yokochi, T., and Nakano, M. (1998). Protective effects of a human 18-kilodalton cationic antimicrobial protein (CAP18)-derived peptide against murine endotoxemia. *Infection and Immunity*, 66, 1861–8.

Krisanaprakornkit, S., Kimball, J. R., and Dale, B. A. (2002). Regulation of human beta-defensin-2 in gingival epithelial cells: The involvement of mitogen-activated protein kinase pathways, but not the NF-κB transcription factor family. *Journal of Immunology*, 168, 316–24.

Lehrer, R., Ganz, T., and Selsted, M. (1991). Defesins: Endogenous antibiotic peptides of animal cells. *Cell*, 64, 229–230.

Lemaitre, B., Nicolas, E., Michaut, L., Reichhart, J.-M., and Hoffmann, J. (1996). The dorsoventral regulatory gene cassette spätzle/Toll/cactus controls the potent antifungal response in Drosophila adults. *Cell*, 86, 973–83.

Leonova, L., Kokryakov, V. N., Aleshina, G., Hong, T., Nguyen, T., Zhao, C., Waring, A. J., and Lehrer, R. I. (2001). Circular minidefensins and posttranslational generation of molecular diversity. *Journal of Leukocyte Biology*, 70, 461–4.

Lillard Jr., J. W., Boyaka, P. N., Chertov, O., Oppenheim, J. J., and McGhee, J. R. (1999). Mechanisms for induction of acquired host immunity by neutrophil peptide defensins. *Proceedings of the National Academy of Sciences USA*, 96, 651–6.

Linzmeier, R., Ho, C. H., Hoang, B. V., and Ganz, T. (1999). A 450-kb contig of defensin genes on human chromosome 8p23. *Gene*, 233, 205–11.

Loury, D., Embree, J. R., Steinberg, D. A., Sonis, S. T., and Fiddes, J. C. (1999). Effect of local application of the antimicrobial peptide IB-367 on the incidence and severity of oral mucositis in hamsters. *Oral Surgery, Oral Medicine, Oral Pathology, Oral Radiology, and Endodontics Online. Access to Oral Surgery, Oral Medicine, Oral Pathology, Oral Radiology, and Endodontics*, 87, 544–51.

Lysenko, E. S., Gould, J., Bals, R., Wilson, J. M., and Weiser, J. N. (2000). Bacterial phosphorylcholine decreases susceptibility to the antimicrobial peptide LL-37/hCAP18 expressed in the upper respiratory tract. *Infection and Immunity*, 68, 1664–71.

Matsushita, I., Hasegawa, K., Nakata, K., Yasuda, K., Tokunaga, K., and Keicho, N.

(2002). Genetic variants of human beta-defensin-1 and chronic obstructive pulmonary disease. *Biochemical and Biophysical Research Communications*, 291, 17–22.

McCray Jr., P. and Bentley, L. (1997). Human airway epithelia express a beta-defensin. *American Journal of Respiratory Cellular and Molecular Biology*, 16, 343–9.

Moon, S. K., Lee, H. Y., Li, J. D., Nagura, M., Kang, S. H., Chun, Y. M., Linthicum, F. H., Ganz, T., Andalibi, A., and Lim, D. J. (2002). Activation of a Src-dependent Raf-MEK1/2-ERK signaling pathway is required for IL-1alpha-induced upregulation of beta-defensin 2 in human middle ear epithelial cells. *Biochimcal and Biophysical Acta*, 1590, 41–51.

Moser, C., Weiner, D. J., Lysenko, E., Bals, R., Weiser, J. N., and Wilson, J. M. (2002). β-Defensin 1 contributes to pulmonary innate immunity in mice. *Infection and Immunity*, 70, 3068–72.

Mukae, H., Iiboshi, H., Nakazato, M., Hiratsuka, T., Tokojima, M., Abe, K., Ashitani, J., Kadota, J., Matsukura, S., and Kohno, S. (2002). Raised plasma concentrations of alpha-defensins in patients with idiopathic pulmonary fibrosis. *Thorax*, 57, 623–8.

Niyonsaba, F., Someya, A., Hirata, M., Ogawa, H., and Nagaoka, I. (2001). Evaluation of the effects of peptide antibiotics human beta-defensins-1/-2 and LL-37 on histamine release and prostaglandin D(2) production from mast cells. *European Journal of Immunology*, 31, 1066–75.

Nizet, V., Ohtake, T., Lauth, X., Trowbridge, J., Rudisill, J., Dorschner, R. A., Pestonjamasp, V., Piraino, J., Huttner, K., and Gallo, R. L. (2001). Innate antimicrobial peptide protects the skin from invasive bacterial infection. *Nature (London)*, 414, 454–7.

Nobuhara, K. K., Fauza, D. O., DiFiore, J. W., Hines, M. H., Fackler, J. C., Slavin, R., Hirschl, R., and Wilson, J. M. (1998). Continuous intrapulmonary distension with perfluorocarbon accelerates neonatal (but not adult) lung growth. *Journal of Pediatric Surgery*, 33, 292–8.

Ochoa, M. T., Stenger, S., Sieling, P. A., Thoma-Uszynski, S., Sabet, S., Cho, S., Krensky, A. M., Rollinghoff, M., Nunes Sarno, E., Burdick, A. E., Rea, T. H., and Modlin, R. L. (2001). T-cell release of granulysin contributes to host defense in leprosy. *Nature Medicine*, 7, 174–9.

O'Connell, B. C., Xu, T., Walsh, T. J., Sein, T., Mastrangeli, A., Crystal, R. G., Oppenheim, F. G., and Baum, B. J. (1996). Transfer of a gene encoding the anticandidal protein histatin 3 to salivary glands. *Hum Gene Therapy*, 7, 2255–61.

Ong, P. Y., Ohtake, T., Brandt, C., Strickland, I., Boguniewicz, M., Ganz, T., Gallo, R. L., and Leung, D. Y. (2002). Endogenous antimicrobial peptides and skin infections in atopic dermatitis. *New England Journal of Medicine*, 347, 1151–60.

Panyutich, A. V., Hiemstra, P. S., van Wetering, S., and Ganz, T. (1995). Human neutrophil defensin and serpins form complexes and inactivate each other. *American Journal of Respiratory Cellular and Molecular Biology*, 12, 351–7.

Panyutich, A. V., Panyutich, E. A., Krapivin, V. A., Baturevich, E. A., and Ganz, T. (1993). Plasma defensin concentrations are elevated in patients with septicemia or bacterial meningitis. *Journal of Laboratory and Clinical Medicine*, 122, 202–7.

Peschel, A., Otto, M., Jack, R., Kalbacher, H., Jung, G., and Götz, F. (1999). Inactivation of the dlt operon in Staphylococcus aureus confers sensitivity to defensins, protegrins, and other antimicrobial peptides. *Journal of Biological Chemistry*, 274, 8405–10.

Putsep, K., Carlsson, G., Boman, H. G., and Andersson, M. (2002). Deficiency of antibacterial peptides in patients with morbus Kostmann: an observation study. *Lancet*, 360, 1144–9.

Schaller-Bals, S., Schulze, A., and Bals, R. (2002). Increased levels of antimicrobial peptides in tracheal aspirates of newborn infants during infection. *American Journal of Respiratory and Critial Care Medicine*, 165, 992–5.

Schittek, B., Hipfel, R., Sauer, B., Bauer, J., Kalbacher, H., Stevanovic, S., Schirle, M., Schroeder, K., Blin, N., Meier, F., Rassner, G., and Garbe, C. (2001). Dermcidin: A novel human antibiotic peptide secreted by sweat glands. *Nature Immunology*, 2, 1133–7.

Schmidtchen, A., Frick, I. M., Andersson, E., Tapper, H., and Bjorck, L. (2002). Proteinases of common pathogenic bacteria degrade and inactivate the antibacterial peptide LL-37. *Molecular Microbiology*, 46, 157–68.

Schutte, B. C. and McCray Jr., P. B., (2002). [Beta]-defensins in lung host defense. *Annual Review of Physiology*, 64, 709–48.

Schutte, B. C., Mitros, J. P., Bartlett, J. A., Walters, J. D., Jia, H. P., Welsh, M. J., Casavant, T. L., and McCray, P. B., Jr. (2002). Discovery of five conserved beta -defensin gene clusters using a computational search strategy. *Proceedings of the National Academy of Sciences USA*, 99, 2129–33.

Scott, M. G., Davidson, D. J., Gold, M. R., Bowdish, D., and Hancock, R. E. (2002). The human antimicrobial peptide LL-37 is a multifunctional modulator of innate immune responses. *Journal of Immunology*, 169, 3883–91.

Selsted, M. E., Harwig, S. S., Ganz, T., Schilling, J. W., and Lehrer, R. I. (1985). Primary structures of three human neutrophil defensins. *Journal of Clinical Investigation*, 76, 1436–9.

Sharma, O. P. (2002). Sarcoidosis and other autoimmune disorders. *Current Opinion in Pulmonary Medicine*, 8, 452–6.

Singh, P., Jia, H., Wiles, K., Hesselberth, J., Liu, L., Conway, B., Greenberg, E., Valore, E., Welsh, M., Ganz, T., Tack, B., and McCray, P. J. (1998). Production of beta-defensins by human airway epithelia. *Proceedings of the National Academy of Sciences USA*, 95, 14961–6.

Singh, P. K., Tack, B. F., McCray Jr., P. B., and Welsh, M. J. (2000). Synergistic and additive killing by antimicrobial factors found in human airway surface liquid. *American Journal of Physiology–Lung Cellular and Molecular Physiology*, 279, L799–805.

Skerlavaj, B., Romeo, D., and Gennaro, R. (1990). Rapid membrane permeabilization and inhibition of vital functions of Gram-negative bacteria by bactenecins. *Infection and Immunity*, 58, 3724–30.

Smith, J., Travis, S., Greenberg, E., and Welsh, M. (1996). Cystic fibrosis airway epithelia fail to kill bacteria because of abnormal airway surface fluid. *Cell*, 85, 229–36.

Soong, L., Ganz, T., Ellison, A., and Caughey, G. (1997). Purification and characterization of defensins from cystic fibrosis sputum. *Inflammation Research*, 46, 98–102.

Sorensen, O. E., Follin, P., Johnsen, A. H., Calafat, J., Tjabringa, G. S., Hiemstra, P. S., and Borregaard, N. (2001). Human cathelicidin, hCAP-18, is processed to the antimicrobial peptide LL-37 by extracellular cleavage with proteinase 3. *Blood*, 97, 3951–9.

Steinberg, D. A., Hurst, M. A., Fujii, C. A., Kung, A. H., Ho, J. F., Cheng, F. C., Loury, D. J., and Fiddes, J. C. (1997). Protegrin-1: a broad-spectrum, rapidly microbicidal peptide with in vivo activity. *Antimicrobial Agents and Chemotherapy*, 41, 1738–42.

Stenger, S., Hanson, D. A., Teitelbaum, R., Dewan, P., Niazi, K. R., Froelich, C. J., Ganz, T., Thoma-Uszynski, S., Melian, A., Bogdan, C., Porcelli, S. A., Bloom, B. R., Krensky, A. M., and Modlin, R. L. (1998). An antimicrobial activity of cytolytic T cells mediated by granulysin. *Science*, 282, 121–5.

Tang, Y.-Q., Yaun, J., Osapay, G., Osapay, C., Tran, D., Miller, C., Quellette, A., and Selsted, M. (1999). A cyclic antimicrobial peptide produced in primate leukocytes by the ligation of two truncated alpha-defensins. *Science*, 286, 498–502.

Territo, M. C., Ganz, T., Selsted, M. E., and Lehrer, R. (1989). Monocyte-chemotactic activity of defensins from human neutrophils. *Journal of Clinical Investigation*, 84, 2017–20.

Tsutsumi-Ishii, Y. and Nagaoka, I. (2002). NF-kappa B-mediated transcriptional regulation of human beta-defensin-2 gene following lipopolysaccharide stimulation. *Journal of Leukocyte Biology*, 71, 154–62.

Tzou, P., Reichhart, J. M., and Lemaitre, B. (2002). Constitutive expression of a single antimicrobial peptide can restore wild-type resistance to infection in immunodeficient Drosophila mutants. *Proceedings of the National Academy of Sciences USA*, 99, 2152–7.

Valore, E. V., Park, C. H., Quayle, A. J., Wiles, K. R., McCray Jr., P. B., and Ganz, T. (1998). Human beta-defensin-1: An antimicrobial peptide of urogenital tissues. *Journal of Clinical Investigation*, 101, 1633–42.

van 't Hof, W., Veerman, E. C., Helmerhorst, E. J., and Amerongen, A. V. (2001). Antimicrobial peptides: Properties and applicability. *Biological Chemsitry*, 382, 597–619.

van Wetering, S., Mannesse-Lazeroms, S. P., Dijkman, J. H., and Hiemstra, P. S. (1997). Effect of neutrophil serine proteinases and defensins on lung epithelial cells: Modulation of cytotoxicity and IL-8 production. *Journal of Leukocyte Biology*, 62, 217–26.

van Wetering, S., Mannesse-Lazeroms, S. P., Van Sterkenburg, M. A., Daha, M. R., Dijkman, J. H., and Hiemstra, P. S. (1997). Effect of defensins on interleukin-8 synthesis in airway epithelial cells. *American Journal of Physiology–Lung Cellular and Molecular Physiology*, 272, L888–96.

van Wetering, S., Mannesse-Lazeroms, S. P., van Sterkenburg, M. A., and Hiemstra, P. S. (2002). Neutrophil defensins stimulate the release of cytokines by airway epithelial cells: Modulation by dexamethasone. *Inflammation Research*, 51, 8–15.

van Wetering, S., Sterk, P. J., Rabe, K. F., and Hiemstra, P. S. (1999). Defensins: Key players or bystanders in infection, injury, and repair in the lung? *Journal of Allergy and Clinical Immunology*, 104, 1131–8.

Vatta, S., Boniotto, M., Bevilacqua, E., Belgrano, A., Pirulli, D., Crovella, S., and Amoroso, A. (2000). Human beta defensin 1 gene: Six new variants. *Human Mutation*, 15, 582–3.

Wang, Y., Agerberth, B., Lothgren, A., Almstedt, A., and Johansson, J. (1998). Apolipoprotein A-I binds and inhibits the human antibacterial/cytotoxic peptide LL-37. *Journal of Biological Chemistry*, 273, 33115–8.

Whitsett, J. A. (2002). Intrinsic and innate defenses in the lung: Intersection of pathways regulating lung morphogenesis, host defense, and repair. *Journal of Clinical Investigation*, 109, 565–9.

Wilmott, R., Fiedler, M., and Stark, J. (1998). Host defense mechanisms. In *Disorders of the Respiratory Tract in Children*, ed. V. Chernick and T. Boat, pp. 238–64. Philadelphia: Saunders.

Wilson, C. L., Ouellette, A. J., Satchell, D. P., Ayabe, T., Lopez-Boado, Y. S., Stratman, J. L., Hultgren, S. J., Matrisian, L. M., and Parks, W. C. (1999).

Regulation of intestinal alpha-defensin activation by the metalloproteinase matrilysin in innate host defense. *Science,* 286, 113–7.

Yanagihara, K., Kadoto, J., and Kohno, S. (2001). Diffuse panbronchiolitis–pathophysiology and treatment mechanisms. *International Journal of Antimicrobial Agents,* 18, S83–7.

Yang, D., Chen, Q., Schmidt, A. P., Anderson, G. M., Wang, J. M., Wooters, J., Oppenheim, J. J., and Chertov, O. (2000). LL-37, the neutrophil granule- and epithelial cell-derived cathelicidin, utilizes formyl peptide receptor-like 1 (FPRL1) as a receptor to chemoattract human peripheral blood neutrophils, monocytes, and T-cells. *Journal of Experimental Medicine,* 192, 1069–74.

Yang, D., Chertov, O., Bykovskaia, S., Chen, Q., Buffo, M., Shogan, J., Anderson, M., Schroder, J., Wang, J., Howard, O., and Oppenheim, J. (1999). Beta-defensins: Linking innate and adaptive immunity through dendritic and T cell CCR6. *Science,* 286, 525–8.

Zanetti, M., Gennaro, R., and Romeo, D. (1995). Cathelicidins: A novel protein family with a common proregion and a variable C-terminal antimicrobial domain. *FEBS letters,* 374, 1–5.

Zeidler, C. and Welte, K. (2002). Kostmann syndrome and severe congenital neutropenia. *Seminars in Hematology,* 39, 82–8.

Zhang, H., Porro, G., Orzech, N., Mullen, B., Liu, M., and Slutsky, A. S. (2001). Neutrophil defensins mediate acute inflammatory response and lung dysfunction in dose-related fashion. *American Journal of Physiology–Lung Cellular and Molecular Physiology,* 280, L947–54.

Zhu, Q. Z., Hu, J., Mulay, S., Esch, F., Shimasaki, S., and Solomon, S. (1988). Isolation and structure of corticostatin peptides from rabbit fetal and adult lung. *Proceedings of the National Academy of Sciences USA,* 85, 592–6.

Index